| | |
|---|---|
| $F_c$ | allowable compressive stress (psi, ksi, kPa, MPa) |
| $F_t$ | allowable tensile stress (psi, ksi, kPa, MPa) |
| $F_y$ | yield stress (psi, ksi, kPa, MPa) |
| GN | giganewton (a unit of force, $10^9$ newtons) |
| GPa | gigapascal (a unit of stress, $10^9$ pascals) |
| h | depth, often cross-sectional (in, ft, mm, m); building height (ft, m) |
| in | inch (a unit of length) |
| I | moment of inertia ($in^4$, $mm^4$, $m^4$) |
| $I_x$ | moment of inertia with respect to an x-axis ($in^4$, $mm^4$, $m^4$) |
| $I_{xx}$ | moment of inertia with respect to the x-centroidal axis ($in^4$, $mm^4$, $m^4$) |
| $I_y$ | moment of inertia with respect to a y-axis ($in^4$, $mm^4$, $m^4$) |
| $I_{yy}$ | moment of inertia with respect to the y-centroidal axis ($in^4$, $mm^4$, $m^4$) |
| k | kip (a unit of force, $10^3$ lb) |
| kg | kilogram (a unit of mass, $10^3$ grams) |
| kip | kilopound (a unit of force, $10^3$ lb) |
| k-ft | kip-feet (a unit of moment) |
| kN | kilonewton (a unit of force, $10^3$ newtons) |
| kN·m | kilonewton-meter (a unit of moment) |
| kPa | kilopascal (a unit of stress, $10^3$ pascals) |
| ksi | kips per square inch (a unit of stress) |
| K | end conditions factor in column buckling calculations |
| lb | pound (a unit of force) |
| lb-ft | pound-feet (a unit of moment) |
| L | length (in, ft, mm, m) |
| LL | live load |
| m | meter (a unit of length) |

(continued inside back cover)

**Third Edition**

# Elementary Structures for Architects and Builders

## Ronald E. Shaeffer

*School of Architecture*
*Florida A & M University*
*Consulting Engineer, Tallahassee, Florida*

**PRENTICE HALL**
*Upper Saddle River, New Jersey      Columbus, Ohio*

**Library of Congress Cataloging-in-Publication Data**

Shaeffer, R. E.
    Elementary structures for architects and builders / Ronald E.
Shaeffer. — 3rd ed.
      p.    cm.
    Includes index.
    ISBN 0-13-348954-X
    1. Structural analysis (Engineering)   2. Structural design.
I. Title.
TA645.S479     1998
624.1′71—dc21              97-5166
                      CIP

Cover photo: © T. Tracy/FPG International
Editor: Ed Francis
Production Editor: Alexandrina Benedicto Wolf
Editorial/Production Supervision: WordCrafters Editorial Services, Inc.
Cover Design Coordinator: Julia Zonneveld Van Hook
Cover Designer: Raymond Hummons
Production Manager: Deidra Schwartz
Marketing Manager: Danny Hoyt
Illustrations on pages 17, 99, 131, 143, 144, 162, 163, 164, 178, 291, 343, and 367
   by Patrick L. Pinnell, AIA

This book was set in Times Roman by TCSystems, Inc., and was printed
and bound by Quebecor Printing/Book Press. The cover was printed by
Phoenix Color Corp.

Appendix H table courtesy, American Forest & Paper Association, Washington, D.C.
Appendix J table extracted with permission from the *Manual of Steel Construction (ASD),*
ninth edition, published by the American Institute of Steel Construction, Inc.

Prentice-Hall International (UK) Limited, *London*
Prentice-Hall of Australia Pty. Limited, *Sydney*
Prentice-Hall of Canada, Inc., *Toronto*
Prentice-Hall Hispanoamericana, S. A., *Mexico*
Prentice-Hall of India Private Limited, *New Delhi*
Prentice-Hall of Japan, Inc., *Tokyo*
Simon & Schuster Asia Pte. Ltd., *Singapore*
Editora Prentice-Hall do Brasil, Ltda., *Rio de Janeiro*

*This work is dedicated to*
*Jane*
*Joseph*
*Kristin*

*Prose requires the use of grammar;*
*POETRY requires an understanding of grammar.*

*Building requires the use of structure;*
*ARCHITECTURE requires an understanding of structure.*

# Preface

This beginning text has been written for students of architecture, building construction, and the related technologies. It is intended to provide the material for a first course in structures, treating the essential topics in statics and mechanics of materials and providing an introduction to structural analysis. The presentation is basically quantitative and will be most effective when used in conjunction with a book emphasizing a qualitative approach to structural behavior.

It is assumed that the student has a background in materials and methods of construction from prior coursework or individual experience. Chapter 5 provides a very brief review of the essential characteristics of a few structural materials but is not sufficient in depth or scope.

A minimal background in calculus and physics has been assumed in writing this material. Most of the derivations of the equations have been placed in the appendices, as they are usually not absolutely essential to the use of the equations themselves. Better students, however, will gain additional understanding and insight by consulting these derivations as they are referenced.

The organization of this edition remains essentially unchanged from the previous two. A substantive content change accomplishes some of the intent of the author's 1980 text, *Building Structures: Elementary Analysis and Design.* That book is the "great-grandfather" of this one and was published using 100% SI units. This edition uses approximately 50% SI units and 50% customary units. It is hoped that instructors will be able to use either system with a minimum of confusion. I do not recommend using a mix of units in the first course in structures because it might hinder the proper understanding of "first principles" that is so essential to the student's future study and practice. During the time the United States is changing to the new units system, I prefer to teach the first course exclusively in SI units and the second course (timber and steel design at our school) using only customary units. Succeeding courses use both.

Part of Chapter 2 has been restructured to proceed more slowly when introducing nonconcurrent force systems and the finding of reactions. I sometimes find out, semesters later, that a student's difficulty with the overall subject matter stems from the lack of a good grasp of free-body concepts and rotational equilibrium. The new approach should help overcome some of this.

Near the end of Chapter 2 a new section has been added involving sloping members and the forces in gable roof construction. This is intended to address arch action and the development of horizontal thrust in a less abstract manner.

I am grateful to Professors Bill Epps, Ken Livingston, and Joel Weinstein for critically reviewing the first edition for Prentice Hall and making several very helpful suggestions. I also wish to thank Professors Ken Dunker and David Glasser for suggestions and error finding. I also appreciate the invaluable feedback from the following reviewers of this edition: Charles W. Graham and Ronald J. Kruhl, Texas A & M University, and Charles Scribner of Vermont Technical College. As previously stated, please inform me of errors that you find; don't assume that I've already been notified. Thank you in advance for your suggestions and assistance!

*R. E. Shaeffer*

# Contents

**1 OVERVIEW 1**

1-1 Definition of Structure    1
1-2 Structure of Buildings    1
1-3 Structural Planning and Design    3
1-4 Types of Loads    4
1-5 Types of Stress    4
1-6 Structural Forms in Nature    5
1-7 Structural Forms in Buildings    8
1-8 Cost    10
1-9 Building Codes    11
1-10 Accuracy of Computations    16


**2 STATICS 19**

2-1 Introduction    19
2-2 Forces    19
2-3 Components and Resultants    20
2-4 Equilibrium of Concurrent Forces    30
2-5 Moments and Couples    40
2-6 Ideal Support Conditions    47
2-7 Rigid-Body Concept and Free-Body Diagrams    51
2-8 Equilibrium of Cantilevered Members    52
2-9 Equilibrium of Simple Beams and Frames    58
2-10 Two-Force Members    66
2-11 Shed and Gable Members    77
2-12 Stability and Determinacy    83
2-13 Pinned Frames    86
2-14 Simple Cable Statics    91
2-15 Conclusion and Procedure    105

## 3 STRUCTURAL PROPERTIES OF AREAS     107

3-1 Introduction
3-2 Centroids     107
3-3 Moment of Inertia     118
3-4 Parallel Axis Theorem     127
3-5 Radius of Gyration     134

## 4 STRESS AND STRAIN     139

4-1 Types of Stress     139
4-2 Basic Connection Stresses     141
4-3 Strain     141
4-4 Stress versus Strain     147
4-5 Stiffness     148
4-6 Total Axial Deformation     151
4-7 Thermal Stresses and Strains     153
4-8 Allowable Stress Design and Limit States Design     158

## 5 PROPERTIES OF STRUCTURAL MATERIALS  161

5-1 Introduction     161
5-2 Nature of Wood     161
5-3 Concrete and Reinforced Concrete     165
5-4 Structural Steel     166
5-5 Masonry and Reinforced Masonry     168
5-6 Creep     168

## 6 SHEAR AND MOMENT   169

6-1 Definitions and Sign Conventions     169
6-2 Shear and Moment Equations     171
6-3 Significance of Zero Shear     182
6-4 Load, Shear, and Moment Relationships     185
6-5 Uniformly Varying Loads     193

## 7 FLEXURAL STRESSES   199

7-1 Introduction     199
7-2 Flexural Strain     200
7-3 Flexural Stress     201
7-4 Section Modulus     218
7-5 Lateral Buckling and Stability     226

## 8 SHEARING STRESSES     229

8-1 Nature of Shearing Stresses     229
8-2 Diagonal Tension and Compression     230

8-3 Basic Horizontal Shearing Stress Equation    231
8-4 Horizontal Shearing Stresses in Timber Beams    240
8-5 Horizontal Shearing Stresses in Steel Beams    244


## 9  DEFLECTION AND INDETERMINATE BEAMS  249

9-1 Introduction    249
9-2 Moment-Area Method    250
9-3 Principle of Superposition    260
9-4 Use of Deflection Formulas    262
9-5 Superposition and Indeterminate Structures    265
9-6 Theorem of Three Moments    271
9-7 Loading Patterns    280


## 10  BEAM DESIGN AND FRAMING  283

10-1 Introduction    283
10-2 Shape of Beam Cross Sections    284
10-3 "Ideal" Beams    284
10-4 Properties of Materials    286
10-5 Tributary Area    287
10-6 Framing Direction    289
10-7 Selecting Wood Beams    292
10-8 Selecting Steel Beams    312
10-9 Design Aids for Wood and Steel    331


## 11  ELASTIC BUCKLING OF COLUMNS  333

11-1 Columns as Building Structural Elements    333
11-2 Column Failure Modes    334
11-3 The Euler Theory    335
11-4 Influence of Different End Conditions    342
11-5 Intermediate Lateral Bracing    353
11-6 Limits to the Applicability of the Euler Equation    362
11-7 Eccentric Loading and Beam-Columns    363


## 12  TRUSSES  365

12-1 Introduction    365
12-2 Analysis by Joint Equilibrium    368
12-3 Method of Sections    379
12-4 Special Types of Trusses    386

## APPENDICES

**A**  Derivation of Basic Flexural Stress Equation        389

**B**  Derivation of Basic Horizontal Shearing Stress Equation        393

**C**  Derivation of Euler Column Buckling Equation        396

**D**  Weights of Selected Building Materials        398

**E**  Properties of Selected Materials        399

**F**  Properties of Areas        401

**G**  Proof of Moment-Area Theorems        403

**H**  Allowable Stresses and Modulus of Elasticity Values for Selected Structural Sawn Lumber        407

**I**  Wood Section Properties        409

**J**  Properties of Selected Steel Sections        411

**K**  Shear, Moment, and Deflection Equations        415

**L**  Introduction to the SI Metric System        418

## ANSWERS TO PROBLEMS        423

## INDEX        445

# 1

# Overview

## 1–1 DEFINITION OF STRUCTURE

The word *structure* has many meanings. Dictionaries usually define it in very general terms, such as the following: "the organization or interrelation of all the parts of a whole; manner of construction." Structures or structured things exist almost everywhere, and any definition will apply more aptly to some than to others. Without confining our use of the word to buildings or other engineered objects, we find that almost everything has structure. It is very difficult to think of anything that is totally without structure. Certainly, every material object has a basic molecular structure, if nothing else. Even outer space, closer to a true vacuum than anything we know, is somewhat defined by the relatively few objects in it. It has been suggested that electrical discharge in the form of lightning has no structure. If we narrow our general terms slightly, we say that lightning seems to behave as if it has no structure. However, it has direction, and this in itself indicates the presence of some structure.

Even intangible things such as thoughts, emotions, and social relationships frequently have definite patterns. Almost everything we do or think has a structure. It is important for us to realize that in this book and our related studies, we are dealing with a very specific and narrow use of the term.

## 1–2 STRUCTURE OF BUILDINGS

Considering only the engineering essentials, the structure of a building can be defined as the assemblage of those parts that exist for the purpose of maintaining shape and stability. Its primary purpose is to resist any loads applied to the building and to transmit those to the ground.

In terms of architecture, the structure of a building is and does much more than that. It is an inseparable part of the building form and to varying degrees is a generator of that form. Used skillfully, the building structure can establish or reinforce orders and rhythms among the architectural volumes and planes. It can be visually dominant or recessive. It can develop harmonies or conflicts. It can be both confining and emancipating, and unfortunately in some cases, it cannot be ignored. It is physical.

1

A structural system is engineered to maintain the architectural form. Therefore, structures for buildings must be rational in terms of their adherence to the fundamental principles of science. Artists can sometimes generate shapes that seemingly ignore any consideration of natural forces, but architects cannot. The principles and tools of physics and mathematics provide the basis for differentiating between rational and irrational forms of construction.

There are at least three items that must be present in the structure of a building:

Stability

Strength and stiffness

Economy

Taking the first of the three requirements, it is obvious that *stability* is needed to maintain shape. An unstable building structure implies unbalanced forces or a lack of equilibrium and a consequent acceleration of the structure or its pieces. (The nature of structural stability is covered in more detail in Section 2–12.)

The requirement of *strength* means that the materials selected to resist the stresses generated by the loads must be adequate. Indeed, a "factor of safety" is usually provided so that under the anticipated loads, a given material is not stressed to a level even close to its rupture point. The material property called *stiffness* is considered with the requirement of strength (i.e., the structure designed must be of sufficient strength *and* stiffness). Stiffness is different from strength in that it involves how much a structure strains or deflects under load. A material that is very strong but lacking in stiffness will deform too much to be of value in resisting the forces applied.

*Economy* of a building structure refers to more than just the cost of the materials used. Construction economy is a complicated subject involving raw materials, fabrication, erection, and maintenance. Design and construction labor costs and the costs of energy consumption must be considered. Speed of construction and the cost of money (interest) are also factors. In most design situations, more than one structural material requires consideration. Competitive alternatives almost always exist, and the choice is seldom obvious.

Apart from these three primary requirements, several other factors are worthy of emphasis. First, the structure or structural system must relate to the building's *function*. It should not be in conflict in terms of form. For example, a linear function demands a linear structure, and therefore it would be improper to roof a bowling alley with a dome. Similarly, a theater must have large, unobstructed spans but a fine restaurant probably should not. **Stated simply, the structure must be appropriate to the function it is to shelter.**

Second, the structure must be *fire-resistant*. It is obvious that the structural system must be able to maintain its integrity at least until the occupants are safely out. Building codes specify the number of hours for which certain parts of a building must resist the heat without collapse. The structural materials used for those elements must be inherently fire-resistant or be adequately protected by fireproofing materials. The degree of fire resistance to be provided will depend on a number of items, including the use and occupancy load of the space, its dimensions, and the location of the building.

Third, the structure should *integrate* well with the building's circulation systems. It should not be in conflict with the piping systems for water and waste, the ducting systems for air, or (most important) the movement of people. It is obvious that the various building systems must be coordinated as the design progresses. One can design in a sequential step-by-step manner within any one system, but the design of all of them should move in a parallel manner toward completion. Spatially, all the various parts of a building are interdependent.

Fourth, the structure must be *psychologically safe* as well as physically safe. A high-rise frame that sways considerably in the wind might not actually be dangerous but may make the building uninhabitable just the same. Lightweight floor systems that are too "bouncy" can make the users very uncomfortable. Large glass windows, uninterrupted by dividing mullions, can be quite safe but will appear very insecure to the occupant standing next to one 40 floors above the street.

Sometimes the architect must make deliberate attempts to increase the apparent strength or solidness of the structure. This apparent safety may be more important than honestly expressing the building's structure, because the untrained viewer cannot distinguish between real and perceived safety.

## 1–3   STRUCTURAL PLANNING AND DESIGN

The building designer needs to understand the behavior of physical structures under load. An ability to intuit or "feel" structural behavior is possessed by those having much experience involving structural analysis, both qualitative and quantitative. The consequent knowledge of how forces, stresses, and deformations build up in different materials and shapes is vital to the development of this "sense."

Beginning this study of forces (statics) and stresses and deformations (mechanics of materials) is most easily done through quantitative methods. These two subjects form the basis for all structural planning and design and are very difficult to learn in the abstract.

In most building design efforts, the initial structural planning is done by the architect. Ideally, the structural and mechanical consultants should work side by side with the architect from the conception of a project to the final days of construction. In most cases, however, the architect must make some initial assumptions about the relationships to be developed between the building form and the structural system. A solid background in structural principles and behavior is needed to make these assumptions with any reasonable degree of confidence. The shape of the structural envelope, the location of all major supporting elements, the directionality (if any) of the system, the selection of the major structural materials, and the preliminary determination of span lengths are all part of the structural process.

Structural design, on the other hand, is done by both the architect and the engineer. The preliminary determination of the size of major structural elements, providing a check on the rationality of previous assumptions, is done by the architect and/or the engineer. Final structural design, involving a complete analysis of all the parts and components, the working out of structural details, and the specifying of structural materials is almost always done by the structural engineer.

Of the two areas, structural planning is far more complex than structural design. It involves the previously mentioned "feeling for structure" or intuition that comes through experience. Structural design can be learned from lectures and books, but it is likely that structural planning cannot. Nevertheless, some insight and judgment can be developed from a minimal background in structural analysis and design. If possible, this should be gained from an architectural standpoint, emphasizing the relationship between the quantities and the resulting qualities wherever possible, rather than from an engineering approach.

This study of quantitative structures can be thorough enough to permit the architect to do completely the analysis for smaller projects, although such depth is not absolutely necessary. At the very least it should provide the knowledge and vocabulary necessary to work with the consulting engineer. It must be remembered that the architect receives much more education that is oriented toward creativity than does the engineer, and therefore needs to maintain control over the design. It is up to the architect to ask intelligent questions and suggest viable alternatives. If handicapped by structural ignorance, some of the design decisions will, in effect, be made by others.

## 1–4   TYPES OF LOADS

In general, loads that act on building structures can be divided into two groups: those due to gravitational attraction and those resulting from other natural causes and elements. Gravity loads can be further classified into two groups: live load and dead load. Building *live loads* include people and most movable objects within the structure or on top of it. Snow is a live load. So is a grand piano, a safe, or a waterbed. *Dead loads*, on the other hand, generally include the immovable objects in a building. The walls (both interior and exterior), floors, mechanical and electrical equipment, and the structural elements themselves are examples of dead loads.

Natural forces not due to gravity that act on buildings are provided by wind and earthquakes. Wind load is a lateral load that varies in intensity with the building's height, location, and shape. Earthquakes can generate very large lateral forces that impact buildings at grade level and act in a "shaking" manner upon the superstructure. (Hurricanes, tornadoes, and earthquakes present special design problems, and local building codes often require certain types of resistive construction.) This basic text will deal primarily with gravity loading.

## 1–5   TYPES OF STRESS

A fundamental concept in structural analysis is that the structure as a whole and each of its elements are in a state of *equilibrium*. This means there are no unbalanced forces acting on the structure or its parts at any point. All forces counteract one another, and this results in equilibrium. When all the forces acting on a given element in the same direction are summed algebraically, the net effect is zero and there will be no acceleration. The object does respond to the forces internally, however. It is pushed or pulled and otherwise deformed; internal stresses of varying types and magnitudes accompany these deformations.

These stresses are named by their action or behavior (i.e., tension, compression, shear, and bending). *Tensile* and *compressive* stresses which act through the axis or center of mass of an object are evenly distributed over the resisting area and result in all the material fibers being stressed to like amounts. *Shearing* stresses and, more important, *bending* stresses are not uniform and usually result in a few fibers of material being deformed to their limit while others remain unstressed or nearly so. Bending is, by far, the structurally least efficient way to carry loads.

Assuming for the moment that we have a material equally strong in tension, compression, shear, and bending, it would be best to load it in tension to achieve its maximum structural capacity. Compressive forces, if applied to a long slender structure, can cause buckling as illustrated in Figure 11–2(b). Buckling always occurs under less load than would be required to fail the materials in true compression (i.e., crushing). Of course, materials are not equal in strength when loaded in different ways. Some materials have almost no tensile strength, and generalizations are very difficult to make. As explained in succeeding chapters, shearing stresses will cause tension and compression; and bending is actually a combination of shear, tension, and compression. Because of the previously mentioned uneven distribution of stress intensity, however, bending is always the most damaging load that can be applied to any structural material.

## 1–6  STRUCTURAL FORMS IN NATURE

Some of the most sophisticated and efficient structures are found in plants, animals, and animal houses. Through adaptation to specific environments over time, natural forms may be refined until they are nearly perfect responses to a given set of forces. Countless examples of this type of form response or form resistance, some less successful than others, may be found all around us. Only a few are cited here, to provide a representative sample. Natural forms can be very complicated in terms of structural analysis, and the reader should not be discouraged at being unable to understand them right away.

As an educational exercise, one may wish to select two or three plant or animal structures and record some preliminary thoughts or ideas about them. What forces act on them? What type of stresses are developed? What parts are strong, weak, stiff, or flexible; and why? The same forms could then be analyzed several months or a year later after completing some formal education in structures. In some cases an object that appears simple and straightforward at first glance becomes quite complex as we learn more about structural behavior.

The egg is one of the classic examples of good structure both in terms of form and material. It is a thin shell which is very strong in compression when loaded uni-

"Structure" is the assembly of parts that maintain the stability of a building. The desire to make the distinction between "working" and secondary elements, and the increasing ability to make it correctly, underlay the origins and growth of modern architecture. The primitive hut's essential structural naturalness was held up as a model of ideal clarity for architecture by the Abbe Laugier in his 1753 *Essay on Architecture;* Le Corbusier's Maisons Dom-ino of 1914, conceived as basic housing, followed Laugier's thinking by identifying slab, column, and footing as the minimum essential structure. Note that the actual structure, revealed in plan and sections, is disguised in the desire for a forceful diagram; the "slab" is a system of beams and joists.

formly. It is doubly curved, which provides some resistance to compression buckling, a problem with all thin shells. In contrast to its strength under uniform loads, the egg is virtually defenseless against point loads. In this case, of course, the development of resistance to point loads would be totally self-defeating.

The scallop shell sketched in Figure 1–1 is considerably stronger than the eggshell. It is much more of a permanent structure and is subject to much greater loads. It is also doubly curved but is much thicker, to provide some resistance to impact loads from predators. Of greater significance, however, is the fluting or small undulations in the surface of the shell. This greatly stiffens the shell and enables it to withstand large loads without buckling. Any type of folding or ribbing of a surface (convolution) adds stiffness, and this principle is used frequently in man-made structures—from building roofs to guard rails.

The ordinary blade of grass provides an interesting example of a form that changes shape constantly over its length. Its cross-section goes from a very strong and stiff tube at the bottom through a V or arc shape at midheight and finally to a very flexible flat shape at the tip. As illustrated in Figure 1–2(b) the V shape is sometimes further refined by a stiffening rib.

A blade of grass acts much like a cantilever beam sticking vertically out of the ground. It deflects when subjected to lateral loads (such as the wind) but resists failure by having its greatest strength located at the bottom, where it is needed to resist bending.

The common spider web is an ingenious tensile structure, light yet strong and easily maintained. It is very redundant, possessing many extra members, and parts of it can be completely torn away without the rest collapsing. Because every member is tensile, the structure is extremely efficient in terms of its self-weight. It is quite flexible and its highly elastic nature is well suited to the impact loads it must sustain.

One of the most sophisticated natural structures is the walnut shell. It has a double curvature and its surface has many convolutions. Figure 1–3 compares half of the shell to a dome. In general, a dome tends to thrust outward at the bottom edge as the load tries to flatten it. This circular bottom edge must be prevented from moving outward or it will develop numerous vertical cracks. One way to contain this edge is with a tension ring around the bottom. The walnut shell provides this in the form of a thickened tapered edge. This thick edge ring also helps to maintain the boundary shape against loads applied in that plane.

As if this were not enough protection for the meat inside, the interior is crossed by several paperlike tensile diaphragms which help to maintain the overall

**Figure 1–1**   Scallop shell.

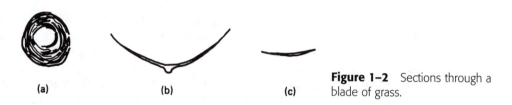

**Figure 1–2** Sections through a blade of grass.

(a)                    (b)                    (c)

spherical shape of the shell. It is not surprising that great force must be applied to fail such a structure.

Because of the success of many natural forms, they are often copied in the design of buildings. Sometimes this is done without much thought and even less skill, and the resulting design is most unfortunate. Success is more likely when we borrow the *principles* of "resistance to loads through form" from nature (rather than the forms themselves) and apply those principles to suit the needs of a particular design problem.

## 1–7  STRUCTURAL FORMS IN BUILDINGS

There are several basic structural elements found in buildings, each of which embodies a different type of structural behavior. The more complicated forms are made up of combinations of the basic ones or are extensions of the same concepts. The basic elements and the stresses they develop under load are as follows:

> *Cable:* pure tension
>
> *Post:* compression (and sometimes bending)
>
> *Beam:*[1] bending and shear
>
> *Truss:* tension and compression
>
> *Arch:* compression (and usually bending)
>
> *Shell:* membrane (tension and/or compression evenly distributed through the shell thickness)

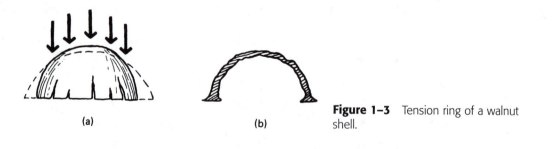

**Figure 1–3** Tension ring of a walnut shell.

(a)                                  (b)

---

[1] *Beam* is a generic term. The name applies to (in order of decreasing size and load capacity):girder, beam, joist, and purlin. A lintel is a special type of beam that spans an opening in a wall.

Success in the application of natural forms to architecture is more likely when the forms are regarded not as models for copying but as types of solution, demonstrations of principle, or suggestive metaphors. We are now quite accustomed to calling the structural frame a "skeleton," and to seeing the various other systems in buildings as also having purposes rather like those in animal anatomy. It was the principles on display in such places as Georges Cuvier's early nineteenth-century museum, not the literal shapes and assemblies of bones, which inspired the analogy. The increased clarity of thinking which resulted contributed to the development of true, skeleton-frame, skyscraper construction.

A considerable portion of this book is devoted to the analysis of each of these structures, except the last two. A proper discussion of arches and shells rightly belongs in a more advanced treatment of the subject, after the basic concepts are understood and some background has been established. Even so, the beginning student can develop some insight into the behavior of these and the more complicated systems from the chapters that follow.

In Table 1–1 we provide some data on different types of building structures. Some of these are quite conventional, while others are used only under very special circumstances. The table has been restricted to systems or parts of systems that form spans, as opposed to supporting elements, such as columns, bearing walls, or vertical cables. (This separation is somewhat arbitrary, and, as seen in the table, many spanning systems act integrally with their supports.)

The tabulation is not to be considered an exhaustive classification of structural systems, and the sketches, especially, are merely representative of the class of structure listed. The figures for span range and span ratios vary widely in many cases, and the values given can only be considered as guidelines. No unusual loads or support conditions have been considered in this table.

The floor systems in the flat-deck category of Table 1–1 occur more frequently than the structures of other groups by virtue of required spans and ease of construction. Each one is compatible with one or more support systems, and these relationships are shown in Figure 1–4.

# 1–8   COST

Probably the question that is asked most frequently of structural consultants is one involving the relative cost of alternative structural systems. As might be suspected, it is also one of the most difficult to answer.

First, the costs of materials, fabrication, and erection are constantly changing and vary considerably with geographic location. The availability of materials and needed construction trades varies widely and transportation costs can be very high. Sometimes the ambient environmental conditions can override other considerations, e.g., exposed steel framing would generally be inappropriate for a seacoast construction site. As mentioned previously, compatibility with other building systems must also be considered (e.g., it might be costly to select a structural system that causes difficulty in the installation of mechanical ducts).

Second, and much harder to assess, is the ultimate or life-cycle cost of one system compared to another when one considers the effects that each has upon the other building systems. For example, the lowest-cost structural system might result in the greatest amount of unused building volume, which would have to be heated and cooled needlessly. The same choice could result in the highest insurance premiums. On the other hand, a low first cost might be better because the cost of money (interest) to construct the building would be less. The general subject of engineering economy treats these issues, and they will not be examined here. For the time being, it will suffice to point out that a proper determination of true cost is not simple.

On the other hand, what is not often understood by beginning designers is how low the cost of a typical building structural system, per se, really is. If we consider superstructure alone (because the cost of a foundation depends so much upon the individual site conditions), we find that it constitutes about 15 to 20% of the total construction cost. This percentage becomes even less if we consider all the costs of a project, including land, interest, fees, and overhead. The 15 to 20% range must be compared to about 35% for the mechanical and electrical systems and about 20% for the nonbearing partitions and interior finishes. These figures all vary greatly with the building type. For example, a hospital would have a high mechanical cost, driving the structural percentage down. On the other hand, a sports stadium with little interior finish costs and very long spans would have a large portion of its cost in the structure. In any event, the cost of the structural system for the great majority of buildings should probably not be a major design determinant. The beginning designer, at least, can derive greater benefit by concentrating less on structural costs and more on the relationships among structure, form, function, and space. Cost may have to compromise these relationships but should not establish them.

## 1–9    BUILDING CODES

Throughout this text, reference is made to various codes and specifications, which provide data on design loads, allowable stresses in materials, properties and dimensions of standard cross sections, and so on. These documents also frequently spell out standard design procedures, construction tolerances, and factors of safety, and in some cases even provide the appropriate design equations. Most of these are developed for the use of design professionals by the materials industry associations, which have as one of their purposes the promotion of the proper and safe use of their material.

In the United States, the model building codes,[2] such as that provided by Building Officials and Code Administrators International, Inc., the International Conference of Building Officials, and the Southern Building Code Congress International, and the many municipal codes often include and/or make reference to specifications of the materials industries. In such cases these specifications, like the rest of the code, must be followed or the design professional has the responsibility of proving the equality or superiority of a different or new procedure. The intent of all building codes is the protection of the health, safety, and welfare of the public; and while some provisions may seem arbitrary or overly

---

[2] It is noted that the three codes addressed here are scheduled to merge into a single "international" code by the year 2000. The author doubts that this will happen on schedule, but applauds the effort.

## IARACTERISTICS OF SELECTED SPANNING SYSTEMS

| *...mary means of resisting loads* | *Spanning system* | *Usual materials and types* | | *Usual span range ft (m)* | *Typical span/ depth ratio* |
|---|---|---|---|---|---|
| Tension | Cable | Steel with joist or concrete panel deck | | 100–500 (30–150) | DNA |
| Compression | Arch | Timber, glued-laminated | | 60–130 (20–40) | DNA |
| | | Timber truss | | 100–220 (30–70) | DNA |
| | | Steel truss | | 130–330 (40–100) | DNA |
| | | Reinforced concrete, convoluted or ribbed | | 80–220 (25–70) | DNA |
| Bending and shear | Flat deck, floor | Wood | Joist with plywood subfloor | 8–20 (2–6) | 20 |
| | | | Beam with plants | 12–30 (4–9) | 18 |
| | | Steel | Beam w/steel subfloor or concrete slab | 15–50 (5–15) | 22 |
| | | | Bar joist with steel subfloor | 12–60 (4–20) | 22 |
| | | Reinf. concrete | Flat plate w/ or w/o drop panels | 10–20 (3–6) | 30 |
| | | | Beam with flat slab | 15–35 (5–10) | 15 |
| | | | Pan joist | 15–35 (5–10) | 20 |
| | | | Waffle pan | 20–50 (6–15) | 22 |
| | | | Precast plank | 20–40 (6–12) | 38 |
| Tension and compression | Truss | Timber members | | 25–100 (7–30) | 5–12 |
| | | Steel members | | 60–200 (20–60) | 5–15 |

| Typical span/ thickness ratio | Advantages | Disadvantages | Comments |
|---|---|---|---|
| 300+ | Long span | High technology; must provide for wind stability | Roof construction only |
| 35 | Appearance (wood finish) | Large pieces to transport | Roof construction only; usually circular or parabolic in shape |
| 40 | Low technology | Not good for concentrated loads | |
| 40 | Long span | Not good for concentrated loads | |
| 30 | Low maintenance | Slow concentration | |
| DNA | Versatile plan and section shapes; low technology | High noise transmission | Popular for residential construction |
| DNA | Open walls; simple foundations; low technology | High noise transmission | Popular for residential construction |
| DNA | Can have composite action | Limited availability in some areas | |
| DNA | Wide range of available spans | No heavy loads | Deep long-span joists span much farther |
| DNA | Low airborne noise transmission | Short spans; openings limited | Popular for high-rise construction |
| DNA | Low airborne noise transmission; ease of construction | Openings limited | Square or rectangular bays |
| DNA | Low airborne noise transmission | Openings limited | Strong directionality |
| DNA | Low airborne noise transmission; appearance (concrete finish) | Poor integration w/ mechanical system | Can cantilever both directions at a corner |
| DNA | Low airborne noise transmission; rapid construction; high span/depth ratio | Requires repetitive bay sizes | Usually prestressed |
| DNA | Long span | Low span/depth ratio | Popular for residential roof construction |
| DNA | Long span | Low span/depth ratio | Popular for industrial roof construction |

**TABLE 1–1** *(continued)*

| Primary means of resisting loads | Spanning system | Usual materials and types | Usual span range ft (m) | Typical span/ depth ratio |
|---|---|---|---|---|
| Membrane action (tension and compression) | Dome | Reinforced concrete thin shell | 50–150 (15–45) | DNA |
| | | Reinforced concrete (convoluted or ribbed) | 100–300 (30–100) | DNA |
| | | Steel truss | 130–500 (40–150) | DNA |
| Membrane action (compression) | Vault | Reinforced concrete thin shell | 60–160 (20–50) | DNA |
| Bending and shear | Barrel vault and folded plate | Reinforced concrete thin shell | 60–120 (20–35) | 12 |
| Tension and compression | Space frame | Steel members | 60–300 (20–90) | DNA |
| | Cable dome | Steel and fabric | 300–800 (90–250) | DNA |
| Membrane action (tension and compression) | Warped surface | Reinforced concrete thin shell | 60–200 (20–60) | DNA |
| Membrane action (tension) | Fabric | Fiberglass or PVC | 60–300 (20–90) | DNA |
| | Cable net | Steel | 200–400 (60–120) | DNA |
| | Air | Fabric-supported | 150–800 (45–250) | DNA |
| | | Fabric-inflated | 60–180 (20–55) | DNA |

| Typical span/ thickness ratio | Advantages | Disadvantages | Comments |
|---|---|---|---|
| 200 | Low stresses | Slow construction; poor acoustics; no point loads | Strong directionality |
| 40 | Can accept moderate point loads | Slow construction | Popular form for sports stadia |
| 60 | Long span | Requires skin for stability | Popular form for sports stadia |
| 175 | Many shapes possible | Slow construction; openings limited; poor acoustics; no point loads | Dominant forms; usually circular or parabolic in shape; roof construction only |
| 200 | Many shapes possible | Slow construction; openings limited | Dominant forms; roof construction only; strong directionality |
| DNA | High span/depth ratio; long span | High technology | Roof construction only |
| DNA | Lightweight | High technology | Roof construction only |
| 200 | Many shapes possible; low stress | Slow construction; openings limited | Dominant forms; roof construction only |
| 600+ | Long span; rapid construction; many shapes possible | High technology; needs flexible skin system | Dominant forms; roof construction only |
| 1000+ | Long span; rapid construction; low air pressure | High noise transmission; needs air locks; fabric deterioration | Roof construction only |
| 30 | No air locks needed; rapid construction | High noise transmission; high pressure; fabric deterioration | Roof construction only |

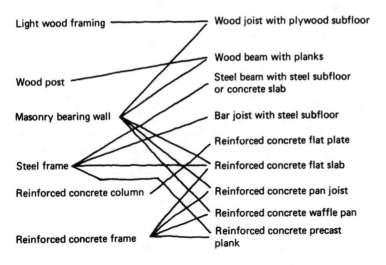

Light wood framing — Wood joist with plywood subfloor

Wood beam with planks

Wood post — Steel beam with steel subfloor or concrete slab

Masonry bearing wall — Bar joist with steel subfloor

Reinforced concrete flat plate

Steel frame — Reinforced concrete flat slab

Reinforced concrete column — Reinforced concrete pan joist

Reinforced concrete waffle pan

Reinforced concrete frame — Reinforced concrete precast plank

**Figure 1–4** Construction compatibility between support and deck systems.

restrictive, they cannot be taken lightly. Indeed, the provisions of a code will generally represent minimum acceptable standards and not design ideals. As mentioned previously, there are cases where the design professional has the cause and responsibility to be more stringent than the code. A clearly written and rational building code can be of real assistance to the designer, and it is well to remember that far fewer failures occur in jurisdictions where good codes are present and enforced.

## 1–10 ACCURACY OF COMPUTATIONS

One of the largest inconsistencies of structural analysis and design procedures is in the determination of a proper accuracy level for computations. The author believes that sometimes it is unfortunate that we possess the machine capability to rapidly produce answers to many decimal places. In structural engineering such precision is seldom, if ever, necessary. In most cases, any values written with more than three significant digits are misleading as to the actual accuracy level and can generate an undeserved sense of confidence.

For most purposes, the levels of precision given in Table 1–2 will probably suffice and may become "too accurate" if not cut off at three significant digits. In this book the writer has attempted to round off quantities to the appropriate level except where clarity would be reduced by so doing.

Structural elements acting mainly in simple compression, like vertical columns and round arches, have a more widespread history of use—and consequently of popular understanding and acceptance—than other elements acting complexly or in pure tension. When the great engineer Gustave Eiffel designed the controversial three-hundred-meter-high Tower for the 1889 Paris World's Fair, he traditionalized its complex, innovative, leaning legs with huge, reassuring "arches in simple compression," which apparently actually help with lateral wind pressure. The American engineer George Washington Ferris responded by inventing his Wheel for the 1893 World's Columbian Exposition in Chicago. In rivalry, and implicit criticism, of the Eiffel Tower, the Ferris Wheel displayed unmistakably its own structural innovation, the hanging of the Wheel from a central hub by cables purely in tension. (Most "Ferris Wheels" since the original have returned to the greater visual reassurance seemingly offered by spokes in compression.)

**TABLE 1–2 ACCEPTABLE COMPUTATION ACCURACY LEVELS**

| Quantity | Sufficient precision |
|---|---|
| Member lengths | Nearest quarter foot, tenth of a meter |
| Cross-sectional dimension | Nearest tenth of an inch, millimeter |
| Force | Nearest tenth of a unit (lb or kips),[a] tenth of a kN |
| Moment | Nearest tenth of a unit (lb-ft or kip-ft),[a] tenth of a kN · m |
| Stress | Nearest unit (psi or ksi, kPa or MPa)[a] |
| Angles | Nearest degree |
| Deflections | Nearest tenth of an inch, millimeter |
| Temperature | Nearest degree |

[a]Whether one works in pounds or kips depends on the magnitude of the loads and forces involved. One kip = 1000 lb. In the metric system, loads and forces are usually expressed in kN; stresses are in kPa or MPa. One MPa = 1000 kPa.

# 2

---

# Statics

---

## 2–1 INTRODUCTION

Statics is one part of a more general subject called mechanics. *Mechanics* involves the study of forces and the effects of those forces on the bodies on which they act. When the forces acting on a body are balanced such that no acceleration is taking place, a state of equilibrium exists. The subject of *statics* is limited to forces acting on bodies in equilibrium.

The branch of mechanics that treats unbalanced systems of forces involving acceleration is called *dynamics*. A third area, which deals with the physical deformations and internal effects in bodies caused by forces, is called *mechanics of materials* or, somewhat incorrectly, *strength of materials*. Mechanics of materials provides the theory behind most of the procedures used in structural analysis and design and, as such, is covered extensively in succeeding chapters of this book.

The subject matter treated in statics rarely poses any difficulty for the majority of students. Much of it is repetitious, merely repeating what has already been covered in a physics course. In some topical sequences, little emphasis is placed on statics, and a thorough understanding of loads and their reactive forces is never really achieved. The reader is cautioned that statics forms the basis for all structural analysis, and all professionals in the field consider it the most important part of any study of quantitative structures.

## 2–2 FORCES

For our purposes, a *force* is a push or pull provided by one object upon another. It can act at a point, such as a concentrated column load on a spread footing, or be distributed, such as the uniform dead load of a slab carried by a beam.

Force is a *vector* quantity, meaning that it has both magnitude and direction. If we let the magnitude of a force be represented graphically by the length of an arrow, the *direction* can be established from its line of action, as shown in Figure 2–1. The line of action extends finitely in front of and behind the force. Each force also has a *sense*, which becomes the sign (+ or −) in algebraic computations. Sense is represented by the arrowhead in Figure 2–1. It is important not to confuse sense and

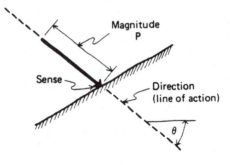

**Figure 2–1** Description of a force.

direction, and it may be helpful to remember that for each direction there are two possible senses. For example, if a certain gravity load acts vertically downward, its direction is vertical and its sense is down.

In statics, we consider only external forces exerted by one body upon another. Internal forces are those exerted by one part of a body upon another part of the same body. The body must be "cut open," in a figurative manner, and the forces made *external* before such forces can be examined or quantified.

## 2–3  COMPONENTS AND RESULTANTS

When two or more forces act at one point on a body (concurrent forces), it may be convenient to replace such forces by a single force which will have the same external effect on the body. This replacement force is called a *resultant*. Figure 2–2 shows a system of two forces being replaced by a resultant. Graphically, it is represented as the diagonal of a parallelogram, which includes the two given forces as adjacent sides. More than two concurrent forces could be treated by successive parallelograms.

It should be apparent that, by following the reverse procedure, one could resolve a given force $R$ into two components along any two lines of action. Figure 2–3 illustrates three such sets of components. The force $R$ could be replaced by any one of these pairs without any change in the action of the particle at $O$. The most useful pair is that shown in Figure 2–3(b), which are called the *rectangular components*.

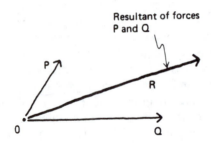

**Figure 2–2** Resultant of two forces.

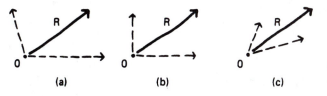

**Figure 2-3**   Different sets of components for the same force R.

(a)  (b)  (c)

The force $F$ in Figure 2–4 may be replaced by two forces, which act in the horizontal ($x$) and vertical ($y$) directions, by noting that

$$F_x = F \cos \theta \tag{2-1}$$

and

$$F_y = F \sin \theta \tag{2-2}$$

where $\theta$ is the angle made by the parent force with the $x$ axis. Conversely, if given the two forces $F_x$ and $F_y$, one could find the magnitude of their resultant by using the *Pythagorean theorem*:

$$F = \sqrt{(F_x)^2 + (F_y)^2} \tag{2-3}$$

The direction of this resultant could be found from the fact that

$$\tan \theta = \frac{F_y}{F_x} \tag{2-4}$$

*go through same point* (handwritten)

The resultant of a system of several concurrent forces can be determined by first resolving each force into its rectangular components as shown in Figure 2–5. These components can then be treated algebraically as indicated by Equations 2–3a and 2–4a.

$$F = \sqrt{(\Sigma F_x)^2 + (\Sigma F_y)^2} \tag{2-3a}$$

$$\tan \theta = \frac{\Sigma F_y}{\Sigma F_x} \tag{2-4a}$$

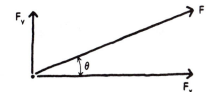

**Figure 2-4**   Rectangular force components.

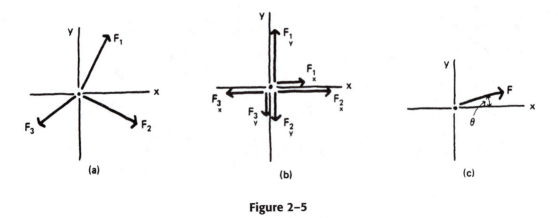

**Figure 2–5**

---

**EXAMPLE
2-1**

Determine the resultant of the three forces shown in Figure 2–6(a).

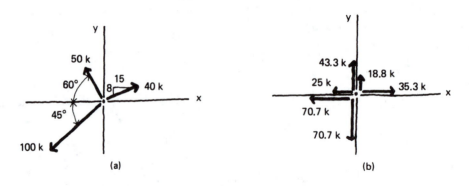

**Figure 2–6**

### Solution
Resolve each force into its rectangular components by using Equations 2–1 and 2–2.
For the 50-kip force,

$$F_x = 50(0.500) = 25.0 \text{ kips}$$
$$F_y = 50(0.866) = 43.3 \text{ kips}$$

For the 40-kip force,

$$F_x = 40\left(\frac{15}{17}\right) = 35.3 \text{ kips}$$

$$F_y = 40\left(\frac{8}{17}\right) = 18.8 \text{ kips}$$

For the 100-kip force:

$$F_x = 100(0.707) = 70.7 \text{ kips}$$

$$F_y = 100(0.707) = 70.7 \text{ kips}$$

These components can then be algebraically summed in their respective directions to get the net components in Figure 2–7(a). Substituting into Equations 2–3a and 2–4a, we obtain

$$F = \sqrt{(\Sigma F_x)^2 + (\Sigma F_y)^2}$$

$$= \sqrt{60.4^2 + 8.6^2}$$

$$= 61.0 \text{ kips}$$

$$\tan \theta = \frac{\Sigma F_y}{\Sigma F_x}$$

$$= \frac{8.6}{60.4}$$

$$\theta = 8°$$

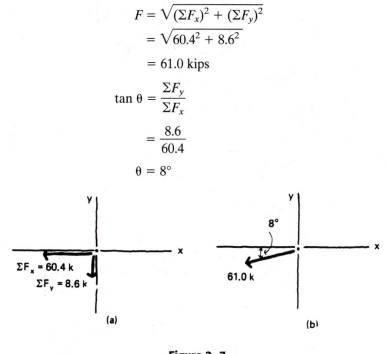

(a)

(b)

**Figure 2–7**

---

**EXAMPLE 2–2M**

Determine the resultant of the three forces shown in Figure 2–8M(a).

*Solution*

Resolve each force into its rectangular components by using Equations 2–1 and 2–2. For the 2-kN force,

$$F_x = 20(0.866) = 17.3 \text{ kN}$$

$$F_y = 20(0.500) = 10.0 \text{ kN}$$

For the 30-kN force,

$$F_x = 30\left(\frac{1}{2.24}\right) = 13.4 \text{ kN}$$

$$F_y = 30\left(\frac{2}{2.24}\right) = 26.8 \text{ kN}$$

For the 100-kN force,

$$F_x = 100\left(\frac{4}{5}\right) = 80 \text{ kN}$$

$$F_y = 100\left(\frac{3}{5}\right) = 60 \text{ kN}$$

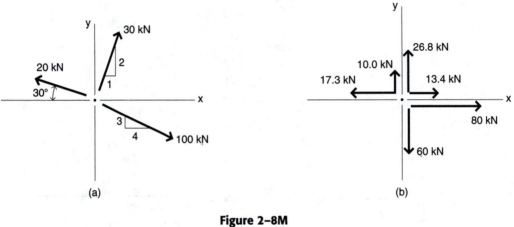

(a)                                    (b)

**Figure 2–8M**

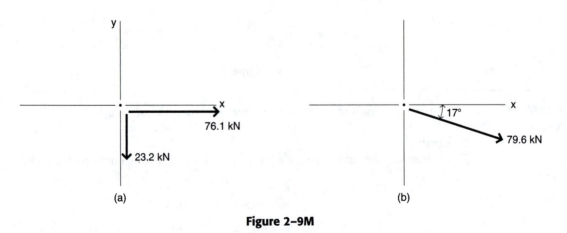

(a)                                    (b)

**Figure 2–9M**

These components can be algebraically summed in their respective directions to get the net components in Figure 2–9M(a). Substituting into Equations 2–3a and 2–4a we obtain

$$F = \sqrt{(\Sigma F_x)^2 + (\Sigma F_y)^2}$$
$$= \sqrt{76.1^2 + 23.2^2}$$
$$= 79.6 \text{ kN}$$

$$\tan \theta = \frac{\Sigma F_y}{\Sigma F_x}$$

$$= \frac{23.2}{76.1}$$

$$\theta = 17°$$

## PROBLEMS

**2–1.** Determine the magnitude, sense, and direction of the resultant of the two forces in Figure 2–10.

**2–1M.** Determine the magnitude, sense, and direction of the resultant of the two forces in Figure 2–10M.

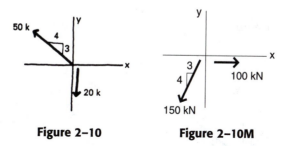

**Figure 2–10**          **Figure 2–10M**

**2–2.** Determine the magnitude, sense, and direction of the resultant of the concurrent force system in Figure 2–11.

**2–2M.** Determine the magnitude, sense, and direction of the resultant of the concurrent force system in Figure 2–11M.

**2–3.** Determine the magnitude, sense, and direction of the resultant of the three concurrent forces in Figure 2–12.

**2–3M.** Determine the magnitude, sense, and direction of the resultant of the three concurrent forces in Figure 2–12M.

**2–4.** Determine the magnitude, sense, and direction of the resultant of the concurrent force system in Figure 2–13.

**2–4M.** Determine the magnitude, sense, and direction of the resultant of the concurrent force system in Figure 2–13M.

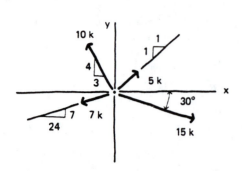

**Figure 2–11**

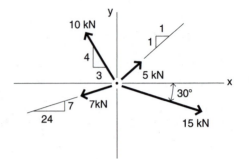

**Figure 2–11M**

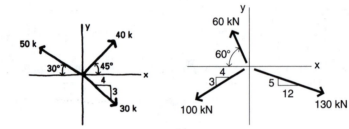

**Figure 2–12**

**Figure 2–12M**

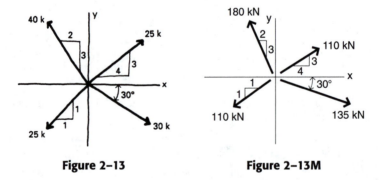

**Figure 2–13**

**Figure 2–13M**

**2–5.** Determine the magnitude, sense, and direction of the resultant of the three forces acting at the top of the pole in Figure 2–14.

**2–5M.** Determine the magnitude, sense, and direction of the resultant of the three forces acting at the top of the pole in Figure 2–14M.

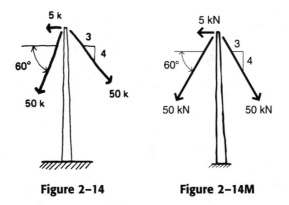

**Figure 2–14**                    **Figure 2–14M**

By using the fact that the opposite sides of a parallelogram are equal, we see that a force triangle (Figure 2–15) may be formed instead of the parallelogram of Figure 2–2. The components $P$ and $Q$ form a head-to-tail arrangement by transposing the force $P$. The resultant $R$ that closes the triangle does not follow this head-to-tail order. This same procedure may be used to determine graphically the resultant of more than two concurrent forces. The forces shown in Figure 2–16(a) may be taken in any order to form a head-to-tail pattern. The force needed to close the polygon is the resultant of the system. Reversing the head-to-tail sequence will give the resultant the proper sense. Figure 2–17 shows how the rectangular components of the forces $P$, $Q$, and $T$ will add to have the same net effect as the components of the resultant force $R$. Figure 2–17(d) shows two polygons (with coincident or overlapping lines), which could be formed by the rectangular components of the forces.

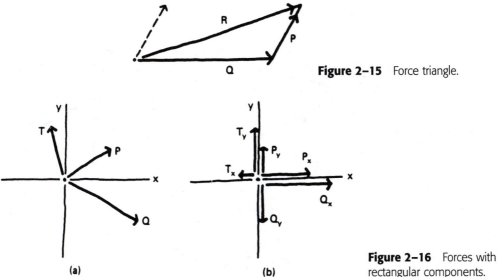

**Figure 2–15**  Force triangle.

**(a)**                              **(b)**

**Figure 2–16**  Forces with rectangular components.

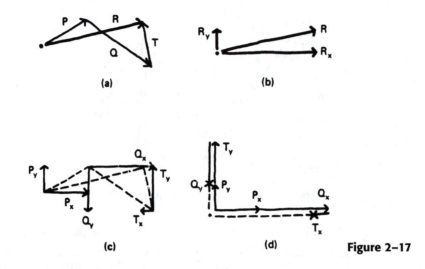

**Figure 2–17**

The dashed resultants are, of course, the same as the components $R_x$ and $R_y$ shown in Figure 2–17(b).

The force that is equal in magnitude, but opposite in sense, to the resultant is called the *equilibrant*. It is, as its name suggests, the one force that is needed to put a given system in equilibrium. It negates or balances the effects of the other forces. Figure 2–18(b) shows the equilibrant $E$ of the previous three-force system. Graphically, the equilibrant will close a polygon of forces by continuing the head-to-tail relationship. In fact, each of the forces shown in Figure 2–19 is the equilibrant of the other three.

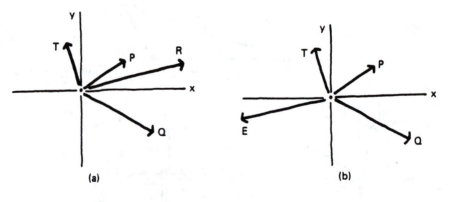

**Figure 2–18** Resultant and equilibrant.

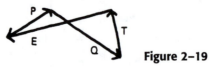

**Figure 2–19**

**EXAMPLE
2–3**

Graphically determine the resultant of the force system in Example 2–1.

**Solution**

The accuracy of any graphical solution depends upon the care and skill provided by the analyst and the scale of the drawings. Large figures will always yield a smaller percentage of error. Even so, one can seldom achieve the accuracy inherent in an algebraic solution (if such precision is necessary). In this case, the force scale is given in Figure 2–20. The answers shown in Figure 2–20 were obtained by measuring directly from the drawing with a scale and protractor.

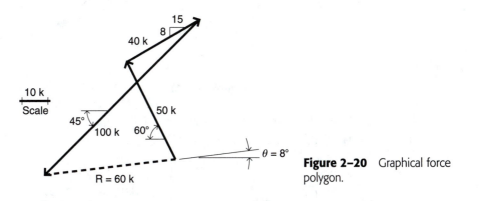

**Figure 2–20** Graphical force polygon.

**EXAMPLE
2–4M**

Graphically determine the resultant of the force system in Example 2–2M.

**Solution**

See remarks in Example 2–3. In this case, the force scale is given in Figure 2–21M. The answers shown in Figure 2–21M were obtained by measuring directly from the drawing with a scale and protractor.

## PROBLEMS

**2–6.** Work Problem 2–2 graphically.

**2–6M.** Work Problem 2–2M graphically.

**2–7.** Work Problem 2–3 graphically.

**2–7M.** Work Problem 2–3M graphically.

**2–8.** Graphically determine the equilibrant of the force system in Figure 2–13.

**2–8M.** Graphically determine the equilibrant of the force system in Figure 2–13M.

**2–9.** Graphically determine the equilibrant of the force system in Figure 2–22.

**2–9M.** Graphically determine the equilibrant of the force system in Figure 2–22M.

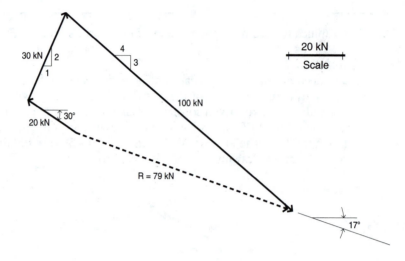

**Figure 2–21M**   Graphical force polygon.

**2–10.** Work Problem 2–5 graphically.

**2–10M.** Work Problem 2–5M graphically.

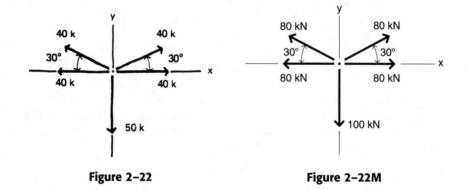

**Figure 2–22**                                              **Figure 2–22M**

# 2–4   EQUILIBRIUM OF CONCURRENT FORCES

As stated in Section 2–1, equilibrium is a state of balance created by opposing forces which act on a body in such a manner that their combined net effect is zero. *Newton's first law* states, in effect, that when a body is at rest (or is moving with a constant velocity in a straight line), the resultant of the force system acting on the body is equal to zero.

If we consider, for the time being, only those force systems which are concurrent, such that there is no tendency for rotation to occur, then equilibrium can be established by setting

$$\Sigma F = 0 \qquad\qquad (2\text{–}5)$$

To ensure force equilibrium in all directions, it is usually easier to work with the two rectangular component equations:

$$\Sigma F_x = 0 \qquad\qquad (2\text{--}5a)$$

$$\Sigma F_y = 0 \qquad\qquad (2\text{--}5b)$$

(These two equations will suffice for coplanar systems; however, a third one in the z direction would be necessary for three-dimensional situations.)  *same point*

The simplest structural systems are those that are concurrent and coplanar, and the two equations above will enable us to analyze such systems for two unknowns. Consider the load $W$ suspended by the two ropes in Figure 2–23(a). The system is a concurrent one of only three forces, $A$ and $B$ and the load $W$, as shown in Figure 2–23(b).

Assuming that the load $W$ is known and the angles of the cables $A$ and $B$ are known, the forces $A$ and $B$ can be readily determined by using the $x$ and $y$ components. With reference to Figure 2–23(c) and using the standard convention that forces upward and to the right are positive in sense, we get

$$\Sigma F_x = 0 \qquad\qquad \Sigma F_y = 0$$

$$B_x - A_x = 0 \qquad\qquad B_y + A_y - W = 0$$

If the components are then expressed in terms of their parent forces $A$ and $B$, the two equations can be solved simultaneously.

Problems of this type are not limited to tensile members, and the cables at $A$ and $B$ could just as well have been rigid bars capable of taking tension or compression. A vector indicating compression will have its arrowhead pointing toward the joint. Where the sense of the force carried by such a member is uncertain, it must be assumed and then verified as the answer is obtained. **A negative sign accompanying the answer will mean an incorrect sense assumption.**

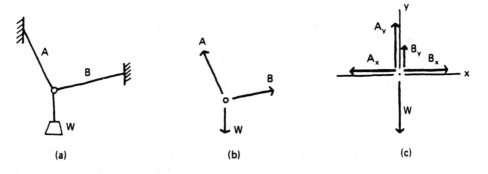

(a)                          (b)                          (c)

**Figure 2–23**

**EXAMPLE**     Determine the magnitude and sense of the forces in the members *A* and *B* in Fig-
**2–5**         ure 2–24.

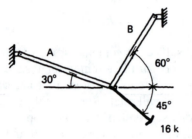

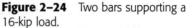

**Figure 2–24**   Two bars supporting a
16 k   16-kip load.

### Solution

If we assume that both bars are in tension, we get the concurrent forces shown in
Figure 2–25(a). Their components appear as in Figure 2–25(b). Writing the two *x*
and *y* equations from the components, we get

$$\Sigma F_x = 0 \qquad\qquad\qquad\qquad \Sigma F_y = 0$$

$$B_x + 0.707(16) - A_x = 0 \qquad\qquad B_y + A_y - 0.707(16) = 0$$

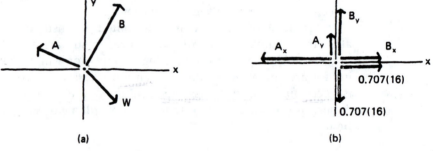

(a)                                              (b)

**Figure 2–25**

In terms of forces *A* and *B*,

$$0.500B - 0.866A + 11.3 = 0$$

$$0.866B + 0.500A - 11.3 = 0$$

Since we have two equations and two unknowns, we can use a simultaneous
solution. This can be done (by obtaining one unknown in terms of the other or) by
addition/subtraction as follows. Multiplying the first equation by 0.866/0.500 will
give us *like* coefficients for *B*. We can then subtract the second equation from it.

$$0.866B - 1.5A + 19.6 = 0$$
$$-(0.866B + 0.5A - 11.3) = 0$$
$$\overline{\phantom{-(0.866B + 0.5A}-2.0A\ \ + \ \ 30.9 = 0}$$
$$A = +15.5$$

Substituting this into any of the equations will give

$$B = +4.2$$

The plus signs indicate that our sense assumptions were correct.

$$A = 15.5 \text{ kips tension}$$

$$B = 4.2 \text{ kips tension}$$

---

**EXAMPLE
2–6M**

Determine the magnitude and sense of the forces in the members $A$ and $B$ in Figure 2–26M.

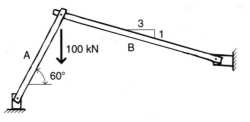

**Figure 2–26M**   Two bars supporting a 100-kN load.

### Solution

If we assume that both bars are in compression, we get the concurrent forces shown in Figure 2–27M(a). Writing the two $x$ and $y$ equations from the components, we get

$$\Sigma F_x = 0 \qquad\qquad\qquad \Sigma F_y = 0$$
$$A_x - B_x = 0 \qquad A_y + B_y - 100 = 0$$

Noting that the hypotense of the 1-on-3 slope is 3.16, we can express the components in terms of sine and cosine values of $A$ and $B$:

$$0.500A - 0.949B = 0$$

$$0.866A + 0.316B - 100 = 0$$

One way to solve the two simultaneous equations is to find $A$ in terms of $B$. From the first equation, we get

$$A = \frac{0.949}{0.500}B = 1.90B$$

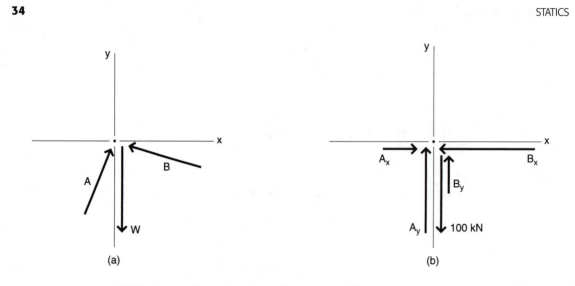

(a)

(b)

**Figure 2–27M**

Substituting into the second equation,

$$0.866(1.90B) + 0.316B - 100 = 0$$
$$B = +51.0$$

and then

$$A = (1.90)(51.0) = +96.9$$

The plus signs indicate that our sense assumptions were correct.

$$A = 96.9 \text{ kN compression}$$
$$B = 51.0 \text{ kN compression}$$

**EXAMPLE 2–7**    Determine the magnitude and sense of the forces in bars $A$ and $B$ in Figure 2–28.

### Solution
For purposes of illustration, we shall assume that bar $A$ is in tension and bar $B$ is in compression. Tension force arrows will act away from the point of concurrency and compression arrows will act in toward it (Figure 2–29). (After an inspection of the horizontal force components, the reader should be able to ascertain that these sense assumptions cannot both be correct.)

$$\Sigma F_x = 0 \qquad\qquad \Sigma F_y = 0$$
$$-B_x - A_x = 0 \qquad\qquad B_y + A_y - 100 = 0$$
$$-0.800B - 0.707A = 0 \qquad 0.600B + 0.707A - 100 = 0$$

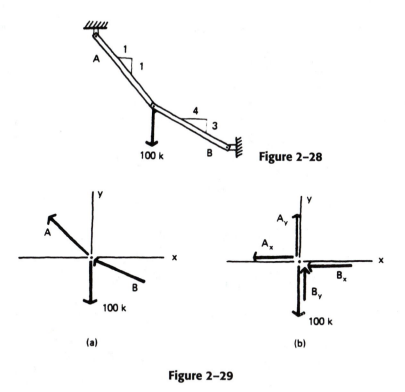

**Figure 2–28**

(a)                                        (b)

**Figure 2–29**

Solving for $A$ and $B$, we get $A = +566$ kips and $B = -500$ kips. The minus sign indicates that bar $B$ is *actually in tension,* not compression as we had assumed.

$$A = 566 \text{ kips tension}$$

$$B = 500 \text{ kips tension}$$

These types of problems can also be solved graphically. In each case, the three forces, load plus two unknowns, are in equilibrium. This means that each one is the equilibrant of the other two, and we must be able to form a head-to-tail triangle. The only magnitude given is that of the applied load, but the directions of all three lines of action are known. Given the three forces, for example, of Figure 2–23(b) and drawing the load vector to scale, only two possible triangles can be formed. These are shown in Figure 2–30. If we place the sense arrows on the triangles in head-to-tail fashion as in Figure 2–31, we see that either triangle will give us the correct results. In each case, the sense arrows point away from the original point of concurrency, indicating tensile forces in both members. The magnitudes are simply measured with the force scale used to draw $W$. As mentioned previously, the accuracy of graphical techniques depends on one's drafting skills and the size of the drawing. Small polygons will usually not give satisfactory results.

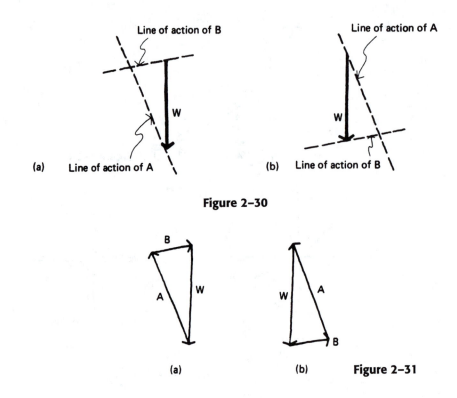

**Figure 2–30**

(a)                                      (b)                    **Figure 2–31**

---

**EXAMPLE
2–8**

Use graphical techniques to check the results of Example 2–7.

**Solution**

The answers in Figure 2–32(a) were obtained by measuring the polygon and are less than 5% off the values obtained in the algebraic solution. Placing the arrows so they act on the original joint, as in Figure 2–32(b), we see that both arrows represent tension.

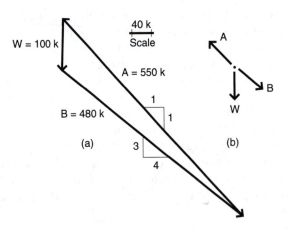

**Figure 2–32**   Force polygon for Example 2–7

**EXAMPLE
2–9**

Use graphical techniques to check the results of Example 2–6M.

**Solution**

The answers in Figure 2–33M(a) were obtained by measuring the polygon and are very close to those obtained algebraically. Placing the arrows so they act on the original joint, as in Figure 2–33M(b), we see that both arrows represent compression.

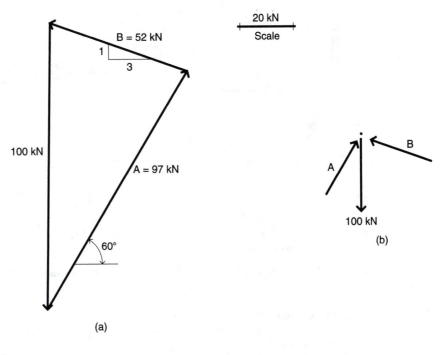

(a)

(b)

**Figure 2–33M**  Force polygon for Example 2–6M.

## PROBLEMS

**2–11.** Determine the magnitude and sense of the forces in cables $A$ and $B$ of Figure 2–34.

**2–11M.** Determine the magnitude and sense of the forces in cables $A$ and $B$ of Figure 2–34M.

**2–12.** Determine the magnitude and sense of the forces in bars $A$ and $B$ of Figure 2–35. Use the algebraic method and then check your results graphically.

**2–12M.** Determine the magnitude and sense of the forces in bars $A$ and $B$ of Figure 2–35M. Use the algebraic method and then check your results graphically.

**2–13.** Determine the magnitude and sense of the forces in bars $A$ and $B$ in Figure 2–36.

**2–13M.** Determine the magnitude and sense of the forces in bars $A$ and $B$ in Figure 2–36M.

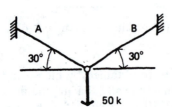

Figure 2–34

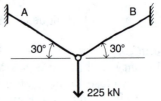

Figure 2–34M

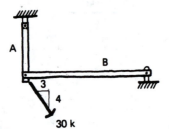

Figure 2–35

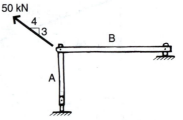

Figure 2–35M

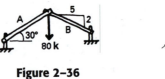

Figure 2–36

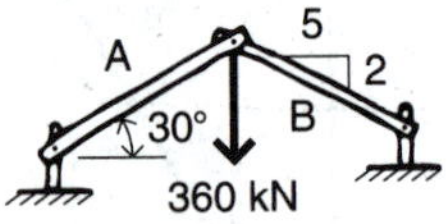

Figure 2–36M

**2–14.** Algebraically determine the forces in members *A* and *B* of the structure in Figure 2–37. (*Note:* The small triangle is another way to represent a *hinge* or *pin* connection.)

**2–14M.** Algebraically determine the forces in members *A* and *B* of the structure in Figure 2–37M. (*Note:* The small triangle is another way to represent a *hinge* or *pin* connection.)

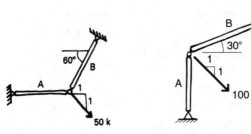

Figure 2–37

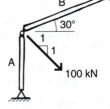

Figure 2–37M

**2–15.** Solve Problem 2–14 graphically.

**2–15M.** Solve Problem 2–14M graphically.

**2–16.** Algebraically determine the forces in the members of the frame in Figure 2–38. Check your results using a large graphical polygon.

**2–16M.** Algebraically determine the forces in the members of the frame in Figure 2–38M. Check your results using a large graphical polygon.

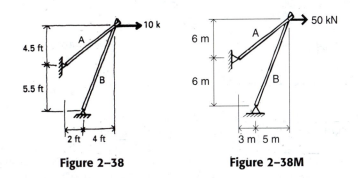

**Figure 2–38**            **Figure 2–38M**

**2–17.** Graphically determine the forces in the bars *A* and *B* of the structure in Figure 2–39.

**2–17M.** Graphically determine the forces in the bars *A* and *B* of the structure in Figure 2–39M.

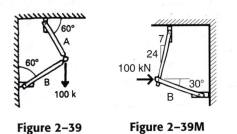

**Figure 2–39**            **Figure 2–39M**

**2–18.** Algebraically determine the forces in the members *A* and *B* for the frame in Figure 2–40. (If one of your sense assumptions was incorrect, can you rationalize the correct sense by comparing this structure to the one in Figure 2–34?)

**2–18M.** Algebraically determine the forces in the members *A* and *B* for the frame in Figure 2–40M. (If one of your sense assumptions was incorrect, can you rationalize the correct sense by comparing this structure to the one in Figure 2–34M?)

**2–19.** With reference to Figure 2–34, what will be the limiting value of the forces in cables *A* and *B* as the two angles labeled 30° approach zero?

**2–19M.** With reference to Figure 2–34M, what will be the limiting value of the forces in cables *A* and *B* as the two angles labeled 30° approach zero?

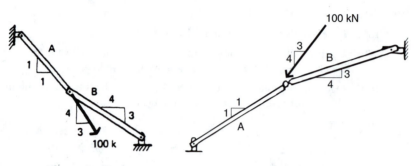

Figure 2–40                    Figure 2–40M

## 2–5   MOMENTS AND COUPLES

A *moment* is a tendency to rotate or twist. When a force acts on a certain object, that object has a tendency to move in the direction of the force. Such motion is called *translation*. Figure 2–41 shows the force *P* having a line of action that passes through the particle at *C*. Under the action of the force, the particle will tend to move along that line of action. However, the particles at *A* and *B* do not lie along the line of action. Under the influence of the force, they will not only tend to translate, but will also tend to rotate (i.e., the force *P* has moment with respect to those two points).[1] Indeed, a force tends to cause rotation, having moment about every point that does not lie along its line of action.

The *magnitude* of the moment of a force acting about somepoint is defined as the product of the force and the perpendicular distance from its line of action to the point. Such a perpendicular distance is often called a *moment arm* and is, in effect, a lever arm. The units of moment are force times distance, and in structural analysis this is usually expressed in pound-feet (lb-ft) or kip-feet (kip-ft). In the SI system, this is expressed in newton-meters (N·m) or kilonewton-meters [kN·m]. Moment also has sense; it either acts clockwise or counterclockwise.

Most moments from forces tend to bend structural objects; for example,

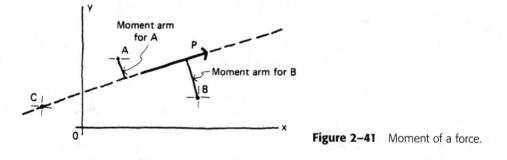

**Figure 2–41**   Moment of a force.

---

[1] The phrases "tend to move" or "tend to rotate" are used because in statics there are generally other forces present that act to balance out or prevent the actual motion.

when two children sit on a seesaw, they bend the board, and how much it is bent depends upon the location (distance) and the weight (force) of a child. The effects of bending moments and their importance are covered in Chapters 6 and 7.

To obtain the magnitude of the moment of a force, with respect to a certain point, it is sometimes easier to work with the components of the force rather than with the force itself. Figure 2–42(a) shows a force $P$ which has a moment arm of distance $d$ with respect to the origin $O$. The magnitude of this moment is $P$ times $d$, and it is clockwise in sense. This value may also be found by using the components shown in Figure 2–42(b). The moment of a force with respect to a point is equal to the algebraic sum of the moments of the components of the force taken with respect to that same point. This means that the moment of $P$ with respect to the point $O$ can also be found as the net effect of $P_y(x)$ counterclockwise and $P_x(y)$ clockwise. (Credit for this observation is given to Pierre Varignon (1654–1722), a French mathematician).

For algebraic purposes, we shall let counterclockwise moments be positive and clockwise moments be negative. (There is no consistency among writers in the mechanics field for this sign convention, and this selection is quite arbitrary. It is only important that one be consistent throughout a given analysis.)

To illustrate *Varignon's theorem*, let us assign values to the quantities as shown in Figure 2–43. The components in Figure 2–43(b) are found as functions of the 20° angle, and the net moment about the origin $O$ of those components is

$$M_o = +P_y(2 \text{ ft}) - P_x(3 \text{ ft})$$
$$= +17.1 \text{ kips}(2 \text{ ft}) - 47.0 \text{ kips}(3 \text{ ft})$$
$$= -107 \text{ kip-ft}$$

or

$$M_o = 107 \text{ kip-ft} \circlearrowright$$

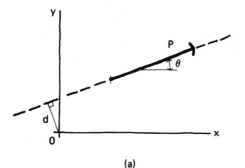 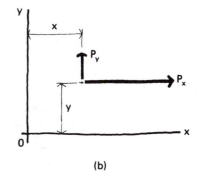

(a)          (b)

**Figure 2–42**  Varignon's theorem.

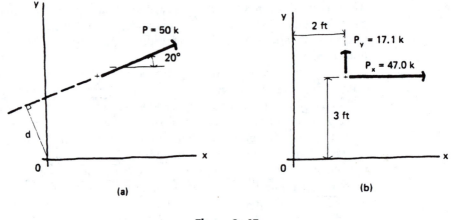

**Figure 2–43**

Alternatively, we can obtain the same value by determining the perpendicular moment arm, distance $d$. Figure 2–44 illustrates one possible way to quantify this distance by first using the triangle *KLM*. With $d$ established as 2.14 ft, the moment of $P$ about the origin is

$$M_o = -P(d)$$
$$= -50 \text{ kips } (2.14 \text{ ft})$$
$$= -107 \text{ kip-ft}$$

or

$$M_o = -107 \text{ kip-ft } \circlearrowright$$

*which agrees with previous answer*

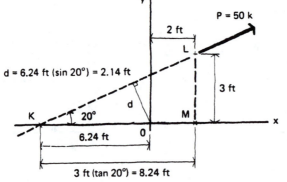

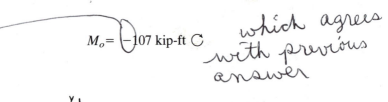

**Figure 2–44**  Determining the moment arm.

The reader should notice that the moment of the force $P$, being a l... perpendicular distance to the line of action, is independent of the loca... force along its line of action. For example, if we relocate $P$ so that it acts at i... $K$ in Figure 2–42, then, using Varignon's theorem, we get

$$M_o = P_x(0 \text{ ft}) - P_y(6.24 \text{ ft})$$
$$= -17.1 \text{ kips}(6.24 \text{ ft})$$
$$= -107 \text{ kip-ft}$$

or

$$M_o = 107 \text{ kip-ft} \circlearrowright$$

This technique of moving a force along its line of action until one of its components passes through the desired moment center can be a valuable shortcut for many moment calculations.

In statics, we frequently encounter a special kind of tendency to rotate, provided by a force system called a couple. A *couple* consists of a pair of equal and opposite parallel forces—two forces that are equal in magnitude, opposite in sense, and have parallel lines of action. It produces only rotation, no translation. It is pure moment. Its magnitude is given by the product of one of the forces and the perpendicular distance between the forces, and its sense is clockwise or counterclockwise. The most important characteristic of a couple is that its effect (magnitude and sense of the moment) is the same with respect to every point in its plane. Its tendency to cause rotation of a point is a constant and is *independent* of the distance between the couple and the point. The student should verify arithmetically that each couple in Figure 2–43M has the same 20 kN·m clockwise effect upon the point $O$. Given the dimensions to *any other point* in the plane of Figure 2–45M, we could show that the same 20 kN·m rotation acts *there* also.

From the standpoint of static equilibrium, the effect of a couple can only be negated or balanced by a second couple of identical magnitude but opposite sense.

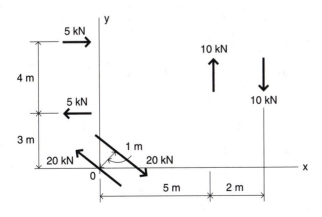

**Figure 2–45M**  Identical couples.

The equilibrant couple must lie in the same (or parallel) plane. The moment of a couple cannot be balanced by a single force because the value of moment for a force is dependent upon the location of the moment center. Similarly, the action of a single force cannot be negated by a couple, for a couple provides no translation.

An understanding of what a couple is enables us to look more closely at what happens inside a bending member (beam). In Chapter 1, mention was made of the inefficiency of bending as a means to carry load. Not only are the stresses distributed very unevenly within a beam, but the small structural depth of most beams necessitates large couple forces to maintain equilibrium. Figure 2–46 shows a beam with a single concentrated load carried between two bearing walls. The beam itself is assumed to be weightless. Let us cut through the beam near the load to examine the internal forces. Isolating the left-hand portion of the beam from the rest of the system, we will see the upward reaction $R$ provided by the wall, as in Figure 2–47(a). An internal downward force called $V$ in Figure 2–47(b) develops as a response to $R$. This force $V$ is provided by the other part of the beam and, for vertical equilibrium, would have to be exactly equal to $R$.

These two forces constitute a couple of magnitude $V$ times $x$ or ($R$ times $x$), which can only be balanced by an opposing couple. In other words, the beam portion might be in force (translational) equilibrium ($\Sigma F = 0$), but it is not in moment (rotational) equilibrium. The second or opposing couple can only be provided by the other beam part upon the cut face, as shown in Figure 2–47(c). Because the moment arm of this opposing couple is limited by the beam depth, the forces $Q$ will be quite large. For moment equilibrium ($\Sigma M = 0$), $Q$ times $y$ must equal $V$ times $x$. The two forces $Q$, of course, are equal according to the definition of a couple. This also assures force equilibrium in the horizontal direction.

The preceding explanation introduced the third equation of static equilibrium, which joins the previous two discussed in Section 2–4 dealing with concurrent forces. Since a beam involves forces that are not concurrent, all three equations were involved.

$$\Sigma F_x = 0 \qquad\qquad (2\text{–}5a)$$

$$\Sigma F_y = 0 \qquad\qquad (2\text{–}5b)$$

$$\Sigma M = 0 \qquad\qquad (2\text{–}6)$$

These are the only three equations of statics that are applicable to coplanar structures.

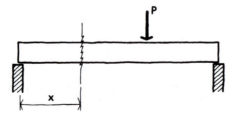

**Figure 2–46** Simple beam.

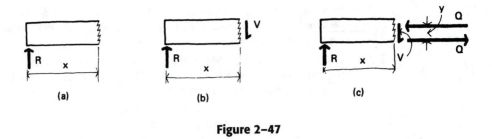

**Figure 2–47**

# PROBLEMS

**2–20.** Determine the magnitude and sense of the moment of the force in Figure 2–48 with respect to points $O, A$, and $B$. In other words, find $M_O$, $M_A$, and $M_B$. Denote the sense in each case by a curved arrow.

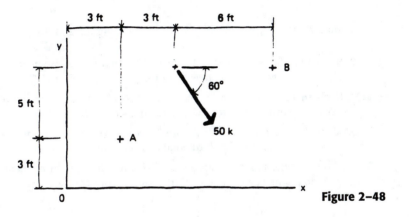

**Figure 2–48**

**2–20M.** Determine the magnitude and sense of the moment of the force in Figure 2–48M with respect to points $O, A$, and $B$. In other words, find $M_O$, $M_A$, and $M_B$. Denote the sense in each case by a curved arrow.

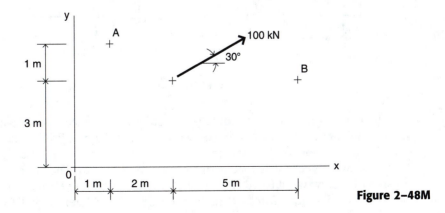

**Figure 2–48M**

**2–21.** Referring to Figure 2–49, determine the magnitude and sense of the moment of the 160-kip force with respect to each of the points $O, A, B$, and $C$. Denote the sense in each case by a curved arrow.

**2–21M.** Referring to Figure 2–49M, determine the magnitude and sense of the moment of the 150-kN force with respect to each of the points $O, A, B$, and $C$. Denote the sense in each case by a curved arrow.

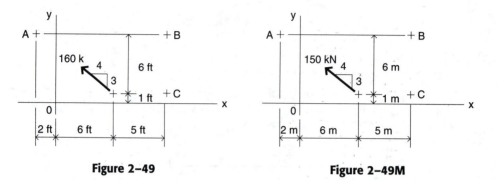

Figure 2–49                    Figure 2–49M

**2–22.** Show that the two forces in Figure 2–50(a) have a net moment of 120 kip-ft counterclockwise with respect to each of the points $O, A, B$, and $C$.

**2–22M.** Show that the two forces in Figure 2–50M(a) have a net moment of 20 kN counterclockwise with respect to each of the points $O, A, B$, and $C$.

**2–23.** Show that the four forces in Figure 2–50(b) have a net moment of zero with respect to points $O$ and $A$ and a third point of your choice.

**2–23M.** Show that the four forces in Figure 2–50M(b) have a net moment of zero with respect to points $O$ and $A$ and a third point of your choice.

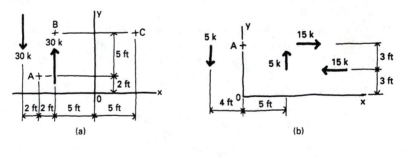

(a)                    (b)

Figure 2–50

Sometimes it is computationally convenient to replace a force with a force and a couple, which have the same effect upon the member. Equilibrium is maintained if we proportion the new force and couple correctly. For example, Figure 2–51(a) shows an eccentric load acting on the top of a pier. The effect on the pier consists of a downward force ($P$) and a clockwise moment ($P$ times $e$). In Figure 2–51(b), two collinear forces $P$ (shown dashed) have been added to the pier.

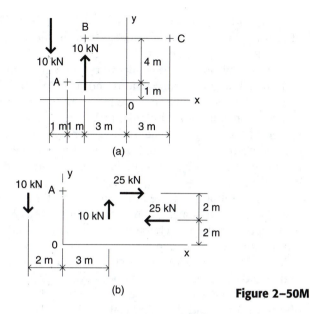

(a)

(b)

**Figure 2–50M**

These forces cancel each other and we have, in effect, added zero to the system. However, the two forces *P* which are circled constitute a clockwise couple of magnitude *P* times *e*. The replacement system is shown in Figure 2–51(c) with the couple represented as a moment arc of value *P(e)*. This new concentric force and couple have the same action upon the pier as did the original eccentric load. Almost all problems involving eccentric loads can be conceptually simplified using such equivalent systems.

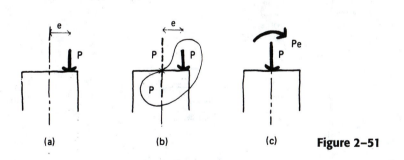

(a)  (b)  (c)  **Figure 2–51**

## 2–6  IDEAL SUPPORT CONDITIONS

The actual support conditions and member connections in real building structures are quite complicated. The proper design of connections, if all the various forces were considered, would be a lengthy and exacting process. An accurate determination of the amount of friction and slip, for example, which occurs under load at a given beam-to-column joint would be almost impossible. Nevertheless, the behavior of the supports can have a critical effect on the structure proper, and the designer cannot ignore the end conditions of any member.

In actual practice, it is necessary to make certain idealized simplifications regarding the nature of connections and supports. The designer must be aware of the fact that these simplifications are false and may or may not approximate the actual construction conditions. Judgment must be used to increase the factors of safety involved in the design whenever it is suspected that an assumed condition departs markedly from the actual one. It is too easy to forget that, while it is simple to draw frictionless rollers or fully rigid, inflexible connections, they are literally impossible to fabricate.

The symbols presented in Figures 2–52 through 2–55 for various ideal support conditions are standard and universally accepted. What is not universally accepted is the determination of characteristics needed, by actual field connections and conditions, for the symbols to be valid representations. Despite this, some attempt has been made in the following discussion to indicate a few types of real connections that would correspond to the various symbols.

The *hanger* can take no compression and is assumed to provide a single force of known sense and direction (Figure 2–52). The *rod* or *angles* are assumed to be

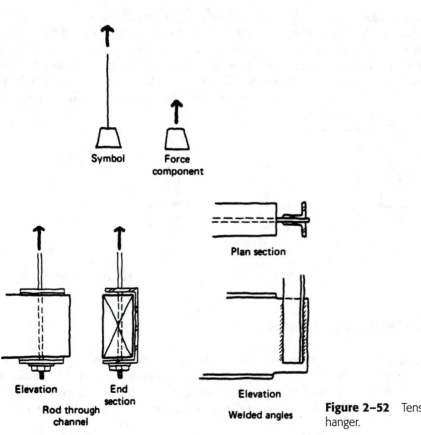

**Figure 2–52**  Tension support or hanger.

long and slender, having no resistance to compression buckling and negligible bending resistance. If the tension member is placed at an angle, its force can be represented as two rectangular components, but these components are *not independent*. They are related to each other by the direction of their resultant, which acts along the member.

The *pin* can take one force of unknown sense and direction (Figure 2–53). In theory, it offers no resistance to rotation (moment). The single force is usually represented by two rectangular components, since this will cover any possible line of action. The two components are independent as to both magnitude and sense and will constitute two unknowns.

The *roller* provides a single force of known direction (Figure 2–54). Its sense is unknown (i.e., it is assumed that the member cannot "lift off" a roller). Like the tension hanger, if the roller or link is on a slope, its force is usually represented as two rectangular components. These two components are *dependent* both as a magnitude and sense and constitute but one unknown. Once either component is deter-

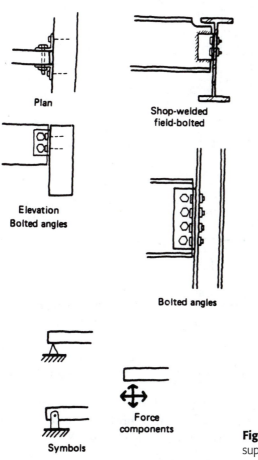

Plan

Shop-welded
field-bolted

Elevation
Bolted angles

Bolted angles

Force
components

Symbols

**Figure 2–53** Pin or hinged supports.

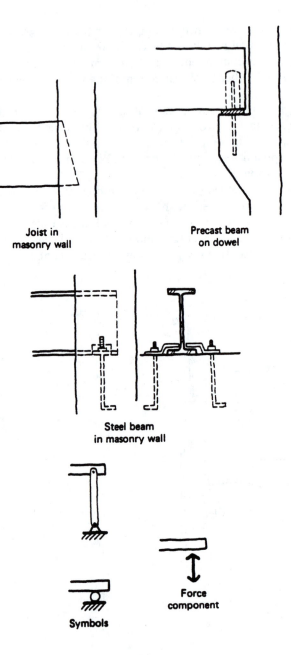

Joist in
masonry wall

Precast beam
on dowel

Steel beam
in masonry wall

Force
component

Symbols

**Figure 2–54** Roller and link
supports.

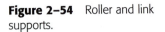

mined, the other is known by trigonometry. The pin and roller are called *simple supports*.

The *link support* is really more of a separate structural member than a support, but it acts very much like a roller. To provide a force of known direction, the link must have a pin (or hinge) at both ends and carry no load in between. Some-

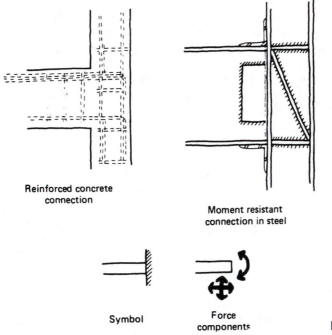

Reinforced concrete
connection

Moment resistant
connection in steel

Symbol

Force
components

**Figure 2–55**

times called a *strut*, it is a special case of a larger group of structural elements, described later (Section 2–8) as two-force members.

The *fixed* or *rigid* or *built-in support* can take one force of unknown sense and direction (Figure 2–55). In theory, it also has full moment resistance (of either sense) and will not change its angle of attachment. Like the pin, the single force is usually represented as two completely independent components. A fixed connection has the potential of three unknowns (i.e., two independent components and a couple). This couple is sometimes called a *moment reaction* or *fixed-end moment*.

## 2–7   RIGID-BODY CONCEPT AND FREE-BODY DIAGRAMS

Before looking further into the subject of static equilibrium, the concept of the *rigid body* should be introduced. In statics, it becomes convenient, if not necessary, to ignore the small deformations and displacements which take place when a member is loaded. To do this, we pretend that the materials used for all structural elements and supports are rigid, having the property of infinite stiffness. We assume that members do not stretch, compress, or bend in any way and that their geometry, therefore, remains constant. This is, of course, never true. Even though structural materials are very stiff, they all deform slightly even under small loads. However, the assumption that structural bodies are rigid greatly simplifies many situations in terms of static equilibrium and, in most cases, introduces an insignificant amount of error. An example of the type of minute change, which is generally ignored, is the shortening of span that takes place when a beam deflects into an arc. Depending on the type of loading, minor changes in the upward reactions at the supports would also occur. Not only would such changes be insignificant if expressed in percentage terms, but they would also be difficult to consider quantitatively because, like other deformations, they vary with the load.

Once all of the external forces have been resolved in terms of statics, the stresses and strains within the various elements of the structure are examined. At this point, it is critical that material deformations are *not* ignored. The rigid-body concept is useful only for the determination of external forces. (It is generally valid but can require modification when applied to the statics of more complicated structural problems.)

While the idea of a structure that is rigid for some analytical operations but not for others may seem incongruous to the novice, a little experience will quickly provide the rationale and judgment behind this concept.

In structural analysis, the principles of statics are used to determine reactive forces, which are responses to the applied loads. These reactions always develop the appropriate magnitudes, senses, and directions, such that the end result is one of equilibrium. In other words, under the combined action of the loads and reactions, each element of the structure has zero tendency to translate and zero tendency to rotate.

Determining the needed reactive forces is made easier if the analyst makes a sketch of the structure or element, showing all the forces involved (known and unknown). Such a sketch is called a *free-body diagram* (FBD), and most structural designers consider making such a diagram the first step in any statics problem. A free-body diagram shows the body in isolation or cut "free" from everything adjacent to it. The effects of all such removed objects are shown as forces acting at the appropriate locations. We have already used these diagrams in this chapter without calling them by name. Figure 2–23(b) is, in effect, a free-body diagram of the central joint. The portion of beam shown in Figure 2–47(c) is another FBD. The examples that follow in Section 2–8 will utilize an FBD.

## 2–8  EQUILIBRIUM OF CANTILEVERED MEMBERS

A *cantilever* is a structural element, often a beam, which has one fixed end and one free end. They are the simplest of structures that involve nonconcurrent force systems. To prove the static equilibrium of a two-dimensional, nonconcurrent force system, three equations are needed:

$$\Sigma F_x = 0$$

$$\Sigma F_y = 0$$

$$\Sigma M = 0$$

The fact that the forces do not pass through a common point means that this third equation, "sum of the moments about any point must be equal to zero," is available to us. As explained in Section 2–6, a fixed-end or moment-resistant connection usually develops three reactive effects: two independent component reactions and one couple or moment reaction. In free-body diagrams these reactions are represented by straight and curved arrows, respectively. A sense must be assumed for each unknown reaction when an FBD is drawn. **(A negative sign on an answer value will mean an incorrect sense assumption.)**

Once a complete FBD for a cantilever is drawn, the fixed-end reactions are found by using the two force equations, followed by a moment equation, written using the fixed end as the moment center. The results are checked by writing a second moment equation using a different moment center. The examples that follow will illustrate this approach.

---

**EXAMPLE 2–10M**

Determine the reactions for the cantilever beam in Figure 2–56M.

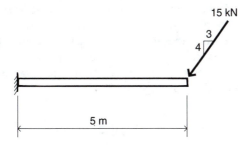

15 kN

3
4

5 m

**Figure 2–56M**  Cantilever beam.

### Solution

The free-body diagram complete with assumed senses is shown in Figure 2–57M. Note that the sloped load was resolved into its rectangular components.

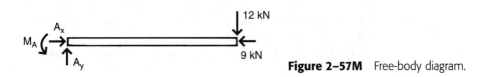

$A_x$

$M_A$

$A_y$

12 kN

9 kN

**Figure 2–57M**  Free-body diagram.

(Notice that the couple or moment reaction at $A$ is given the name $M_A$. To be consistent with the component reactions, it probably should be called $A_m$, but $M_A$ is the commonly accepted nomenclature.)

Solving for the horizontal reaction, we get

$$\Sigma F_x = 0$$

$$A_x - 9 = 0 \qquad A_x = 9 \text{ kN}$$

$$\rightarrow$$

Solving for the vertical reaction, we get

$$\Sigma F_y = 0$$

$$A_y - 12 = 0 \qquad A_y = 12 \text{ kN} \uparrow$$

Lastly we sum the moments about the point $A$ to solve for $M_A$. (This moment center is particularly appropriate because it eliminates the newly found values of $A_x$ and $A_y$. If by chance these values were found in error, they would not influence the answer we will get for the moment reaction. In general, it is good procedure to select a moment center that will eliminate previously found values if possible.)

$$\Sigma M_A = 0$$

$$M_A - 12(5) = 0 \qquad M_A = 60 \text{ kN} \cdot \text{m} \,($$

(Notice here the difference between the two terms $M_A$. The first one is in the equation statement, which is in the form of a *command*. The equation says, in effect, "Add up all the moments about the point $A$ and set that sum equal to zero." The second $M_A$ is the *name* of a moment reaction that the wall provides, acting on the end of the beam.)

Notice that each answer provides magnitude, direction, and sense.

Finally, a check can be achieved by taking moments about some point other than $A$. We can use the end $B$, for example.

$$\overset{?}{\Sigma M_B = 0}$$

$$M_A - A_y(5) = 0$$

$$60 - 12(5) = 0$$

$$\overset{\checkmark}{0 = 0}$$

(Notice that there is no moment arm for $M_A$ in this equation. After all, it is a couple and a couple has no moment center; it acts equally about every point in its plane. Besides, if it had a moment arm, the term would have units of force times distance squared, and this would have no place in a moment equation where all the terms must have units of moment.)

To complete this example, an answer free-body diagram (AFBD) is provided (Figure 2–58M).

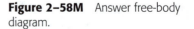

**Figure 2–58M** Answer free-body diagram.

**EXAMPLE 2-11**

Determine the reactions at the wall for the cantilevered frame of Figure 2–59.

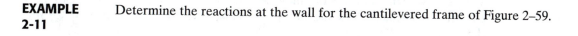

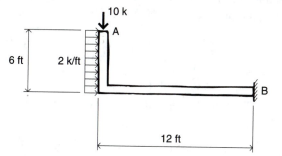

**Figure 2–59** Cantilevered frame.

### Solution

A free-body diagram is provided in Figure 2–60. The uniform load has been replaced by a fictitious resultant for convenience. (Note that the resultant acts through the center of the uniform load.) The two force equations are applied first, followed by a moment equation written for the moment center $B$. This moment center eliminates the values of $B_x$ and $B_y$, and in case they were found incorrectly, the answer for $M_B$ will not be affected.

$$\Sigma F_x = 0$$
$$12 - B_x = 0 \qquad B_x = 12 \text{ kips}$$
$$\leftarrow$$

$$\Sigma F_y = 0$$
$$-10 + B_y = 0 \qquad B_y = 10 \text{ kips}\uparrow$$

$$\Sigma M_B = 0$$
$$-12(3) + 10(12) + M_B = 0$$
$$M_B = -84 \qquad M_B = 84 \text{ kip-ft}^{\curvearrowright}$$

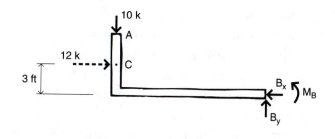

**Figure 2–60** Free-body diagram.

The negative sign on the answer indicates that our sense assumption for $M_B$ was incorrect. It really acts clockwise, so that sense is reflected by the curved arrow with the answer.

A check is achieved by writing a second moment equation about a different moment center, say the point $C$.

$$\overset{?}{\Sigma M_C = 0}$$

$$-B_x(6) + B_y(12) + M_B = 0$$

$$-12(3) + 10(12) + (-84) = 0$$

$$\overset{\checkmark}{0 = 0}$$

(Notice that the value for $M_B$ was plugged in *with* its negative sign!)
Figure 2–61 provides an AFBD.

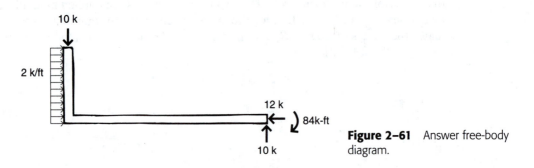

**Figure 2–61**  Answer free-body diagram.

Notice that the AFBD clearly shows that the frame is in equilibrium under the action of three couples.

## PROBLEMS

**2–24.** Determine the reactions for the cantilevered beam of Figure 2–62.

**2–24M.** Determine the reactions for the cantilevered beam of Figure 2–62M.

**2–25.** Determine the reactions for the cantilevered frame of Figure 2–63.

**2–25M.** Determine the reactions for the cantilevered frame of Figure 2–63M.

**2–26.** Determine the reactions for the cantilevered frame of Figure 2–64.

**2–26M.** Determine the reactions for the cantilevered frame of Figure 2–64M.

**2–27.** Determine the reactions for the cantilevered beam carrying a uniformly varying load in Figure 2–65. (Hint: The resultant of the uniformly varying load must act through its center of gravity, i.e., centroid. See Appendix F.)

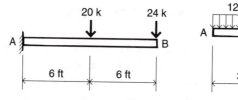

**Figure 2–62**

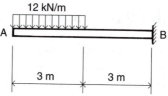

**Figure 2–62M**

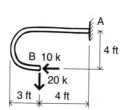

**Figure 2–63**

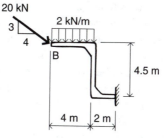

**Figure 2–63M**

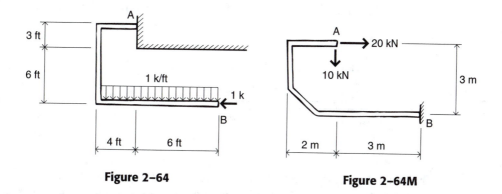

**Figure 2–64**

**Figure 2–64M**

**2–27M.** Determine the reactions for the cantilever in Figure 2–65M, which is carrying a trapezoidal load. (Hint: First separate the load into two parts, a uniform load and a uniformly varying load. The resultant of the triangular portion will act through its center of gravity, i.e., centroid. See Appendix F.)

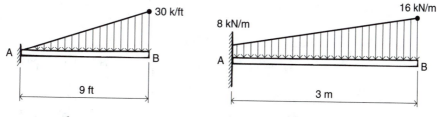

**Figure 2–65**

**Figure 2–65M**

## 2–9 EQUILIBRIUM OF SIMPLE BEAMS AND FRAMES

The word *simple* in statics is a reference to how a structure is supported. When one refers to a simple beam or simple frame, it means that the beam or frame is supported at two places and that one support is a pin and the other is a roller. The procedure for obtaining the reaction components (usually three in number, two at the pin and one at the roller) for such a structure is quite different from the way cantilevers are treated. Whenever practicable, moment equations are written for more than one point, and one of the two force equations is left for use as an independent check on the answers. The judicious selection of moment centers such that previously found values are eliminated from the next equation is a desirable approach. The following examples will serve to illustrate this technique.

It should be noted that regardless of the number of moment centers selected, however, there remain only three independent equations in planar statics, and thus *a maximum of three independent unknowns can be determined*. Other techniques must be used to supplement these three equations whenever the number of independent reaction components exceeds three.

**EXAMPLE 2-12**

Determine the reaction components for the simply supported beam in Figure 2–66.

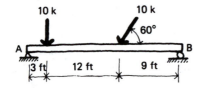

**Figure 2-66**  Beam with two loads.

**Solution**

The free-body diagram is shown in Figure 2–67. The senses of the unknown reactions must be assumed. (**A negative sign on the answer will mean an incorrect sense assumption.**) The load that acts at an angle has been resolved into its rectangular components. The three equations of equilibrium are then used to find the unknown force components.

$$\Sigma F_x = 0$$

$$A_x - 5 = 0 \qquad A_x = 5 \text{ kips}$$
$$\circlearrowleft$$

$$\Sigma M_A = 0$$

$$B_y(24) - 8.67(15) - 10(3) = 0 \qquad B_y = 6.67 \text{ kips}\uparrow$$

$$\Sigma M_B = 0$$

$$-A_y(24) + 10(21) + 8.67(9) = 0 \qquad A_y = 12.0 \text{ kips}\uparrow$$

$$?$$
$$\Sigma F_y = 0$$
$$-10 - 8.67 + 12.0 + 6.67 = 0$$
$$\checkmark$$
$$0 = 0$$

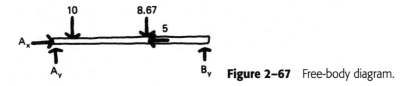

**Figure 2–67** Free-body diagram.

Notice that each answer specifies magnitude, direction, and sense. The AFBD is provided in Figure 2–68.

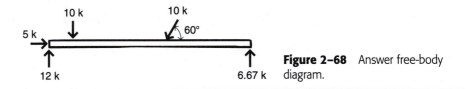

**Figure 2–68** Answer free-body diagram.

**EXAMPLE 2–13M**

Determine the reaction components for the structure in Figure 2–69M.

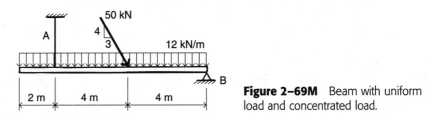

**Figure 2–69M** Beam with uniform load and concentrated load.

**Solution**

The free-body diagram is provided in Figure 2–70M. The uniform load has been re-placed by an equivalent concentrated load acting through the middle of the 10-m long load. The senses of the unknown reactions must be assumed. (**A negative sign on the answer will mean an incorrect sense assumption.**) The load that acts at an angle has been resolved into its rectangular components. The three equations of equilibrium are then used to find the unknown force components.

$$\Sigma F_x = 0$$
$$-B_x - 30 = 0 \qquad B_x = 5 \text{ kN}$$
$$\leftarrow$$

$$\Sigma M_B = 0$$

$$-A_y(8) - 40(4) - 120(5) = 0 \qquad A_y = 95 \text{ kN}\uparrow$$

$$\Sigma M_A = 0$$

$$B_y(8) - 40(4) - 120(3) = 0 \qquad B_y = 65 \text{ kN}\uparrow$$

$$\overset{?}{\Sigma F_y = 0}$$

$$-120 - 40 + 95 + 65 = 0$$

$$\overset{\checkmark}{0 = 0}$$

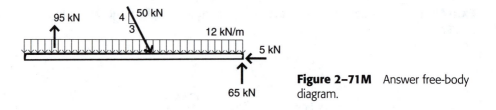

**Figure 2–70M**  Free-body diagram.

Notice that each answer specifies magnitude, direction, and sense. The AFBD is provided in Figure 2–71M.

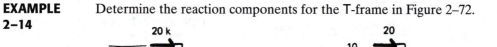

**Figure 2–71M**  Answer free-body diagram.

---

**EXAMPLE 2–14**

Determine the reaction components for the T-frame in Figure 2–72.

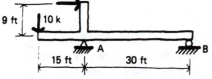

**Figure 2–72**  T-frame.

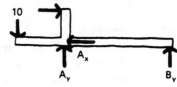

**Figure 2–73**  Free-body diagram.

**Solution**

$$\Sigma F_x = 0$$

$$20 - A_x = 0 \qquad A_x = 20 \text{ kips}$$

$$\leftarrow$$

$$\Sigma M_A = 0$$

$$10(15) - 20(9) + B_y(30) = 0 \qquad B_y = 1 \text{ kip}\uparrow$$

$$\Sigma M_B = 0$$

$$10(45) - 20(9) - A_y(30) = 0 \qquad A_y = 9 \text{ kips}\uparrow$$

$$?$$

$$\Sigma F_y = 0$$

$$-10 + 9 + 1 = 0$$

$$\checkmark$$

$$0 = 0$$

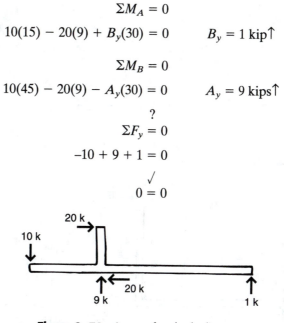

**Figure 2–74**  Answer free-body diagram.

**EXAMPLE 2–15M**

Determine the reaction components for the U-frame in Figure 2–75M.

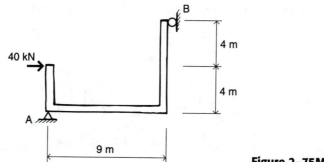

**Figure 2–75M**  U-frame.

**Solution**

Notice that the rigid body concept must be employed so the frame cannot roll downwards and fall off the roller.

$$\Sigma F_y = 0 \qquad A_y = 0$$

$$\Sigma M_A = 0$$

$$-40(4) + B_x(8) = 0 \qquad B_x = 20 \text{ kN}$$

$$\leftarrow$$

$$\Sigma M_B = 0$$

$$40(4) - A_x(8) = 0 \qquad A_x = 20 \text{ kN}$$
$$\leftarrow$$

$$\overset{?}{\Sigma F_x = 0}$$

$$40 - 20 - 20 = 0$$
$$\checkmark$$
$$0 = 0$$

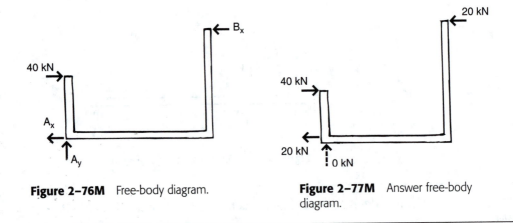

**Figure 2–76M**  Free-body diagram.

**Figure 2–77M**  Answer free-body diagram.

---

**EXAMPLE 2–16**

Determine the reaction components for the beam in Figure 2–78.

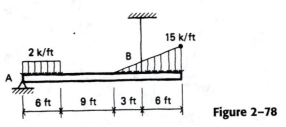

**Figure 2–78**

**Solution**

The uniform load and the uniformly varying load are both converted to equivalent concentrated loads for the purpose of obtaining reactions. Each load shown dashed in Figure 2–79 is placed at the centroid of its load area. (See Appendix F for the triangular load.)

$$\Sigma F_x = 0 \qquad\qquad A_x = 0$$

$$\Sigma M_A = 0$$

$$B_y(18) - 12(3) - 67.5(21) = 0 \qquad\qquad B_y = 80.75 \text{ kips} \uparrow$$

$$\Sigma M_B = 0$$

$$-A_y(18) + 12(15) - 67.5(3) = 0$$

$$A_y = -1.25 \qquad A_y = 1.25 \text{ kips}\downarrow$$

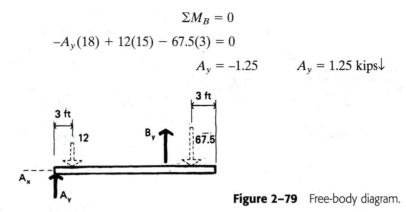

**Figure 2–79** Free-body diagram.

(The negative sign means that the sense of $A_y$ shown on the FBD is incorrect.)

$$\overset{?}{\Sigma F_y = 0}$$

$$A_y - 12 - B_y - 67.5 = 0$$

$$-1.25 - 12 + 80.75 - 67.5 = 0$$

$$\overset{\checkmark}{0 = 0}$$

(Notice that when the answer obtained for $A_y$ was substituted in the checking equation, it was substituted in *with* its negative sign!)

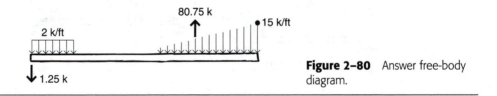

**Figure 2–80** Answer free-body diagram.

## PROBLEMS

**2–28.** Determine the reaction components for the simple beam of Figure 2–81.

**2–28M.** Determine the reaction components for the simple beam of Figure 2–81M.

**2–29.** Determine the reaction components for the frame loaded by wind in Figure 2–82.

**2–29M.** Determine the reaction components for the frame loaded by wind in Figure 2–82M.

**2–30.** Determine the reaction components for the beam in Figure 2–83.

**2–30M.** Determine the reaction components for the beam in Figure 2–83M.

**2–31.** Determine the reaction components for the U-shaped frame of Figure 2–84.

**2–31M.** Determine the reaction components for the U-shaped frame of Figure 2–84M.

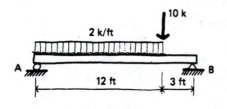

**Figure 2–81**

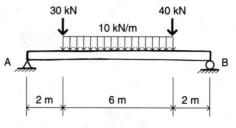

**Figure 2–81M**

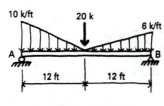

**Figure 2–82**

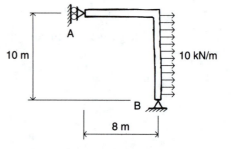

**Figure 2–82M**

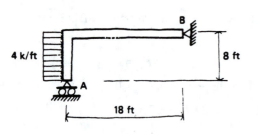

**Figure 2–83**

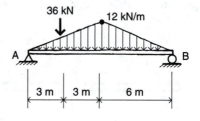

**Figure 2–83M**

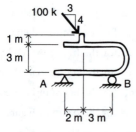

**Figure 2–84**

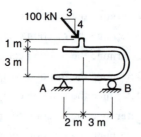

**Figure 2–84M**

**2–32.** Determine the reaction components for the rigid member in Figure 2–85.

**2–32M.** Determine the reaction components for the rigid member in Figure 2–85M.

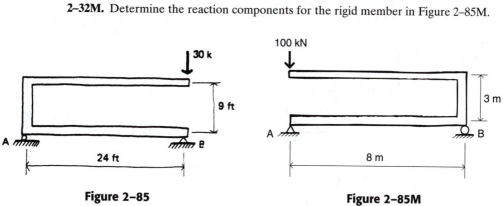

**Figure 2–85**                    **Figure 2–85M**

**2–33.** Determine the reaction components for the rigid body in Figure 2–86.

**2–33M.** Determine the reaction components for the rigid body in Figure 2–86M.

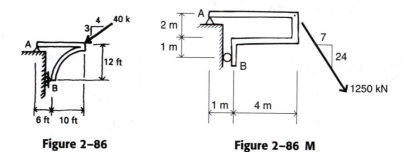

**Figure 2–86**                    **Figure 2–86 M**

**2–34.** Determine the reaction components for the rigid member in Figure 2–87.

**2–34M.** Determine the reaction components for the rigid member in Figure 2–87M.

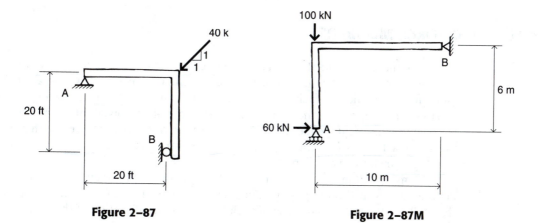

**Figure 2–87**                    **Figure 2–87M**

**2–35.** Determine the reaction components for the beams of Figure 2–88. The connection at B may be assumed to act like a pin in both beams. (*Hint*: The components of the sloped cable force are related to each other by the angle of the cable.)

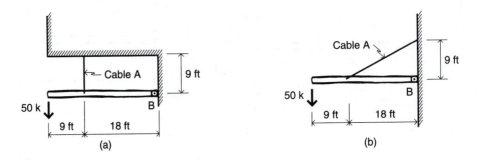

**Figure 2–88**   Beams suspended by cables.

**2–35M.** Determine the reaction components for the beams of Figure 2–88M. (*Hint*: The components of the force in the sloping member *BC* are related to each other by the angle of the member.)

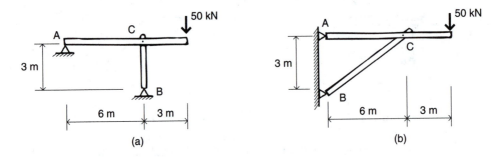

**Figure 2–88M**   Beams supported by struts.

## 2–10   TWO-FORCE MEMBERS

When a structural element is hinged or pinned at each end and carries no load in between, it is called a *two-force member*. Such elements have only two forces acting on them, one applied at each of the two pins. To maintain equilibrium, these forces must be equal, opposite, and collinear. If the forces are resolved into components, the components are dependent on the line of action of the parent force. In this case, the line of action must connect the two pins. In other words, the direction of the two forces is known by inspection.

Member *DB* in Figure 2–89 is a two-force member. Provided that no intermediate load is placed on it, any force acting at *D* or *B* must have the line of action shown. The magnitude and sense of such forces will be a function of the loading on

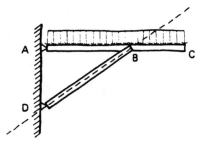

**Figure 2–89** Two-force member.

member *AC*. There are only two possibilities for equilibrium, as shown in Figure 2–90(a) and (b). Any components drawn must be consistent as to sense, as in Figure 2–90(c) or (d). This means that if one of the four components is determined, the others are known automatically. Similarly, if one of the four components is assumed as to sense, the other three sense assumptions are made without choice.

A *cable* is a two-force member of limited sense (i.e., only tension). A roller is the simplest of all two-force members, and the link support is by definition a two-force member. The determination of force direction by two-force members is a very useful tool in statics. It means, for example, that the reactions of the three structures of Figure 2–91 will be identical. In each case, the reaction at *B* (which passes through *C*) will be as shown in Figure 2–92(a). For algebraic determination, it is probably easiest to deal with the dependent components shown in Figure 2–92(b). The components in this case will have the relationship.

$$\frac{B_x}{B_y} = \frac{3}{4}$$

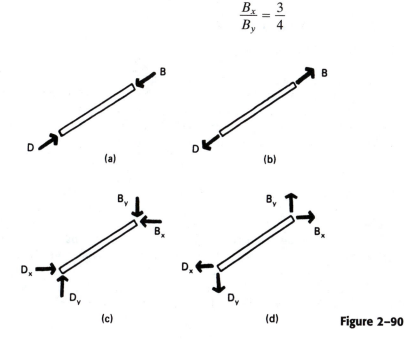

**Figure 2–90**

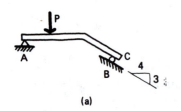

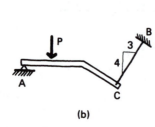

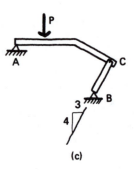

**Figure 2–91**   Three two-force support elements.

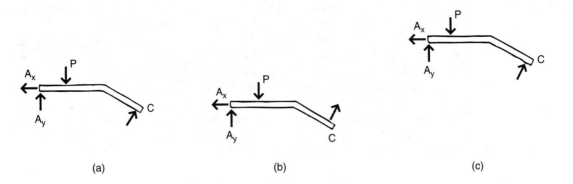

**Figure 2–92**   Free-body diagrams.

Because the equilibrium determination of a two-force member is trivial, it is best to remove it from free-body diagrams just as if it were a support. This was done in the examples and problems of Section 2–4 for the equilibrium of concurrent forces. Each bar or cable functioned as a two-force member and was removed when we sketched an FBD of the point of concurrency.

**EXAMPLE
2–17**

Determine the reaction components at $A$ and $B$ for the structure in Figure 2–93.

**Solution**

The member $BC$ is a two-force member that holds up the bar $AD$. (In this case, it is fairly obvious that $BC$ acts in compression, and the force components will be assumed accordingly. When the sense is not so obvious, it may be guessed as either tension or compression, consistent with the options of Figure 2–90.) The reader should recognize from Figure 2–94(b) that $B_x = C_x$ and $B_y = C_y$. When

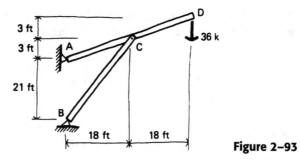

**Figure 2–93**

writing the equations of equilibrium for the FBD of member $AD$, these equivalencies will be used. Also recognize that the two-force member is on a slope of 24 ft vertical on 18 ft horizontal, and therefore $B_x = 3/4 B_y$ or $B_y = 4/3 B_x$.

$$\Sigma M_A = 0$$

$$-B_x(3) + B_y(18) - 36(36) = 0$$

$$\text{But } B_x = \frac{3}{4} B_y:$$

$$-\frac{3}{4} B_y(3) + 18 B_y - 1296 = 0 \qquad B_y = 82.3 \text{ kips} \uparrow$$

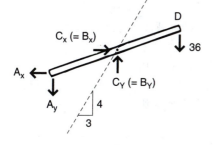

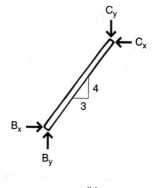

(a)                       (b)

**Figure 2–94**   Free-body diagrams.

If $B_y$ is 82.3, then

$$B_x = \frac{3}{4}(82.3) \qquad B_x = 61.7 \text{ kips}$$
$$\rightarrow$$

The two-force equations can be used to get $A_x$ and $A_y$ from an FBD of member $AD$ or an FBD of the entire structure, as in Figure 2–95.

$$\Sigma F_x = 0$$

$$-A_x + 61.7 = 0 \qquad A_x = 61.7 \text{ kips}$$
$$\leftarrow$$

$$\Sigma F_y = 0$$

$$-A_y + 82.3 - 36 = 0 \qquad A_y = 46.3 \text{ kips} \uparrow$$

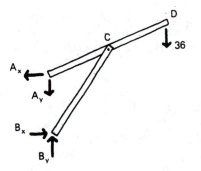

**Figure 2–95**  Free-body diagram of the whole structure.

A check on our work is essential because, in this case, all the answers were made dependent on the first answer for $B_y$. Using Figure 2–95 and taking moments about $B$, we get

$$\overset{?}{\Sigma M_B = 0}$$

$$A_x(21) - 36(36) = 0$$

$$\overset{\checkmark}{61.7(21) - 1296 \approx 0}$$

---

**EXAMPLE 2–18M**

Determine the reaction components at $A$ and $B$ for the structure in Figure 2–96M.

**Solution**

The member $AC$ is a two-force member that keeps the horizontal load from rotating member $DB$ to the right. (In this case it is fairly obvious that $AC$ acts in tension, and the force components will be assumed accordingly. When the sense is not so obvious, it may be guessed as either tension or compression, consistent with the options of Figure 2–90.) The reader should recognize from Figure 2–97M(b) that $C_x = A_x$ and $C_y = A_y$. When writing the equations of equilibrium for the FBD of member $DB$, these equivalencies will be used. Also recognize that the two-force member is on a slope 6 m vertical on 4 m horizontal, and therefore

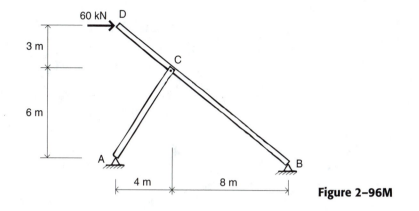

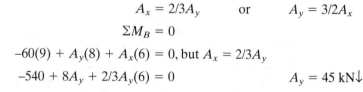

**Figure 2–96M**

$$A_x = 2/3A_y \qquad \text{or} \qquad A_y = 3/2A_x$$

$$\Sigma M_B = 0$$

$$-60(9) + A_y(8) + A_x(6) = 0, \text{ but } A_x = 2/3A_y$$

$$-540 + 8A_y + 2/3A_y(6) = 0 \qquad\qquad A_y = 45 \text{ kN} \downarrow$$

If $A_y$ is 45 $k$N, then

$$A_x = 2/3(45) \qquad\qquad A_x = 30 \text{ kN}$$
$$\leftarrow$$

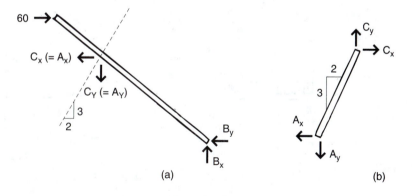

(a)                                      (b)

**Figure 2–97M**   Free-body diagram.

The two-force equations can be used to get $B_x$ and $B_y$ from an FBD of member *DB* or an FBD of the entire structure, as in Figure 2–98M.

A check on our work is essential because, in this case, all the answers were made dependent on the first answer for $A_y$. Using Figure 2–98M and taking moments about *A*, we get

$$?$$
$$\Sigma M_A = 0$$
$$-60(9) + B_y(12) = 0$$
$$\checkmark$$
$$-540 + 45(12) = 0$$
$$0 = 0$$

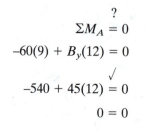

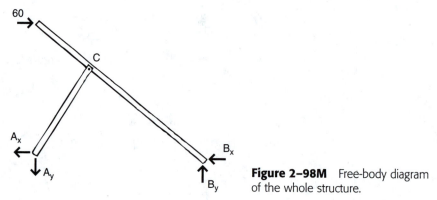

**Figure 2–98M**  Free-body diagram of the whole structure.

---

**EXAMPLE 2–19**

Determine the reaction components at $A$ and $B$ for the structure in Figure 2–99(a).

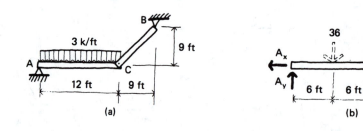

(a)                              (b)

**Figure 2–99**

**Solution**

The two-force member $BC$ acts on a 45° slope, which means that $B_x = B_y$. Using the free-body diagram in Figure 2–99(b),

$$\Sigma M_A = 0$$
$$-36(6) + B_y(12) = 0 \qquad B_y = 18 \text{ kips} \uparrow$$

From the slope of $BC$,

$$B_x = B_y \qquad B_x = 18 \text{ kips}$$
$$\rightarrow$$

$$\Sigma M_B = 0$$
$$-A_y(12) + 36(6) = 0 \qquad A_y = 18 \text{ kips} \uparrow$$
$$\Sigma F_x = 0$$
$$-A_x + B_x = 0 \qquad A_x = 18 \text{ kips}$$
$$\leftarrow$$

Since $B_x = 18$,

$$\overset{?}{\Sigma F_y = 0}$$
$$+ 18 - 36 + 18 = 0$$
$$\overset{\checkmark}{0 = 0}$$

From this Example one can see that the length of the two-force member does not affect the external forces involved; only the slope (or geometry) is important. The line of action of the forces acting on a two-force member always connects the two pins present at each end of the member. It then becomes apparent that the shape of the two-force member between the pins is arbitrary. The shape of the two-force member will *not* affect the statics of the structure. This means that the three structures shown in Figure 2–100 would have the same external reactions as we found in Example 2–19.

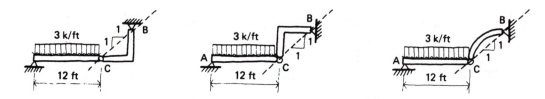

**Figure 2–100**   Three structures that have the same reactions as the structure in Figure 2–99(a).

## PROBLEMS

**2–36.** Determine the reaction components for the structure in Figure 2–101.

**2–36M.** Determine the reaction components for the structure in Figure 2–101M.

**2–37.** Determine the reaction components for the three-hinged arch in Figure 2–102.

**2–37M.** Determine the reaction components for the three-hinged arch in Figure 2–102M.

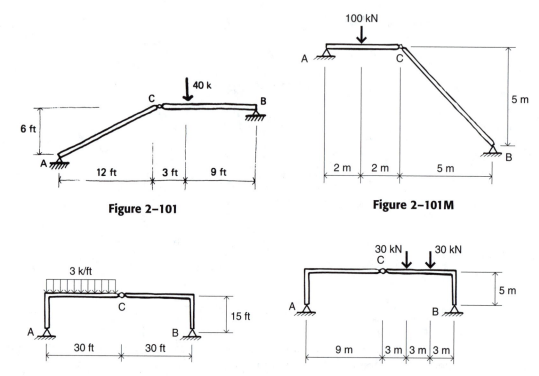

**Figure 2–101**

**Figure 2–101M**

**Figure 2–102**

**Figure 2–102M**

**2–38.** Determine the reaction components for the structure in Figure 2–103.

**2–38M.** Determine the reaction components for the structure in Figure 2–103M.

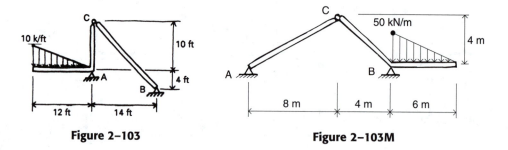

**Figure 2–103**

**Figure 2–103M**

**2–39.** Determine the reaction components for the structure in Figure 2–104.

**2–39M.** Determine the reaction components for the structure in Figure 2–104M.

**2–40.** Determine the reaction components for the structure in Figure 2–105. (*Hint:* Make an FBD of member *BC* to prove that $B_y$ and $C_y$ are both zero.)

**2–40M.** Determine the reaction components for the structure in Figure 2–105M. (*Hint:* Make an FBD of member *BC* to prove that $B_y$ and $C_y$ are both zero.)

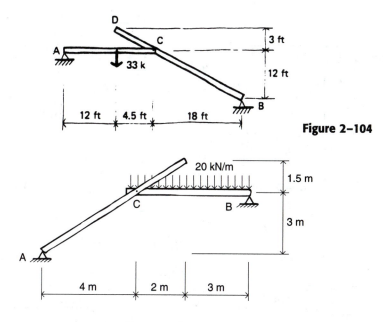

**Figure 2–104**

**Figure 2–104M**

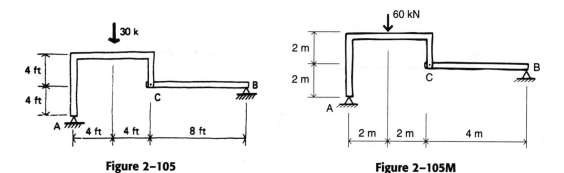

**Figure 2–105**                                                **Figure 2–105M**

**2–41.** With reference to the structure in Figure 2–105, move the load so that it acts in the middle of member *BC*, and then determine the reaction components.

**2–41M.** With reference to the structure in Figure 2–105M, move the load so that it acts in the middle of member *BC*, and then determine the reaction components.

**2–42.** Determine the reaction components for the structure in Figure 2–106.

**2–42M.** Determine the reaction components for the structure in Figure 2–106M.

**2–43.** Determine the reaction components for the structure in Figure 2–107.

**2–43M.** Determine the reaction components for the structure in Figure 2–107M.

**2–44.** Determine the reaction components for the frame in Figure 2–108.

**2–44M.** Determine the reaction components for the frame in Figure 2–108M.

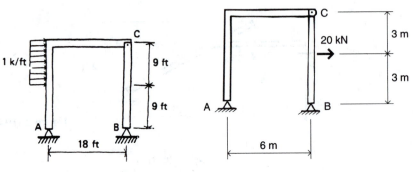

**Figure 2–106**

**Figure 2–106M**

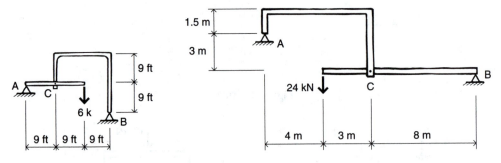

**Figure 2–107**

**Figure 2–107M**

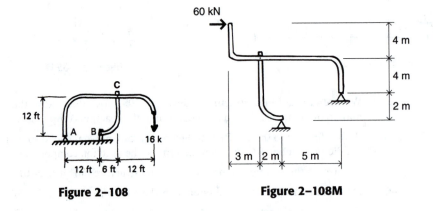

**Figure 2–108**

**Figure 2–108M**

**2–45.** Determine the reaction components for the frame in Figure 2–109.

**2–45M.** Determine the reaction components for the frame in Figure 2–109M.

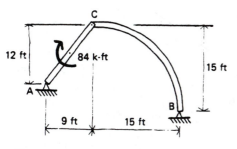

**Figure 2–109**  Structure acted upon by a couple.

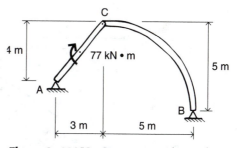

**Figure 2–109M**  Structure acted upon by a couple.

## 2–11 SHED AND GABLE MEMBERS

Beams such as those illustrated in Figure 2–110 occur commonly in roof construction and are usually subject to uniform loads from snow and wind. Loads from snow are uniform along the horizontal projection of the span and are different from dead loads, such as the roof deck, which are uniform along the sloped length. A very steep roof has little projected plan area but may have a high dead load per unit run of horizontal span.

In Figure 2–111, a convenient angle of slope has been chosen to show the difference between load on a slope and load on the projection. Assume that the spacing of the roof beams is such that the snow load is 2 kN per meter of horizontal projection, for a total load of 8 kN. Likewise, take the structural dead load of the roof, including an allowance for beam self-weight, as 2 kN per meter of sloped length, for a total of 10 kN. Both loads are gravity loads and act vertically downward, so each reaction will be 9 kN upward. If the beams are supported equally by girders at each end as in Figure 2–110, there will be no horizontal thrust, only downward load. To illustrate that this is the case, we will break the load into components that act perpendicularly and tangentially to the beam. The snow load, if spread over the sloped length, will be 8 kN/5m or 1.6 kN/m. Adding this to the dead load, the total load per meter of slope will then be 3.6 kN, as in Figure 2–112(a). If this value is broken into components as in Figure 2–112(b), these components will sum to perpendicular and tangential reactions at each end of the span as shown in Figure 2–112(c). The resultant of these two reactions is, of course, vertical as in Figure 2–112(d) and equal to the original computed 9 kN.

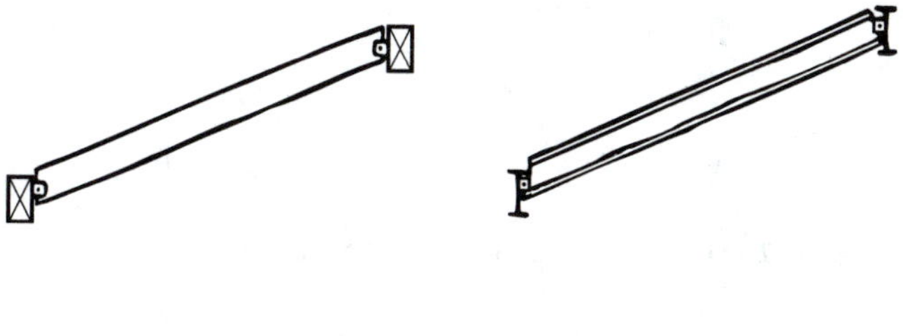

**Figure 2–110**  Sloping beams.

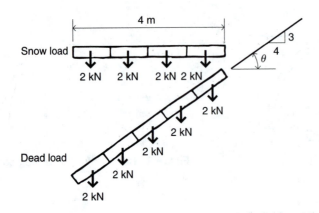

Snow load

2 kN    2 kN    2 kN  2 kN

2 kN

2 kN

2 kN

Dead load

2 kN

2 kN

**Figure 2–111**

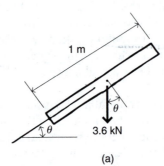

1 m

$\theta$

3.6 kN

(a)

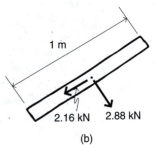

1 m

2.16 kN    2.88 kN

(b)

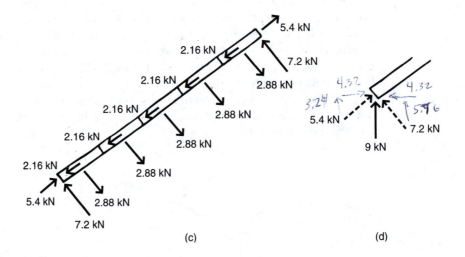

5.4 kN

2.16 kN

7.2 kN

2.16 kN

2.88 kN

2.16 kN

2.88 kN

2.16 kN

2.88 kN

2.16 kN

2.88 kN

5.4 kN        2.88 kN

7.2 kN

(c)

4.32

3.24

4.32

5.4 kN        5.76

7.2 kN

9 kN

(d)

**Figure 2–112**  Sloped beam components.

It is usual to convert loads along the sloped length to loads acting on the horizontal projection. In this example, the dead load along the slope was 2 kN/m. The value we get when converting this to load acting along the horizontal projection will always be larger. From Figure 2–112(a) it can be shown that the relationship is a function of the cosine of the angle made with the horizontal. The load on the horizontal projection will be equal to the load acting along the slope divided by the cosine of that angle. In our case

$$\frac{2 \text{ kN/m}}{\dfrac{4}{5}} = 2.5 \text{ kN/m}$$

The total load per meter of horizontal projection will then be 2 plus 2.5 or 4.5 kN/m. The total load on the beam will be 4 meters (of horizontal projection) times 4.5 kN/m which equals 18 kN. Each vertical reaction will be 9 kN, once again.

The resolution of forces is not so straightforward if the sloping surfaces are configured to carry load by the development of a horizontal thrust. If there is no beam at the top end to provide the vertical relative force, horizontal forces must be present. This situation might be typified by a residential attic space, as illustrated in Figure 2–113, where the ridge member has a negligible capacity. In such cases, the attic floor serves as a tensile tie to resist the outward thrust of the rafters and the structure acts like a simple truss.

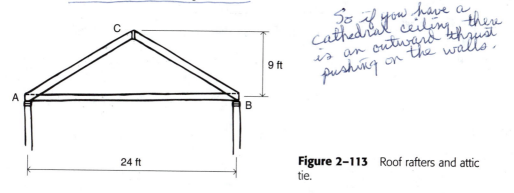

*So if you have a cathedral ceiling there is an outward thrust pushing on the walls.*

**Figure 2–113**   Roof rafters and attic tie.

---

**EXAMPLE 2–20**

Referring to the gable roof of Figure 2–113 and the free-body diagrams of Figure 2–114, solve for the horizontal and vertical forces involved by using the equations of statics.

**Solution**

To find $C_x$, using the rafter on the left,

$$\Sigma M_A = 0$$

$$C_x (9) - 3600 (6) = 0 \qquad C_x = 2400 \text{ lbs}$$

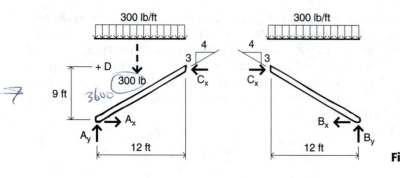

**Figure 2–114**  Free-body diagrams.

To find $A_y$,

$$\Sigma F_y = 0$$

$$A_y - 3600 = 0 \qquad A_y = 3600 \text{ lbs} \uparrow$$

The point $D$ is a convenient moment center to find $A_x$ because it eliminates the two previously found values.

$$\Sigma M_D = 0$$

$$-3600(6) + A_x(9) = 0 \qquad A_x = 2400 \text{ lbs}$$

$$\rightarrow$$

$$\overset{?}{\Sigma F_x = 0}$$

$$A_x - C_x = 0$$

$$2400 - 2400 = 0$$

$$\overset{\checkmark}{0 = 0}$$

The foregoing indicates that the attic floor members would have to be designed for tension as well as bending (which would be generated by the gravity forces due to people and goods in the attic).

The tensile force in the tie is directly proportional to its elevation within the gable structure. That is, if a cathedral ceiling is desired as in Figure 2–115, the force in the tie (collar beam) will be much greater, which in turn will cause a lot of bending in the rafters and, in some cases, connection difficulties.

---

**EXAMPLE
2–21**

Determine the tensile force in collar beam shown in Figure 2–115.
Figure 2–116 provides a free-body diagram for the lefthand rafter of Figure 2–115.

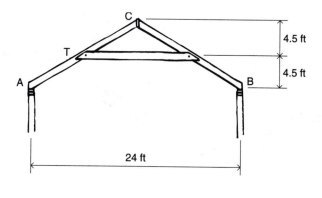

**Figure 2–115** Cathedral ceiling.

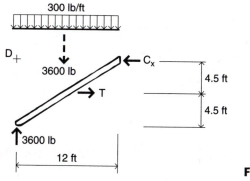

**Figure 2–116** Free-body diagram.

---

### Solution

The vertical reaction remains at 3600 lbs, of course, but the tie force will double.

$$\Sigma M_D = 0$$

$$-3600(6) + T(4.5) = 0 \qquad T = 4800 \text{ lbs}$$

$$\rightarrow$$

---

## PROBLEMS

**2–46.** Determine the load on the horizontal projection (kips/ft) for the rafter of Figure 2–117 if the load along the slope is 1 kip/ft of sloped length.

**2–46M.** Determine the load on the horizontal projection (kN/m) for the rafter of Figure 2–117M if the load along the slope is 6 kN/m of sloped length.

**2–47.** The roof rafters of Figure 2–118 are supported by a partition that runs down the cen-

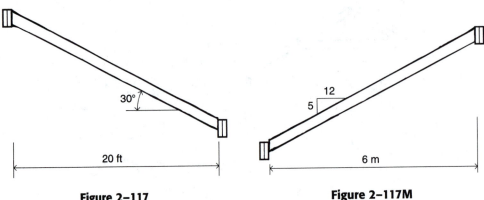

**Figure 2–117**                    **Figure 2–117M**

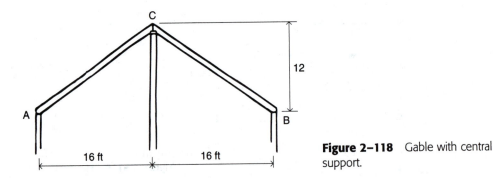

**Figure 2–118**   Gable with central
support.

ter of the building. The total load is 80 lb/ft of sloped rafter length. Determine the
horizontal thrust that each rafter will exert at the top of an outer wall if the central
load-bearing partition is accidentally removed.

**2–47M.** The roof rafters of Figure 2–118M are supported by a partition that runs down the
center of the building. The snow load is 3.6 kN/m of horizontal projection and the
dead load is 2.4 kN/m of sloped rafter. Determine the horizontal thrust that each
rafter will exert at the top of an outer wall if the central load-bearing partition is accidentally removed.

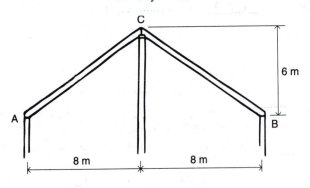

**Figure 2–118M**   Gable with central
support.

**2–48.** With reference to Figure 2–115, determine the force in the collar beam if it is moved so that it is a mere 1 foot down from the apex. Assume the loading is unchanged at 300 lb/ft of horizontal projection.

**2–48M.** The roof rafters of Figure 2–119M carry a total load 1 kN/m on the horizontal projection.
(a) Determine the tension in the collar beam.
(b) Determine the tension in the collar beam if it is relocated to the eave elevation.

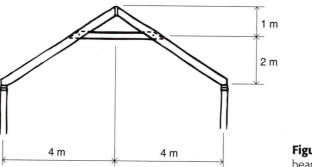

**Figure 2–119M**   Rafters with collar beam.

# 2–12   STABILITY AND DETERMINACY

To be in a state of static equilibrium, a structure must meet the requirements of stability. Loads and reactions bear no meaningful relationship to one another in an unstable structure. Structural stability is accomplished through the geometry of the members and the support (or boundary) conditions present. First, a stable structure is one that will remain at rest under any realistic loading pattern. For example, the simple beam in Figure 2–120(a) is generally unstable. Even though it may remain at rest under a specific load, such as in Figure 2–120(b), it is still judged unstable. Second, the structure must be capable of carrying load without requiring an angular change in its geometry. The structure held in place by cables in Figure 2–121 is unstable because its load-carrying ability depends on a change in geometry—in this case, a small motion to the right. The amount of motion is not important. The fact that such motion must take place before the structure can accept load is important.

It may be helpful to consider the three concurrent force structures in Figure 2–122. The horizontal bars of Figure 2–122(b) are unstable because, as two-force members, they cannot develop the vertical components needed to equilibrate the load $P$. In other words, an angle change is necessary. But if we accept the concept of rigid bodies, such an angle change is impossible because the "rigid" bars cannot elongate to accommodate this change.

Structural stability, for the purposes of equilibrium, is a theoretical concept, and it should be remembered that bodies are considered weightless as well as rigid

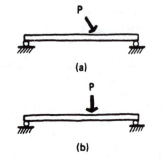

(a)

(b)

**Figure 2–120**

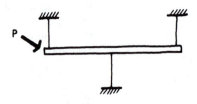

**Figure 2–121**

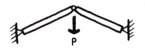

Stable

(a)

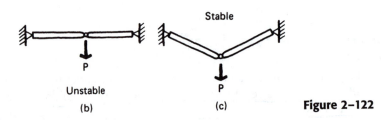

Unstable

(b)

Stable

(c)

**Figure 2–122**

and that rollers and pins are assumed to be frictionless. Examine each of the structures in Figure 2–123 with respect to the previous discussion. By our rules, the two shown in Figure 2–123(a) and (b) are unstable, while those of Figure 2–123(c) and (d) are stable.

Once a structure has satisfied the conditions of stability, it can then be classified as determinate or indeterminate with respect to its reactions. A member or structure is statistically *determinate* if the number of independent reaction components does not exceed the number of applicable independent equations of equilibrium. (This is really just a precise way of stating the familiar axiom, "The number of unknowns cannot exceed the number of equations.") If this is true, then those reaction components can be determined using the techniques of statics alone. If, on the other hand, there exist extra or redundant reaction components, then the structure is said to be statically *indeterminate*. This means that, in order to determine the reactions, the analyst will have to consider more than the basic equilibrium of forces. In general, this means that the deformation or deflection of a structure or member must be examined.

If an indeterminate structure has only one redundant force component, it is

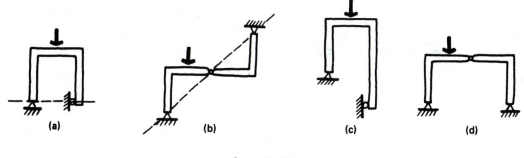

**Figure 2-123**

described as being indeterminate to the first degree; if two redundant components, to the second degree; and so on. Some structures, such as a light timber frame of stud and joist construction, have mostly determinate members. Others, such as the typical cast-in-place reinforced concrete frame, are highly indeterminate.

Basic procedures for analyzing structures with one degree of redundancy are presented briefly in Chapter 9, but a proper treatment of indeterminate structures is beyond the scope of this book. All the examples and problems presented for quantitative solution in this chapter are statistically determinate.

## PROBLEMS

**2–49.** Determine whether or not each of the structures in Figure 2–124 is stable. If stable, then ascertain if it is determinate or indeterminate. If indeterminate, to what degree?

**2–50.** Each of the structures in Figure 2–125 is either unstable or indeterminate. Change or remove *one* of the support conditions in each case so that the resulting structure will be stable and determinate.

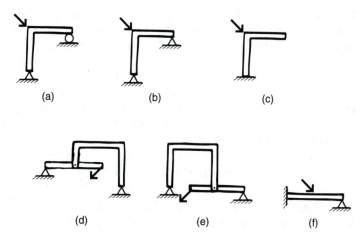

**Figure 2-124**

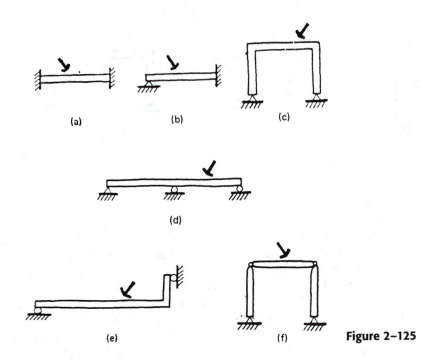

**Figure 2–125**

## 2-13   PINNED FRAMES

In this chapter, we have considered many structures that utilize pinned joints to transmit forces from one member to another. In reality, such joints are seldom really pinned but do have enough flexibility that they can be considered moment-free for the purposes of statics. Even when such joints are continuous and moment-resistant (and the structure is therefore usually indeterminate), it sometimes helps to make the temporary and fictitious assumption of pinned connections. This can aid the designer in doing a preliminary visual analysis of the basic manner in which a completed structure might carry its loads. The effects of continuity can then be imposed on this qualitative analysis to achieve a higher degree of accuracy and understanding. An initial pin-joint assumption can serve as a starting point in the study of many structures.

The statics of any structure is made simpler if there is no moment at the joints. For example, at a joint involving only two members, it is logical to conclude that the force exerted by the first member upon the second is equal and opposite to and collinear with the force exerted by the second member upon the first. This is essential for equilibrium of the pin itself.

Before any member can be designed, it is necessary to determine all the forces acting upon it. A structural analysis of the A-frame in Figure 2–126 would start with a complete determination of all the force components acting on each member. In this case, the external reactions at $A$ and $B$ could be easily obtained using a free-body diagram of the whole structure. Note that there would be no $A_x$ component.

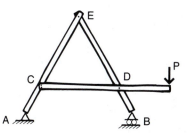

**Figure 2–126**  Simple A-frame.

The joint forces at *C, D,* and *E* can be found by using an FBD for each member, as in Figure 2–127. The forces are assumed to act with arbitrary sense except that, of course, they must act in an opposite manner on the two members meeting at a given joint. The equations of equilibrium can be applied to each member in turn, moving from one free body to another as the various forces become known. An incorrect sense assumption will provide a negative answer as usual, but it must be remembered that any sense change will affect *two* free bodies.

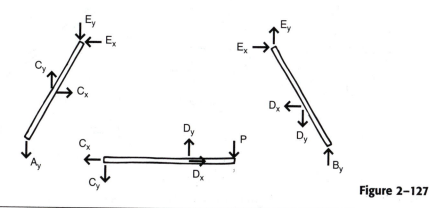

**Figure 2–127**

**EXAMPLE 2–22M**

Determine all the force components acting on each member of the pin-jointed frame in Figure 2–128.M.

**Solution**

The four external reaction components cannot be simply obtained from an FBD of the whole, so the structure will be taken apart as in Figure 2–129M. Sense assumptions are made by attempting to trace the pattern each load might take to the supports. Recognition of the two-force member *CD* is essential to the solution.

FBD I:
$$\Sigma M_A = 0$$
$$-36(3) + F_y(8) = 0$$
$$F_y = 13.5$$

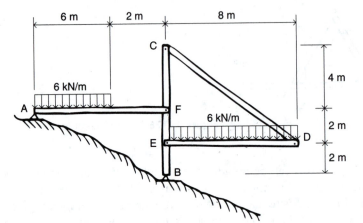

**Figure 2–128M** Schematic section of a hillside dwelling.

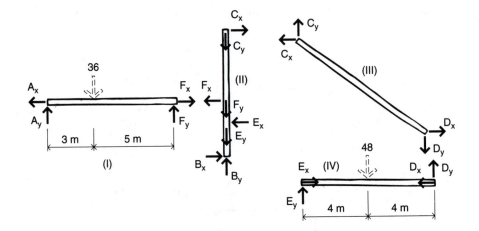

**Figure 2–129M** Free-body diagrams.

FBD 1:

$$\Sigma M_F = 0$$
$$36(5) - A_y(8) = 0$$
$$A_y = 22.5$$

FBD IV:

$$\Sigma M_E = 0$$
$$-48(4) + D_y(8) = 0$$
$$D_y = 24$$

FBD IV:

$$\Sigma M_D = 0$$
$$48(4) + E_y(8) = 0$$
$$E_y = 24$$

FBD III:
$$D_x = \tfrac{4}{3}D_y$$
$$= \tfrac{4}{3}(24)$$
$$D_x = 32$$

FBD III:
$$\Sigma F_x = 0$$
$$- C_x + D_x = 0$$
$$C_x = 32$$

FBD III:
$$\Sigma F_y = 0$$
$$C_y - D_y = 0$$
$$C_y = 24$$

FBD IV:
$$\Sigma F_x = 0$$
$$-E_x + D_x = 0$$
$$E_x = 32$$

FBD II:
$$\Sigma M_B = 0$$
$$- C_x(8) + F_x(4) + E_x(2) = 0$$
$$- 32(8) + 4F_x + 32(2) = 0$$
$$F_x = 48$$

FBD II:
$$\Sigma M_F = 0$$
$$-C_x(4) - E_x(2) + B_x(4) = 0$$
$$- 32(4) - 32(2) + 4B_x = 0$$
$$B_x = 48$$

FBD II:
$$\Sigma F_y = 0$$
$$- C_y - F_y - E_y + B_y = 0$$
$$-24 - 13.5 - 24 + B_y = 0$$
$$B_y = 61.5$$

FBD I:
$$\Sigma F_x = 0$$
$$-A_x + F_x = 0$$
$$A_x = 48$$

Using the entire structure for a check, we obtain

$$\overset{?}{\Sigma F_x = 0}$$
$$48 - 48 = 0$$
$$\checkmark$$
$$0 = 0$$
$$\Sigma F_y \overset{?}{=} 0$$
$$22.5 - 36 - 48 + 61.5 = 0$$
$$\checkmark$$
$$0 = 0$$

The correct answers are provided in Figure 2–130M.

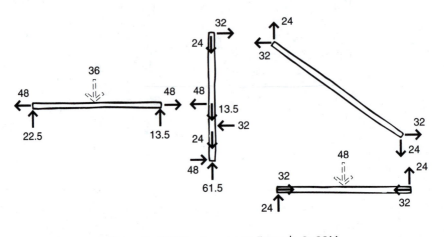

**Figure 2–130M**   Answers to Example 2–22M.

## PROBLEMS

**2–51.** Determine all the force components acting on each member of the frame in Figure 2–131.

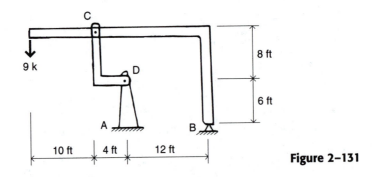

**Figure 2–131**

**2–51M.** Determine all the force components acting on each member of the frame in Figure 2–131M.

**2–52.** Determine all the force components acting on each member of the frame in Figure 2–132.

**2–52M.** Determine all the force components acting on the main carriage member, $AD$, in Figure 2–132M.

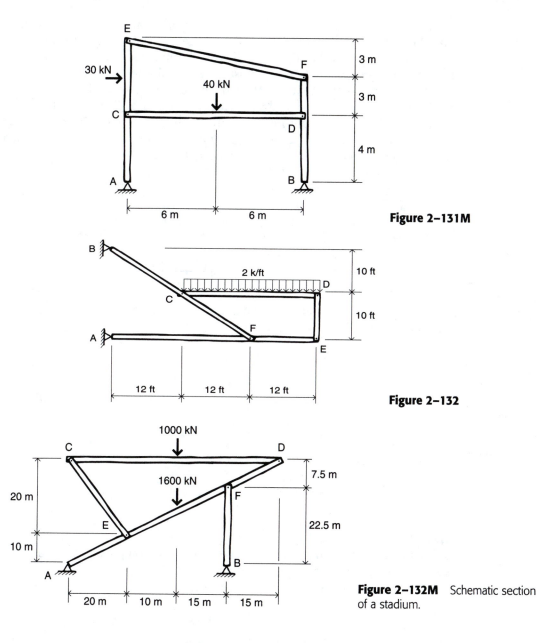

**Figure 2–131M**

**Figure 2–132**

**Figure 2–132M**  Schematic section of a stadium.

## 2–14   SIMPLE CABLE STATICS

A flexible cable will assume a specific geometry when acted upon by one or more point loads. This geometry is dependent on the relative magnitude and location of each load, the length of the cable, and the height of the supports. The designer usually has little control over the loads but can select the length of the cable, thereby controlling the sag, and the support locations. In general, the greater the sag, the less will be the force in the cable. For reasons of equilibrium discussed previously, a taut cable of very slight sag will have to withstand tremendous internal forces.

Analysis of one-way cable systems is made simple by the fact that a cable has effectively zero moment resistance. It acts like a link chain and, if assumed weightless, will take a straight-line geometry between the loads. Each cable segment then acts like a two-force tension member. Each load point is held in concurrent equilibrium by the load and two internal cable forces, one on each side of the load.

If the loads are applied vertically, the horizontal component of the cable force is a *constant* throughout the cable, and only the vertical component varies from segment to segment. Since the vertical component is dependent on the cable slope, the largest tension in the cable will occur where the cable is greatest, usually at the highest cable support.

*slope of the*

---

**EXAMPLE 2–23**

Determine the distance $y$ and the magnitude of the force in each segment of the cable in Figure 2–133.

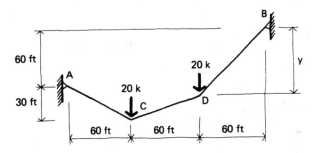

**Figure 2–133**   Cable with third-point loading.

**Solution**

First determine the external reactions. The ratio of the components $A_x$ and $A_y$ will be 2:1, following the slope of the cable segment $AC$.

Using the FBD in Figure 2–134, we obtain

$$\Sigma M_B = 0$$

$$-A_x(60) - A_y(180) + 20(120) + 20(60) = 0$$

$$-2A_y(60) - 180A_y + 2400 + 1200 = 0$$

$$A_y = 12 \qquad A_y = 12 \text{ kips} \uparrow$$

$$A_x = 24 \qquad A_x = 24 \text{ kips}$$

$$\leftarrow$$

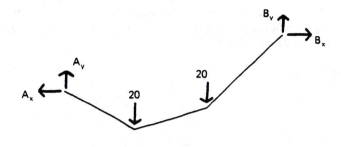

**Figure 2–134** Free-body diagram.

$$\Sigma F_x = 0$$

$$-A_x + B_x = 0$$

$$B_x = 24 \qquad B_x = 24 \text{ kips}$$

$$\rightarrow$$

$$\Sigma M_A = 0$$

$$-20(60) - 20(120) + B_y(180) - 24(60) = 0$$

$$B_y = 28 \qquad B_y = 28 \text{ kips} \uparrow$$

$$?$$

$$\Sigma F_y = 0$$

$$12 - 20 - 20 + 28 = 0$$

$$\checkmark$$

$$0 = 0$$

The distance $y$ must be such that the cable segment $BD$ has a slope that fits the component ratio of $B_y$ to $B_x$. Therefore,

$$\frac{28}{24} = \frac{y}{60} \qquad y = 70 \text{ ft}$$

The cable force in segment $AC$ is

$$T_{AC} = \sqrt{12^2 + 24^2}$$

$$= 26.8 \qquad T_{AC} = 26.8 \text{ kips}$$

The cable force in segment $BD$ is

$$T_{BD} = \sqrt{24^2 + 28^2}$$

$$= 36.9 \qquad T_{BD} = 36.9 \text{ kips}$$

The vertical component of the force in *CD* can be determined by considering its slope or from the vertical equilibrium of either point *C* or *D*, as shown in Figure 2–135. The force in segment *CD* will then be

$$T_{CD} = \sqrt{24^2 + 8^2}$$
$$= 25.3 \qquad\qquad T_{CD} = 25.3 \text{ kips}$$

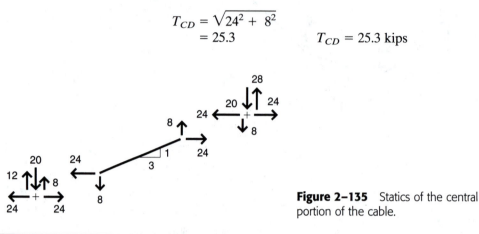

**Figure 2–135** Statics of the central portion of the cable.

---

**EXAMPLE 2–24M**

Determine the distance and the magnitude of the force in each segment of the cable in Figure 2–136M.

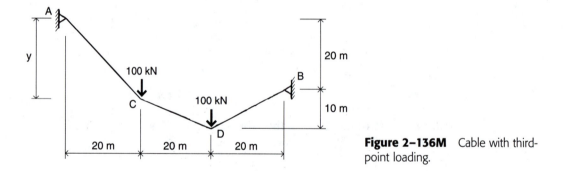

**Figure 2–136M** Cable with third-point loading.

**Solution**

First determine the external reactions. The ratio of the components $B_x$ and $B_y$ will be $2:1$, following the slope of the cable segment *DB*. Using the FBD in Figure 2–137M, we obtain

$$\Sigma M_A = 0$$
$$B_y(60) + B_x(20) - 100(20) - 100(40) = 0$$
$$60B_y + 2B_y(20) - 2000 - 4000 = 0$$
$$B_y = 60 \qquad\qquad B_y = 60 \text{ kN} \uparrow$$
$$B_x = 120 \qquad\qquad B_x = 120 \text{ kN}$$
$$\rightarrow$$

$$\Sigma F_x = 0$$

$$-A_x + B_x = 0$$

$$A_x = 120 \qquad A_x = 120 \text{ kN}$$

$$\Sigma M_B = 0$$

$$+ 120(20) - A_y(60) + 100(20) + 100(40) = 0$$

$$+ 2400 - 60 A_y + 2000 + 4000 = 0$$

$$A_y = 140 \qquad A_y = 140 \text{ kN} \uparrow$$

$$?$$

$$\Sigma F_y = 0$$

$$+ 60 - 100 - 100 + 140 = 0$$

$$\checkmark$$

$$0 = 0$$

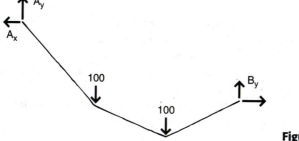

**Figure 2–137M** Free-body diagram.

The distance $y$ must be such that the cable segment $AC$ has a slope that fits the component ratio of $A_x$ and $A_y$. Therefore

$$\frac{140}{120} = \frac{y}{20} \qquad y = 23.3 \text{ m}$$

The cable force in segment $AC$ is

$$T_{AC} = \sqrt{140^2 + 120^2}$$
$$= 198 \qquad T_{AC} = 198 \text{ kN}$$

The cable force in segment $BD$ is

$$T_{BD} = \sqrt{60^2 + 120^2}$$
$$= 134 \qquad T_{BD} = 134 \text{ kN}$$

The vertical component of the force in *CD* can be determined by considering its slope or from the vertical equilibrium of either point *C* or *D*, as shown in Figure 2–138M. The force in segment *CD* will then be

$$T_{CD} = \sqrt{120^2 + 40^2}$$
$$= 126 \qquad\qquad T_{CD} = 126 \text{ kN}$$

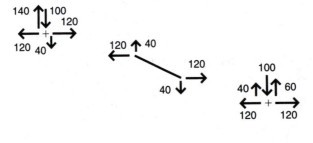

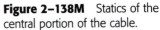

**Figure 2–138M**  Statics of the central portion of the cable.

When the load points get very close together and are of equal magnitude, the load may be considered to be uniform. If it is uniform along the horizontal projection as in Figure 2–139, the cable will assume a parabolic shape. The maximum sag and its location, labeled *y* and *x*, respectively, will depend on the cable length and the relative support heights as before. Given the value of the uniform load, the designer can vary the span and sag and easily check the influence of such changes on the cable force. With a uniformly loaded cable, the lowest point will usually have a slope of zero, and thus the force in the cable at that point will have no vertical component. This means that, referring to Figure 2–140 and assuming known loads and support locations, we can write two independent moment equations, one about *A* and the other about *B*. The equations will have the same two unknowns, *H* and *x*, and can be solved simultaneously. Once *x* is known, the vertical components of each

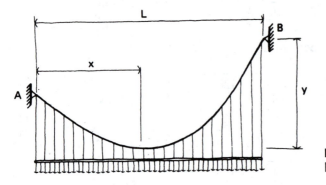

**Figure 2–139**  Cable with a uniform load.

reaction can be obtained from equilibrium in the vertical direction. Since $H$ is constant throughout the cable, the largest cable force will again be at the uppermost support.

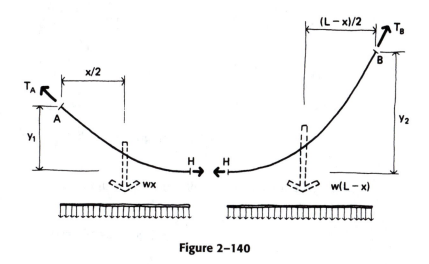

**Figure 2–140**

**EXAMPLE 2–25**

Determine the highest tension in the uniformly loaded cable of Figure 2–141. The free-body diagrams are shown in Figure 2–142.

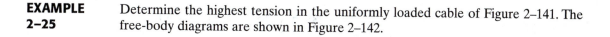

240 ft

12 ft

18 ft

B

A

w = 1 k/ft

**Figure 2–141**

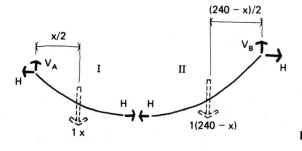

x/2

$(240 - x)/2$

$V_A$

I

II

$V_B$

H

H

H

H

1 x

1(240 − x)

**Figure 2–142** Free-body diagrams.

### Solution

FBD I: $\Sigma M_A = 0$

$$-1x\left(\frac{x}{2}\right) + H(18) = 0$$

$$18H - \frac{x^2}{2} = 0$$

$$36H - x^2 = 0$$

FBD II: $\Sigma M_B = 0$

$$-H(30) + 1(240 - x)\left(\frac{240 - x}{2}\right) = 0$$

Solving the two equations will yield the quadratic

$$\frac{2}{5}x^2 - 288x - 34\,560 = 0$$

$$x = \frac{-b \pm \sqrt{b^2 - 4ac}}{2a}$$

$$= \frac{-288 \pm \sqrt{288^2 - 4(\frac{2}{5})(-34\,560)}}{2(\frac{2}{5})}$$

$$= -825 \text{ and } + 105$$

The maximum tension will occur at point $B$. From previous work,

$$H = \frac{x^2}{36}$$

$$= \frac{105^2}{36}$$

$$= 306$$

FBD II: $\Sigma F_y = 0$

$$V_B = 1(240 - x) = 0$$

$$V_B - 1(240 - 105) = 0 \qquad\qquad V_B = 135$$

$$T_B = \sqrt{135^2 + 306^2}$$

$$= 334 \qquad\qquad T_B = 334 \text{ kips}$$

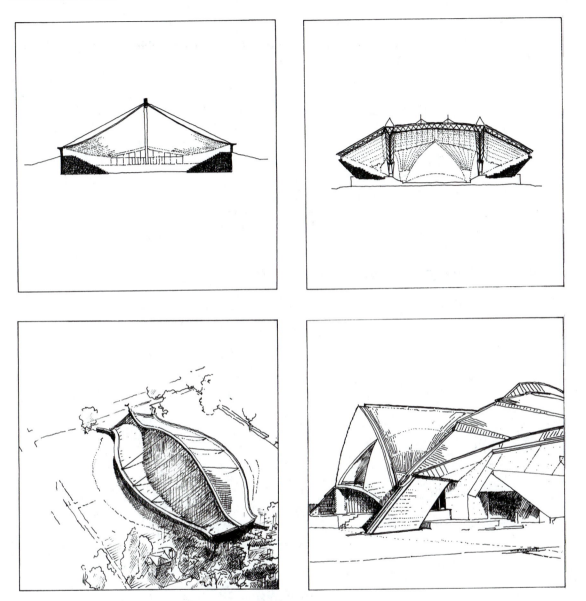

The brilliantly simple structural idea in Eero Saarinen's 1956–58 hockey rink—to use a central concrete "spine" to support, and in turn be laterally stabilized by, catenary steel cables defining the roof enclosure—proved in practice to be inadequate. Additional visible, exterior cables had to be provided, anchoring the spine sideways in both directions to a concrete base associated with the audience seating. By comparison, the paired spines of Fumihiko Maki's gymnasium of 1984 attain their stability integrally, through roof trusses which serve as stabilizing struts, again carrying any lateral loads down to the anchoring mass of the seating. Both architects took general design cues from the basic structural elements; Saarinen found his design's overall architectural character in the sweeping curves of the cables, while Maki drew his forms from the origami-like system of stiffening angular and curving planes.

**EXAMPLE**
**2–26M**
Determine the highest tension in the uniformly loaded cable of Figure 2–143M.

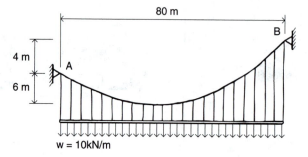

**Figure 2–143M**

**Solution**

FBD I:
$$\Sigma M_A = 0$$
$$-10x\left(\frac{x}{2}\right) + H(6) = 0$$
$$6H - 5x^2 = 0$$

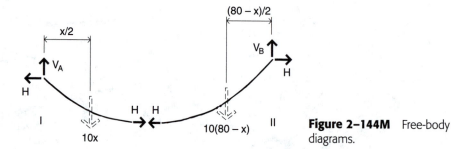

**Figure 2–144M**   Free-body diagrams.

FBD II:
$$\Sigma M_B = 0$$
$$-H(10) + 10(80 - x)\left(\frac{80 - x}{2}\right) = 0$$
$$-H + 3200 - 80x + \frac{x^2}{2} = 0$$

Solving the two equations will yield the quadratic

$$\frac{x^2}{3} + 80x - 3200 = 0$$

$$x = \frac{-b \pm \sqrt{b^2 - 4ac}}{2a}$$

$$= \frac{-80 \pm \sqrt{80^2 - 4(\frac{1}{3})(-3200)}}{2(\frac{1}{3})}$$

$$= -275 \text{ and } 34.9$$

The maximum tension will occur at point $B$. From previous work,

$$H = \tfrac{5}{6}x^2$$
$$= \tfrac{5}{6}(34.9^2)$$
$$= 1015$$

FBD II:
$$\Sigma F_y = 0$$
$$V_B - 10(80 - x) = 0$$
$$V_B - 10(80 - 34.9) = 0$$
$$V_B = 451$$
$$T_B = \sqrt{451^2 + 1015^2}$$
$$= 1110 \qquad\qquad T_B = 1110 \text{ kN}$$

---

**EXAMPLE 2–27**   Determine the value of $H$, the horizontal component of the cable force, in terms of $w$, $L$, and $y$ as defined in Figure 2–145. The two cable supports are on the same level.

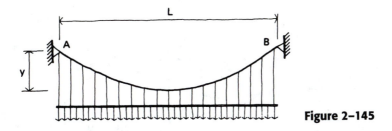

**Figure 2–145**

### Solution
The slope is zero at midspan. Using the free-body diagram in Figure 2–146, we get

$$\Sigma M_B = 0$$

$$\frac{wL}{2}\left(\frac{L}{4}\right) - H(y) = 0$$

$$H = \frac{wL^2}{8y} \qquad\qquad (2\text{–}7)$$

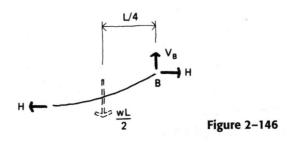

**Figure 2–146**

## PROBLEMS

**2–53.** Determine the maximum sag and the largest tension in the symmetrically loaded cable of Figure 2–147.

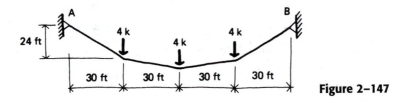

**Figure 2–147**

**2–53M.** Determine the maximum sag and the largest tension in the symmetrically loaded cable of Figure 2–147M.

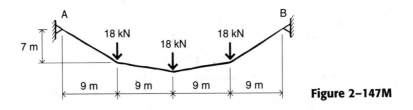

**Figure 2–147M**

**2–54.** One-half of a suspension structure is shown in Figure 2–148. How much tension will be in the tieback cable $AB$, and how much compression in the mast $BC$?

**2–54M.** One-half of a suspension structure is shown in Figure 2–148M. How much tension will be in the tieback cable $AB$, and how much compression in the mast $BC$?

**2–55.** A total of 60 kips must be supported by the cable in Figure 2–149. How much should be placed at each of the two load points to achieve the geometry shown?

**2–55M.** A total of 270 kN must be supported by the cable in Figure 2–149M. How much should be placed at each of the two load points to achieve the geometry shown?

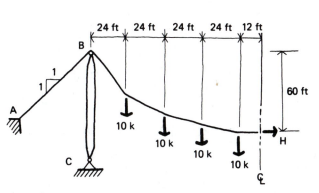

**Figure 2–148**

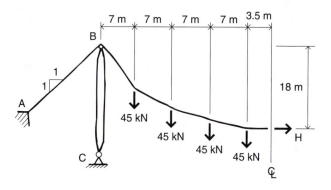

**Figure 2–148M**

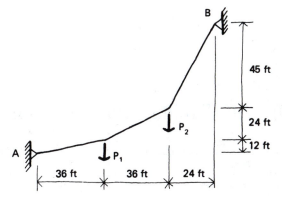

**Figure 2–149**

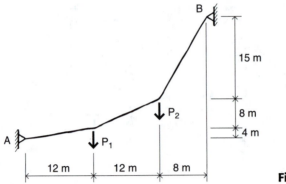

**Figure 2–149M**

**2–56.** With reference to Figure 2–145, if $L = 300$ ft and $w = 1$ kip/ft, determine the maximum tension in the cable when

    **(a)** $y = 300$ ft
    **(b)** $y = 150$ ft
    **(c)** $y = 75$ ft
    **(d)** $y = 30$ ft
    **(e)** $y = 15$ ft

**2–56M.** With reference to Figure 2–145, if $L = 100$ m and $w = 15$ kN/m, determine the maximum tension in the cable when

    **(a)** $y = 100$ m
    **(b)** $y = 50$ m
    **(c)** $y = 25$ m
    **(d)** $y = 10$ m
    **(e)** $y = 5$ m

**2–57.** Determine the maximum tension in the cable of Figure 2–150.

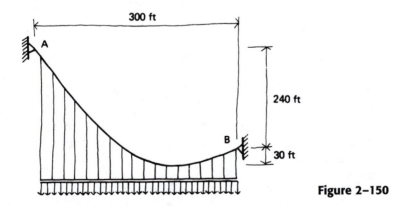

**Figure 2–150**

**2–57M.** Determine the maximum tension in the cable of Figure 2–150M.

**2–58.** The cables shown in Figure 2–151 carry a uniform load (on the horizontal projection) of 4 kips/ft over the entire 240 ft. Determine the required values of $H_L$ and $x$ in order to maintain equilibrium.

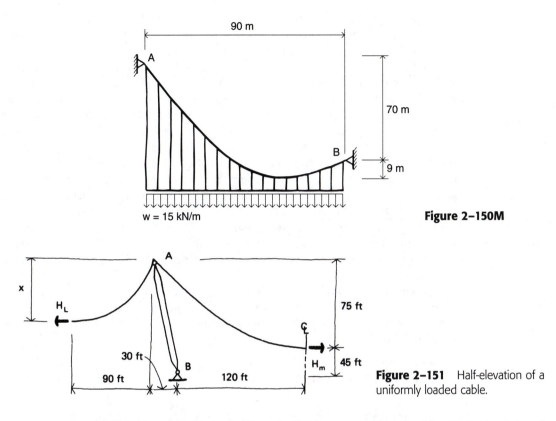

**Figure 2–150M**

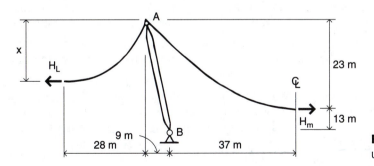

**Figure 2–151**   Half-elevation of a uniformly loaded cable.

**2–58M.** The cables shown in Figure 2–151M carry a uniform load (on the horizontal projection) of 60 kN/m over the entire 74 m. Determine the required values of $H_L$ and $x$ in order to maintain equilibrium.

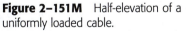

**Figure 2–151M**   Half-elevation of a uniformly loaded cable.

# 2–15   CONCLUSION AND PROCEDURE

The preceding sections have provided an introduction to concepts of static equilibrium in planar force systems. The principles can easily be extended to three-dimensional systems by increasing the number of equilibrium equations to six, one force equation and one moment equation for each of the three coordinate axes. Problem examples have not been included for three-dimensional systems because no new and fundamental concepts would be involved.

Statics is a subject, like most areas of structural analysis, that cannot be learned by reading about it. The few concepts involved are deceptively simple, but any real understanding comes only through practice. The forces of statics have very real effects on all structures, and the designer must become familiar with the action of such forces and the responses (reactions) made by the structures. While the directions and relative magnitudes of reactive forces are far more important to the architect than their actual quantitative values, no purely qualitative approach to the study of physical forces has proven to be sufficient. It is also important not to confine one's study of the subject to text or classroom examples. The beginning student is urged to attempt to conceptualize or sense the forces that provide equilibrium to the objects in his or her immediate environment. Because of the presence of gravity, such forces can be found everywhere. Begin to examine the statics of ordinary objects such as chairs, doors, signposts, bridges, trees, spider webs, playground equipment, and buildings.

Success in quantitative problem analysis comes only through experience, but that experience should be directed so there is a minimum of wasted effort. The writer's own mistakes and those of his students have resulted in the recommendations that follow.

1. Always take time to review the given situation thoroughly. Read the problem carefully, noting what it is that you are to find. What is it that the client is asking you to do?

2. To study forces, always sketch free-body diagrams. Do not attempt to work without them. The mere process of sketching can help you to "see" the forces. If you are a visually minded person, you need a picture.

3. Check your assumptions and your answers as you proceed. Make qualitative estimates. Make graphical polygon checks. Are the results rational? Arithmetic may be neat and precise, but common sense is more valuable.

4. Attempt to work slowly and carefully at the beginning of any analysis. Charging into a problem seems to generate more errors than does racing to finish it.

5. Record your work in neat and orderly fashion. Always include units and senses with numerical answers. State assumptions clearly and make notes explaining your procedure as you work. Always work as though someone were gong to review your records and follow your steps at a later date. Most often, that someone will be you.

# 3

# Structural Properties of Areas

## 3-1 INTRODUCTION

This chapter is devoted to two concepts that have to do with certain properties of cross sections. In structural analysis, it is necessary to consider more than just the number of square inches included within a cross-sectional area. The shape of the section (or how the material is distributed) is equally important.

In this discussion, the term *cross section* is a general one applying to any element or even to a section through an entire structure. It is appropriate to refer to the cross-sectional areas of not only beams and columns, but also trusses, footings, walls, folded plates, segments of shells, and so on.

Two very important concepts are the *centroid* of an area, which is analogous to the center of mass of a volume; and the *moment of inertia* of an area, which is most simply described as a measure of resistance to bending and buckling.

## 3-2 CENTROIDS

The center of mass or center of gravity may be visualized as the location of the resultant of a set of parallel forces. Figure 3–1 illustrates a thin, flat plate of homogeneous material, lying in a horizontal plane. Each element $dA$, located by coordinates $x$ and $y$, will be acted upon by a vertical force due to gravity. The resultant of all these parallel forces will be located at the center of gravity of the plate. Conceptually, this point would be the place where we could attach a vertical string to hang the plate and have the plate remain horizontal. It is dimensioned by $\bar{x}$ and $\bar{y}$ in Figure 3–1.

If we let the thickness approach zero, the plate becomes an area and the center of gravity is then called a *centroid*. The centroid is like an average location of all the small elemental areas, $dA$. Mathematically, it can be located with respect to any reference axes by two equations:

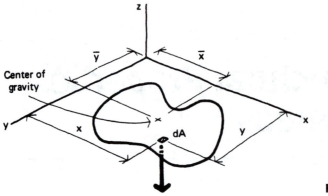

**Figure 3–1**

$$\bar{x} = \frac{\int_0^A x\, dA}{\int_0^A dA} \tag{3–1}$$

$$\bar{y} = \frac{\int_0^A y\, dA}{\int_0^A dA} \tag{3–2}$$

The denominator in each of these expressions is, of course, the total area, and the numerator is called the *first moment* of the area, being a summation of areas times distances. The numerator is also sometimes called the *statical moment* of the area.

The concept of the centroid is used in most of the principles found in the study of mechanics of materials and is, therefore, critical to the understanding of structures. As indicated, the center of gravity is an easier concept to grasp, as it can be experimentally determined. For example, if we cut a shape out of cardboard, the shape will balance on the end of a pencil placed under its center of gravity. If the cardboard is uniform in thickness, the center of gravity will be at the centroid of the area. (This experiment will not always be practical because the centroid, and center of gravity for that matter, do not have to be located physically within the area or on the object. For example, the centroid of a doughnut shape would be at the center of the hole.)

For most structural applications, the areas involved are regular geometric shapes, portions of such shapes, or combinations of them. This means that it is seldom necessary to use the calculus to determine a centroid location. Since the centroid of a regular shape is usually known by inspection, the integrals become algebraic sums of the statical moments and areas involved.

$$\overline{x} = \frac{\Sigma x_i A_i}{\Sigma A_i} \qquad (3\text{--}1a)$$

$$\overline{y} = \frac{\Sigma y_i A_i}{\Sigma A_i} \qquad (3\text{--}2a)$$

These versions of the basic equations are used in the examples that follow. Appendix F will be useful in obtaining the centroids of often-used shapes.

After the centroid of an area has been determined, it is good practice to sketch a rough scale figure of the area, showing the location of the centroid. Gross errors will appear readily in such a sketch.

---

**EXAMPLE 3–1** Determine the centroid location of the T shape in Figure 3–2. (*Note:* Unless otherwise indicated, examples using customary units will have cross-sectional dimensions in inches.)

### Solution
The shape is comprised of two rectangles with their centroids as shown in Figure 3–3. By symmetry, the centroid of the T will lie along the $y$ axis, and $\overline{x}$ must be zero (Figure 3–4). Note that the statical moment of an area will be zero whenever the reference axis passes through the centroid. The distance $\overline{y}$, however, has a value between 3 and 7 in. and can be found by Equation 3–2a.

$$\overline{y} = \frac{\Sigma y_i A_i}{\Sigma A_i}$$

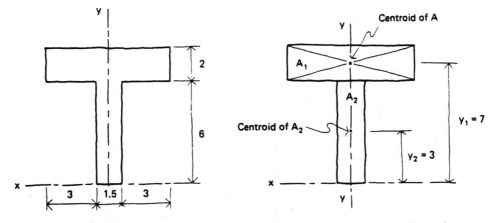

**Figure 3–2** T-Shaped cross section.

**Figure 3–3** Two rectangles in the T shape.

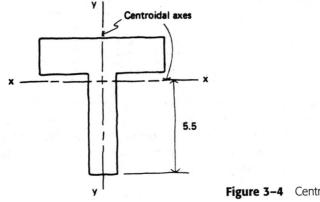

**Figure 3–4** Centroid location.

$$= \frac{y_1 A_1 + y_2 A_2}{A_1 + A_2}$$

$$= \frac{7(2)(7.5) + 3(1.5)(6)}{2(7.5) + 1.5(6)}$$

$$= 5.5 \text{ in.}$$

**EXAMPLE 3–2M**     Find $\bar{x}$ and $\bar{y}$ for the shape in Figure 3–5M. (*Note*: Unless otherwise indicated, examples using SI units will have cross-sectional dimensions in millimeters.)

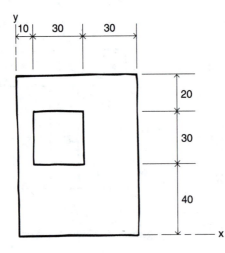

**Figure 3–5M** Rectangular shape with a square hole.

### Solution

Subtract the statical moment and the area of the hole. If we let the subscript 1 represent the area of the rectangle and 2 represent the square hole, we get

$$\bar{x} = \frac{x_1 A_1 - x_2 A_2}{A_1 - A_2}$$

$$= \frac{35(70)(90) - 25(30)(30)}{70(90) - 30(30)}$$

$$= 37 \text{ mm}$$

(See Figure 3–6M). Similarly, for the $y$ direction,

$$\bar{y} = \frac{y_1 A_1 - y_2 A_2}{A_1 - A_2}$$

$$= \frac{45(90)(70) - 55(30)(30)}{90(70) - 30(30)}$$

$$= 43 \text{ mm}$$

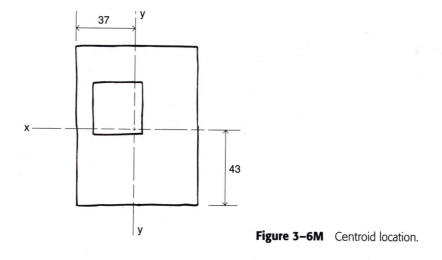

**Figure 3–6M** Centroid location.

**EXAMPLE 3-3M**    Determine the $\bar{y}$ for the section shown in Figure 3–7M.

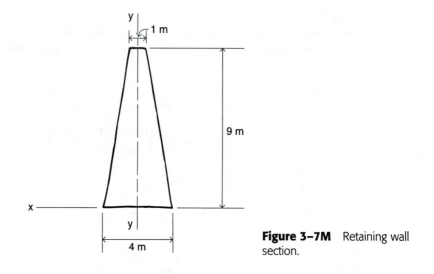

**Figure 3–7M**  Retaining wall section.

## Solution

Assume the area to be made up of two triangles (subscript 2) and one rectangle (subscript 1).

$$\bar{y} = \frac{y_1 A_1 + 2y_2 A_2}{A_1 + 2A_2}$$

$$= \frac{4.5(1)(9) + 2(3)\left(\frac{1}{2}\right)(1.5)(9)}{(1)(9) + 2\left(\frac{1}{2}\right)(1.5)(9)}$$

$$= 3.6 \text{ m}$$

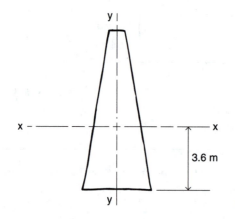

**Figure 3–8M**  Centroid location.

# PROBLEMS

**3–1.** Determine $\bar{y}$ for the cross section in Figure 3–9.

**3–1M.** Determine $\bar{y}$ for the cross section in Figure 3–9M.

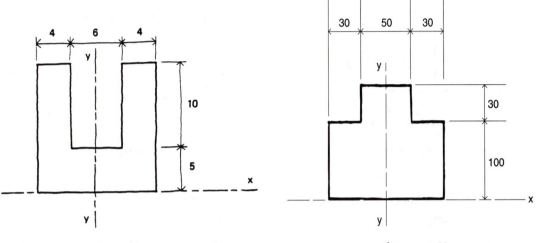

**Figure 3–9**   Channel beam cross section.

**Figure 3–9M**

**3–2.** Determine the centroid location for the symmetrical angle in Figure 3–10.

**3–2M.** Determine the centroid location for the symmetrical angle in Figure 3–10M.

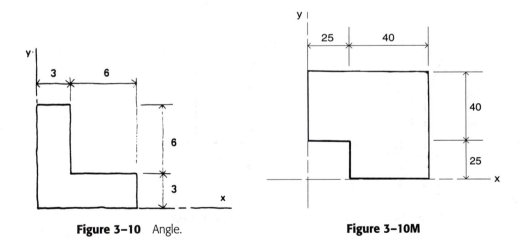

**Figure 3–10**   Angle.

**Figure 3–10M**

**3–3.** Determine the $\bar{y}$ for the shape in Figure 3–11.

**3–3M.** Determine the $\bar{y}$ for the shape in Figure 3–11M.

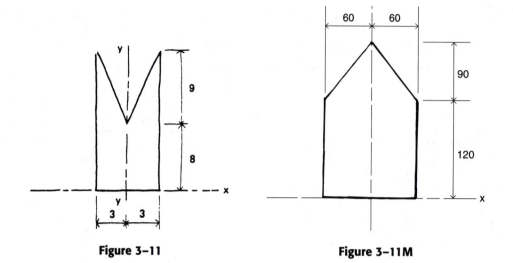

**Figure 3–11**            **Figure 3–11M**

**3–4.** Determine the $\bar{y}$ for the shape in Figure 3–12.

**3–4M.** Determine the $\bar{y}$ for the shape in Figure 3–12M.

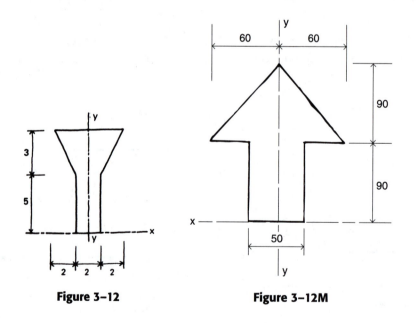

**Figure 3–12**            **Figure 3–12M**

**3–5.** Determine $\bar{y}$ for the shape in Figure 3–13.

**3–5M.** Determine $\bar{y}$ for the shape in Figure 3–13M.

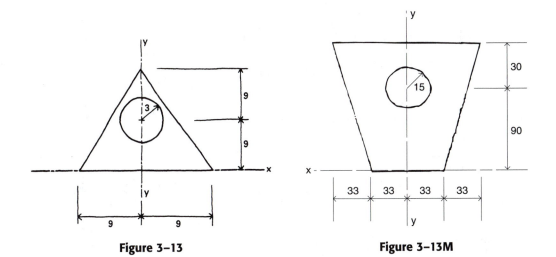

**Figure 3–13**                    **Figure 3–13M**

**3–6.** Figure 3–14 shows a steel beam cross section made up of a C6 × 10.5 channel and a W10 × 12 wide-flange shape. The centroid of the channel has been located by dimension. Other properties of the two parts may be obtained from Appendix J. Determine $\bar{y}$.

**3–6M.** Figure 3–14M shows a steel beam cross section made up of a C150 × 16 channel and a W250 × 18 wide-flange shape. The centroid of the channel has been located by dimension. Other properties of the two parts may be obtained from Appendix J. Determine $\bar{y}$.

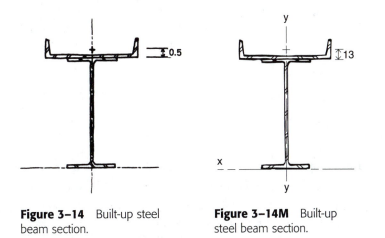

**Figure 3–14**   Built-up steel beam section.

**Figure 3–14M**   Built-up steel beam section.

**3–7.** Determine the centroid location for the group of footing pads shown in Figure 3–15. Each square is 9 ft on a side and each circle is 9 ft in diameter.

**3–7M.** Determine the centroid location for the group of footing pads shown in Figure 3–15M. Each square is 3 m on a side and each circle is 3 m in diameter.

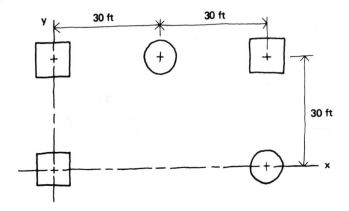

**Figure 3–15**  Footing plan.

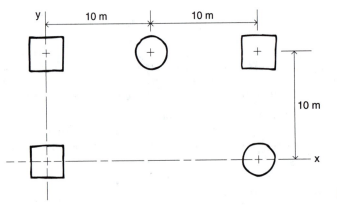

**Figure 3–15M**  Footing plan.

**3–8.** Locate the centroid of the prestressed single-T in Figure 3–16.

**3–8M.** Locate the centroid of the prestressed single-T in Figure 3–16M.

**3–9.** Figure 3–17 shows a cross section through a timber beam taken where a hole 2 in. in diameter was drilled to permit the passage of a pipe. Determine $\bar{y}$ with respect to the $x$ axis, which was a centroidal axis before the hole was drilled. (*Hint*: First determine $\bar{y}$ with respect to the base of the section.)

**3–9M.** Figure 3–17M shows a cross section through a timber beam taken where a hole 50 mm in diameter was drilled to permit the passage of a pipe. Determine $\bar{y}$ with respect to the $x$ axis, which was a centroidal axis before the hole was drilled. (*Hint*: First determine $\bar{y}$ with respect to the base of the section.)

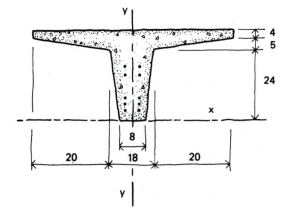

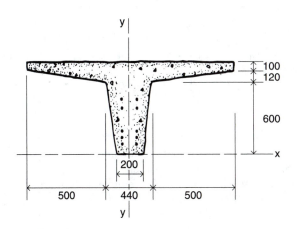

Figure 3–16 Prestressed beam cross section.

Figure 3–16M Prestressed beam cross section.

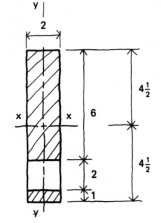

Figure 3–17

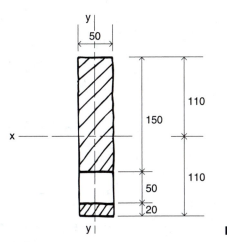

Figure 3–17M

## 3-3   MOMENT OF INERTIA

The *moment of inertia* of an area is a mathematical concept that is used to quantify the resistance of various sections to bending or buckling. It is a shape factor that measures the relative location of material in a cross section in terms of effectiveness. A beam section with a large moment of inertia, or *I* value, will have smaller stresses and deflections under a given load than one with a lesser *I* value. A thin shell will have less tendency to buckle if its surface is shaped so that a large moment of inertia is present. A long, slender column will not buckle laterally if the moment of inertia of its cross section is sufficient.

The concept of moment of inertia is vital to understanding the behavior of most structures, and it is unfortunate that it has no accurate physical analogy or description. Mathematically it is easy to compute, and in structural analysis it is easy to use, but beginning students often have difficulty with its abstract nature.

A moment-of-inertia value can be computed for any shape with respect to any reference axis. Figure 3–18 shows this general situation. The moment of inertia of an area about a given axis is defined as the sum of the products of all the elemental areas and the square of their respective distances to that axis. Thus we get the following two equations from Figure 3–18:

$$I_x = \int_0^A y^2 \, dA \tag{3-3}$$

$$I_y = \int_0^A x^2 \, dA \tag{3-4}$$

An *I* value has units of length to the fourth power, because the distance from the reference axis is squared. For this reason, the moment of inertia is sometimes called the *second moment* of an area. More important, it means that elements that are relatively far away from the axis will contribute substantially more to an *I* value than those which are close. Assuming that each element has the same area *dA*, then one located at twice the distance of another will have four times the moment of inertia.

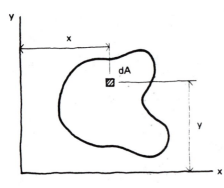

**Figure 3–18**

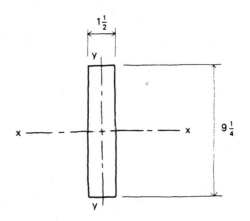

**Figure 3–19** A 2 ×10 joist with actual dimensions in inches.

For structural analysis purposes, usually only two $I$ values are important, the ones that can be computed with respect to the centroidal $x$ and $y$ axes. These are called the *principal axes.* Figure 3–19 shows a simple wood joist of rectangular cross section.

Following the previous discussion, this shape has a much larger moment of inertia about the $x$ axis than it does about the $y$ axis. This confirms what we already know from experience—that a rectangular shape has more resistance to bending if used as a joist than if used as a plank. Laid on its side, as in Figure 3–20(b), so that the load would be parallel to the $x$ axis, the rectangle would deflect and break under relatively low load. Like many structural elements, the rectangle has a strong axis and a weak axis. It is far more efficient to load the cross section so that bending occurs about the strong axis. Figure 3–20 illustrates strong-axis bending and weak-axis bending.

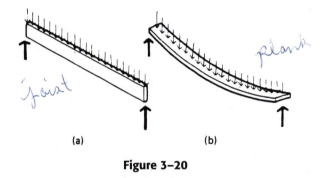

(a)                    (b)

**Figure 3–20**

It may help to understand the concept of moment of inertia if we draw an analogy based upon real inertia due to motion and mass. Imagine the two shapes in Figure 3–21 to be cut out of heavy sheet material and placed on an axle ($xx$) so they could spin about it. The two shapes have equal areas, but the one in Figure 3–21(a) has a much higher moment of inertia ($I_{xx}$) with respect to the axis of spin. It would be much harder to start it spinning, and once moving, much harder to stop. The same principle is involved when a figure skater spins on the ice. With arms held close in, the skater will rotate rapidly, and with arms outstretched (creating increased resistance to spin and more inertia), the skater slows down.

Similarly, a beam section shaped as in Figure 3–21(a), with flanges located far from the centroidal axis, will have far more resistance to bending than the cruciform shape in Figure 3–21(b).

As with centroidal determinations, it is seldom necessary to use the calculus to find moments of inertia for the shapes commonly used in building structures. Most often, $I$ values for regular shapes can be expressed in terms of the section dimensions. For example, the centroidal moment of inertia ($I_{xx}$) of a simple rectangle of dimensions $b$ and $h$ is found, using Figure 3–22, as follows:

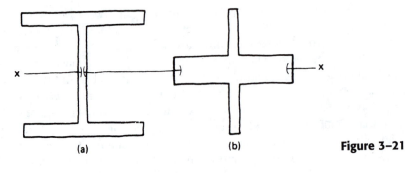

(a)                                        (b)                        **Figure 3–21**

$$I_{xx} = \int_0^A y^2 \, dA$$

$$= \int_{-h/2}^{h/2} y^2 b \, dy$$

$$= b \left[ \frac{y^3}{3} \right]_{-h/2}^{h/2}$$

$$= \frac{b}{3} \left[ \frac{h^3}{8} + \frac{h^3}{8} \right]$$

$$I_{xx \atop \substack{\text{centroid} \\ \text{rectangle}}} = \frac{bh^3}{12}$$

The moment of inertia expressions for several geometric shapes appear in Appendix F and will be used in the example problems. The values with respect to an axis located along an edge or base are also given, as these are sometimes useful.

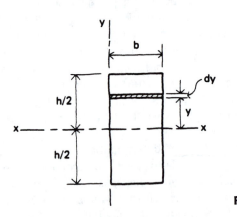

**Figure 3–22**

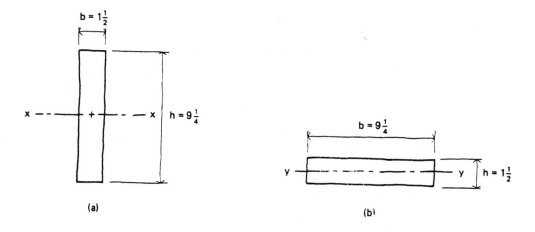

**Figure 3-23**

We can now quantify the previous illustration of how a rectangular joist has greater resistance to bending if used upright as opposed to flat. For the joist in question and referring to Figure 3-23, the centroidal moments of inertia are

$$I_{xx} = \frac{bh^3}{12} \qquad\qquad I_{yy} = \frac{bh^3}{12}$$

$$= \frac{1.5(9.25)^3}{12} \qquad\qquad = \frac{9.25(1.5)^3}{12}$$

$$= 98.9 \text{ in}^4 \qquad\qquad = 2.60 \text{ in}^4$$

The figures indicate that the nominal 2 × 10 section has a strong-axis moment of inertia that is almost 40 times larger than its weak-axis moment of inertia. The strong- and weak-axis moments of inertia for this and other timber rectangular sections may be found in Appendix I. The same values for selected steel beam shapes appear in Appendix J.

**EXAMPLE
3-4M**

Determine the centroidal moments of inertia for the wide-flange shape in Figure 3-24M.

### Solution

The $I_{yy}$ computation is merely the summation of the $I$ values of three rectangles. To compute $I_{xx}$, it will be necessary to subtract the $I$ values of two rectangles of "space," located on either side of the web, from the $I$ value of the enclosing rectangle.

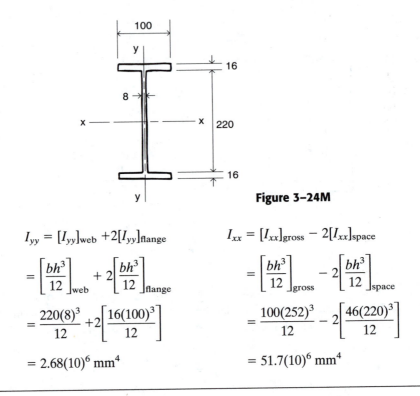

**Figure 3–24M**

$$I_{yy} = [I_{yy}]_{\text{web}} + 2[I_{yy}]_{\text{flange}}$$

$$= \left[\frac{bh^3}{12}\right]_{\text{web}} + 2\left[\frac{bh^3}{12}\right]_{\text{flange}}$$

$$= \frac{220(8)^3}{12} + 2\left[\frac{16(100)^3}{12}\right]$$

$$= 2.68(10)^6 \text{ mm}^4$$

$$I_{xx} = [I_{xx}]_{\text{gross}} - 2[I_{xx}]_{\text{space}}$$

$$= \left[\frac{bh^3}{12}\right]_{\text{gross}} - 2\left[\frac{bh^3}{12}\right]_{\text{space}}$$

$$= \frac{100(252)^3}{12} - 2\left[\frac{46(220)^3}{12}\right]$$

$$= 51.7(10)^6 \text{ mm}^4$$

---

**EXAMPLE 3–5**

Determine the moment of inertia with respect to the $xx$ axis of the 4-in square which has been rotated 45°, as in Figure 3–25.

**Solution**

The section is comprised of two triangles attached at their bases. From Appendix F, for each triangle,

$$I_{xx \atop \substack{\text{base} \\ \text{triangle}}} = \frac{bh^3}{12}$$

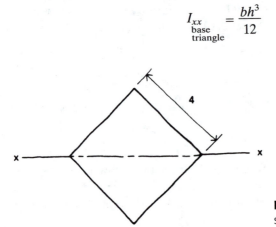

**Figure 3–25** Rotated square section.

The dimensions $b$ and $h$ can be computed because of the 45° relationship.

$$h = 4(\sin 45°)$$
$$= 2.83 \text{ in}$$

$$b = \frac{4}{\sin 45°}$$
$$= 5.66 \text{ in}$$

$$I_x = 2\left[\frac{bh^3}{12}\right]$$

$$= 2\left[\frac{5.66(2.83)^3}{12}\right]$$

$$= 21.4 \text{ in}^4$$

---

**EXAMPLE 3–6M**

Compute the centroidal moments of inertia of the shape in Figure 3–26M.

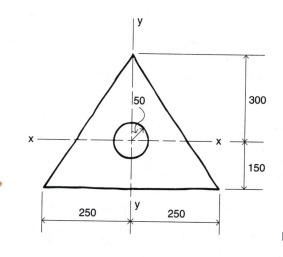

**Figure 3–26M**  Triangle with hole.

### Solution

$I_{xx}$ can be obtained by subtracting the $I$ of the circular hole from the $I$ of the gross triangle. (*Note:* This direct subtraction is only possible because the triangle and the circle share the same $xx$ centroidal axis.) $I_{yy}$ can be computed by subtracting the $I$ of the hole from the larger $I$ computed by adding two values for triangles about a common base. The common base in this case is the vertical $yy$ axis.

$$I_{xx} = [I_{xx}]_{\text{gross}} - [I_{xx}]_{\text{hole}}$$

$$= \left[\frac{bh^3}{36}\right]_{\text{gross}} - \left[\frac{\pi R^4}{4}\right]_{\text{hole}}$$

$$= \frac{500(450)^3}{36} - \frac{\pi(50)^4}{4}$$

$$= 1260(10)^6 \text{ mm}^4$$

$$I_{yy} = 2[I_y]_{\text{base}} - [I_{yy}]_{\text{hole}}$$

$$= 2\left[\frac{bh^3}{12}\right]_{\text{base}} - \left[\frac{\pi R^4}{4}\right]_{\text{hole}}$$

$$= 2\left[\frac{450(250)^3}{12}\right] - \frac{\pi(50)^4}{4}$$

$$= 1170(10)^6 \text{ mm}^4$$

## PROBLEMS

**3–10.** Determine the centroidal moments of inertia for the hollow beam cross section of Figure 3–27.

**3–10M.** Determine the centroidal moments of inertia for the slotted beam cross section of Figure 3–27M.

TRICKY TO GET ~~CAN'T~~ ~~DO~~ Iyy W/O USING PARALLEL AXIS TH. WHICH IS NOT DISCUSSED UNTIL NEXT SECTION

TO GET Iy:
1. FIND Iy GROSS
2. SUBTRACT Iy FOR ABCD
3. ADD Iy FOR SMALL RECT. IN CTR

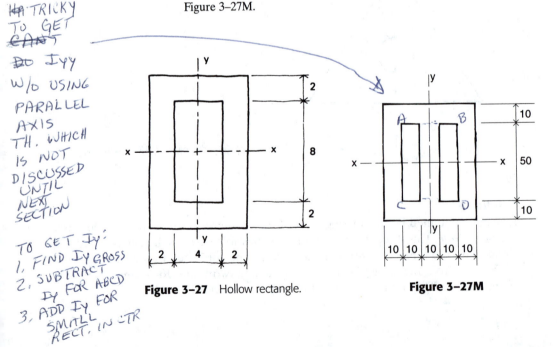

**Figure 3–27** Hollow rectangle.

**Figure 3–27M**

**3–11.** Determine the centroidal moments of inertia for the T-beam cross section of Figure 3–2.

**3–11M.** Determine $I_{yy}$ for the shape in Figure 3–9M.

**3–12.** Determine the $I_{xx}$ value for the tapered section in Figure 3–28.

**3–12M.** Determine the $I_{xx}$ value for the tapered section in Figure 3–28M.

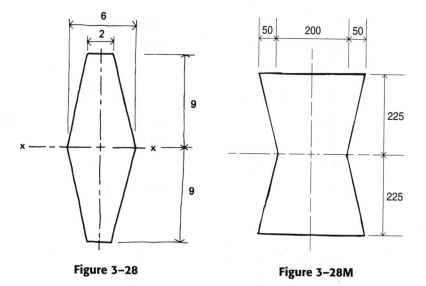

**Figure 3–28**                    **Figure 3–28M**

**3–13.** Determine the $I_{xx}$ value for the precast plank cross section of Figure 3–29. Each hole is 6 in. in diameter.

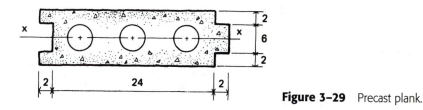

**Figure 3–29**   Precast plank.

**3–13M.** Determine the $I_{xx}$ value for the precast plank cross section of Figure 3–29M. Each hole is 150 mm in diameter.

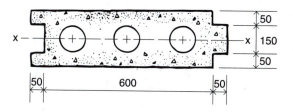

**Figure 3–29M**   Precast plank.

**3–14.** Determine the moment of inertia, with respect to the centroidal $xx$ axis, of the section in Figure 3–30. It is built up from two skins, each ½ in. thick, and six nominal 2 × 6 members. (Actual dimensions are 1½ × 5½ in.)

**3–14M.** Determine the moment of inertia, with respect to the centroidal $xx$ axis, of the section in Figure 3–30M. It is built up from two skins, each 12 mm thick, and six 38 × 139 mm flange pieces.

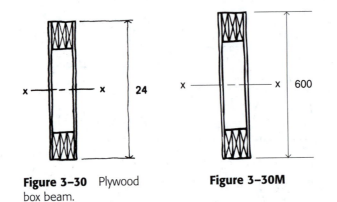

**Figure 3–30**  Plywood box beam.         **Figure 3–30M**

**3–15. (a)** Determine the $I_{yy}$ for the chevron shape in Figure 3–31.
   **(b)** Determine the $I_{xx}$ for the same shape. Note that $\bar{y}$ can be found by inspection.

**3–15M. (a)** Determine the $I_{yy}$ for the inverted chevron shape in Figure 3–31M.
   **(b)** Determine the $I_{xx}$ for the same shape. Note that $\bar{y}$ can be found by inspection.

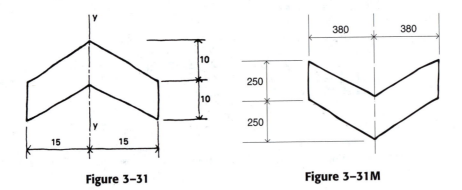

**Figure 3–31**              **Figure 3–31M**

**3–16.** With reference to the T shape of Figure 3–32, first locate the $xx$ centroidal axis by finding $\bar{y}$, and then computer the value of $I_{xx}$.

**3–16M.** With reference to the T shape of Figure 3–32M, first locate the $xx$ centroidal axis by finding $\bar{y}$, and then compute the value of $I_{xx}$.

**3–17M.** Using the centroid location determined for the tapered section in Example 3–3M, determine the $I_{xx}$ value.

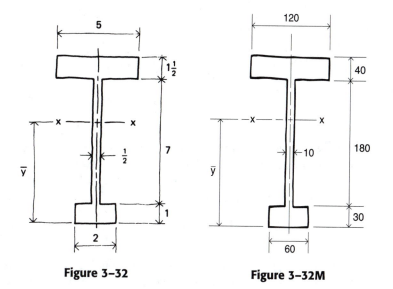

**Figure 3–32**

**Figure 3–32M**

## 3–4 PARALLEL AXIS THEOREM

The addition and subtraction of moment of inertia values for parts of complex shapes can become confusing and lead to errors. The parallel axis theorem provides a simple way to compute the moment of inertia of a shape about any axis parallel to a centroidal one. Its use not only saves time but eliminates most of the confusion. It is easily derived. Assume that we wish to find the moment of inertia of the general shape in Figure 3–33 with respect to an $x'$ axis which is parallel to the centroidal $xx$ one and located $d$ distance away. The general expression will give us

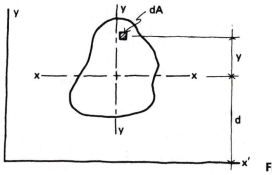

**Figure 3–33**

$$I_{x'} = \int_0^A (y + d)^2 \, dA$$

$$= \int_0^A (y^2 + 2yd + d^2) \, dA$$

$$I_{x'} = \int_0^A y^2 dA + 2d \int_0^A y \, dA + d^2 \int_0^A dA$$

The integral in the second term of this expression is the statical moment of the shape with respect to one of its own centroidal axes and, as such, must be zero-valued. Since the first term is a centrodial moment of inertia for the shape, the parallel axis theorem is reduced to

$$I_{x'} = I_{xx} + Ad^2 \tag{3–5}$$

where $I_{x'}$ = moment of inertia of the shape about a remote $x'$ axis
$\quad\ I_{xx}$ = moment of inertia of the shape about the centroidal $xx$ axis
$\quad\ A$ = area of the shape
$\quad\ d$ = perpendicular distance between the $xx$ and $x'$ axes

In almost all applications of this theorem, the $x'$ or remote axis is actually the centroidal axis of a composite section made up of several geometric shapes. The parallel axis theorem is applied to each of the shapes in turn to find the total or composite moment of inertia. In every case, the axes of the individual shapes *must* be centroidal ones. The following examples will illustrate the theorem's use.

---

**EXAMPLE 3–7**  Determine the moment of inertia about the $xx$ centroidal axis for the shape shown in Figure 3–34.

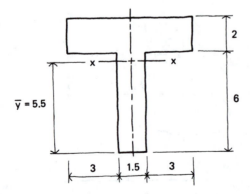

**Figure 3–34**  T-beam of Example 3–1.

## Solution

Since neither rectangle has its centroid coincident with the centroid of the entire section, two applications of the parallel axis theorem will be used. The $d$ distances are as given in Figure 3–35.

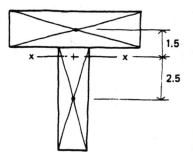

**Figure 3–35** Parallel axis distances for Example 3–7.

$$I_{xx} = [I_{xx} + Ad^2]_{\text{flange}} + [I_{xx} + Ad^2]_{\text{stem}}$$

$$= \left[\frac{bh^3}{12} + Ad^2\right]_{\text{flange}} + \left[\frac{bh^3}{12} + Ad^2\right]_{\text{stem}}$$

$$= \left[\frac{7.5(2)^3}{12} + 7.5(2)(1.5)^2\right] + \left[\frac{1.5(6)^3}{12} + 1.5(6)(2.5)^2\right]$$

$$= 122 \text{ in}^4$$

---

**EXAMPLE 3–8M**    Determine the $I_{yy}$ value for the retaining wall section of Figure 3–7M.

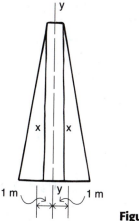

**Figure 3–36M**

### Solution

The $d$ distances are given in Figure 3–36M. The central rectangle has its $yy$ axis coincident with the $yy$ axis of the entire shape, but the triangles do not.

$$I_{yy} = [I_{yy}]_{rectangle} + 2[I_{yy} + Ad^2]_{triangle}$$

$$= \left[\frac{bh^3}{12}\right]_{rectangle} + 2\left[\frac{bh^3}{36} + Ad^2\right]_{triangle}$$

$$= \left[\frac{9(1)^3}{12}\right] + 2\left[\frac{9(1.5)^3}{36} + \frac{1}{2}(9)(1.5)(1)^2\right]$$

$$= 15.9 \text{ m}^4$$

---

**EXAMPLE 3–9**  To get additional bending resistance, a plate is welded to the top flange of a W14 × 22 as shown in Figure 3–37. The wide-flange shape itself has an $I_{xx}$ value of 199 in.$^4$, an area of 6.49 in.$^2$, and is 13.74 in. in actual depth. Determine the centroidal $xx$ moment of inertia for this built-up section.

### Solution

First determine the location of the centroidal $xx$ axis by finding $\bar{y}$. Selecting a reference axis through the bottom edge of the beam, we get

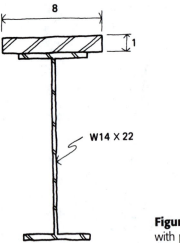

**Figure 3–37**  Wide-flange beam with plate.

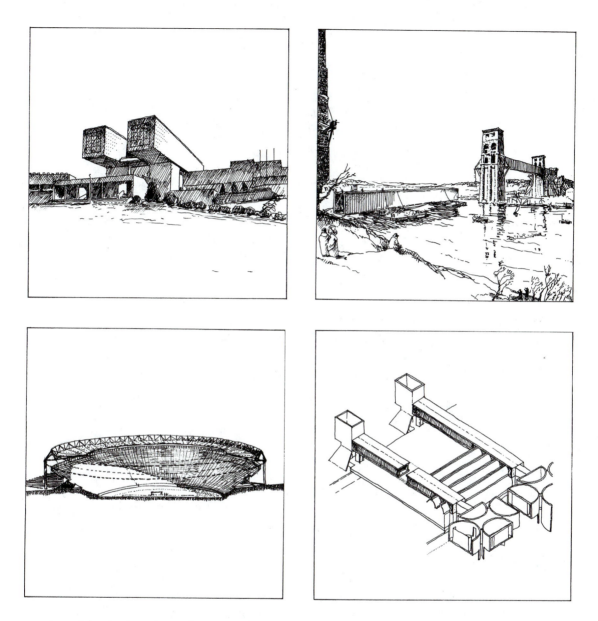

Large tube structures have often proven attractive to architects and engineers for their simultaneous structural and sculptural strengths. With materials concentrated at the outer edges of the volume, a tube attains a high moment of inertia (I) relative to its overall cross-sectional area, and the interior space can often be useful as well. Thus Robert Stephenson's 1849 Britannia Bridge, whose spanning sections were riveted metal tubes sized to carry early railway trains inside them, had a forceful presence in the land- and seascape of the Menai Straits. Arata Isozaki's 1974 Kitakyushu City Museum turns a hilltop into a spectacular new sort of Acropolis for Art. The 1990 renovation of the Rome Olympic Stadium by Massimo Majowiecki uses a space-grid tube, triangular in section, to form an enormous, stiff compression ring, elliptical in overall plan, for carrying the wide canopy over a completely unobstructed view of the playing field. Louis I. Kahn's early scheme (c. 1960) for the Salk Institute Research Building envisioned a system of structural tubes, "servant space," carrying the roof vaults and mechanical systems for the "served space," of the laboratories below.

**131**

$$\bar{y} = \frac{\Sigma y_i A_i}{\Sigma A_i}$$

$$= \frac{6.87(6.49) + 14.24(8)(1)}{6.49 + 8(1)}$$

$$= 10.9 \text{ in}$$

$$I_{xx} = [I_{xx} + Ad^2]_{\text{wide flange}} + [I_{xx} + Ad^2]_{\text{plate}}$$

With reference to Figure 3–38, we get

$$I_{xx} = [199 + 6.49(4)^2] + \left[\frac{8(1)^3}{12} + 8(1)(3.3)^2\right]$$

$$= 391 \text{ in}^4$$

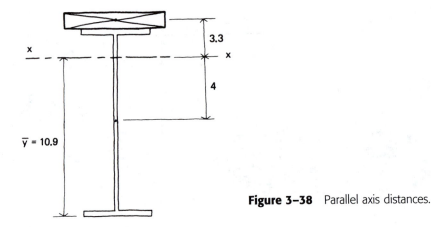

**Figure 3–38**   Parallel axis distances.

## PROBLEMS

**3–18.** Use the parallel axis theorem to solve Problem 3–12.

**3–18M.** Use the parallel axis theorem to solve Problem 3–12M.

**3–19.** Determine the $I_{xx}$ for the channel beam in Figure 3–9.

**3–19M.** Determine the $I_{xx}$ for the section in Figure 3–9M.

**3–20.** Two 2 × 10 joists enclose a 2 × 4 member to make the beam section of Figure 3–39. Determine the centroidal moment of inertia, $I_{xx}$. (Section properties are given in Appendix I.)

**3–20M.** Two 38 × 235 joists enclose a 38 × 89 member to make the beam section of Figure 3–39. Determine the centroidal moment of inertia, $I_{xx}$. (Section properties are given in Appendix I.)

**3–21.** Use the parallel axis theorem to solve Problem 3–15.

**Figure 3–39** Built-up timber beam.

**3–21M.** Use the parallel axis theorem to solve Problem 3–15M.

**3–22.** Knowing that $\bar{y}$ has a value of 6.34 in. in Figure 3–32, determine $I_{xx}$ for the shape by using the parallel axis theorem.

**3–22M.** Knowing that $\bar{y}$ has a value of 160 mm in Figure 3–32M, determine $I_{xx}$ for the shape by using the parallel axis theorem.

**3–23.** Determine the $I_{xx}$ value for the steel box beam of Figure 3–40. It is made of two C10 × 25 channels and two plates as shown. Each channel has an $I_{xx}$ of 91.2 in.$^4$, an area of 7.35 in.$^2$, and is 10.0 in. in depth.

**3–23M.** Determine the $I_{xx}$ value for the steel box beam of Figure 3–40M. It is made of two C250 × 37 channels and two plates as shown. Each channel has an $I_{xx}$ of $38.0(10)^6$ mm$^4$, an area of 4740 mm$^2$, and is 254 mm in depth.

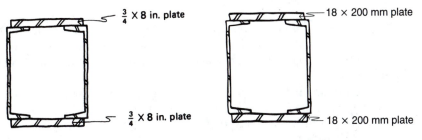

**Figure 3–40** Steel box beam.           **Figure 3–40M**

**3–24.** The 2 × 12 joist of Figure 3–41 has an $I_{xx}$ value of 178 in.$^4$, as given in Appendix I. What will be the percentage decrease in this value if a hole 2 in. in diameter is drilled through it at
(a) location A?
(b) location B?

**3–25M.** The wide-flange beam shape of Figure 3–24M must be cut in half at the $xx$ axis to make two structural T shapes, each 126 mm in depth. Compute the $I_{xx}$ value for one of the T sections. What percentage does this represent of the total $I_{xx}$ previously computed in Example 3–4M?

**3–26.** Locate the centroidal $xx$ axis and determine $I_{xx}$ for the concrete shape in Figure 3–42.

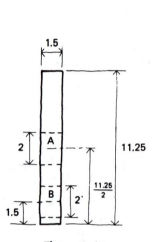

**Figure 3–41**

**Figure 3–42**  Prestressed concrete section.

## 3–5  RADIUS OF GYRATION

The *radius of gyration* is a concept that expresses a relationship between the area of a cross section and a centroidal moment of inertia. It is a shape factor, which measures resistance to bending (or buckling) about a certain axis, and accounts for both $I$ and $A$.

Continuing the concept of rotational inertia discussed in Section 3–3, the radius of gyration ($r$) represents the location of two parallel lines, one on each side of the axis of spin, at which all of the mass of an object could be concentrated with no change in inertia. For a cross section instead of a mass, we say that all the area may be placed in two lines with no change in the moment of inertia. The $r$ value with respect to a centroidal axis, as indicated in Figure 3–43, is a perpendicular distance from the axis to one of the imaginary lines of concentration. For the shapes most fre-

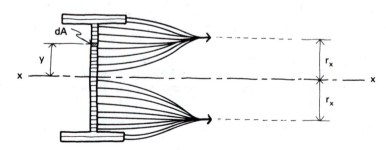

**Figure 3–43**  Concept of radius of gyration.

quently encountered in structural analysis, there are two $r$ values, $r_x$ and $r_y$, one each for the strong and weak axes, respectively.

If, by definition, the moment of inertia for the shape does not change, then from Figure 3–43,

$$I_{xx} = \int_0^A y^2 \, dA = r_x^2 \int_0^A dA$$

If we sum up the elemental areas, we get

$$I_{xx} = r_x^2 A$$

Solving for the radius of gyration, we obtain

$$r_x = \sqrt{\frac{I_{xx}}{A}} \tag{3–6}$$

Similarly for the weak axis,

$$r_y = \sqrt{\frac{I_{yy}}{A}} \tag{3–7}$$

The radius of gyration is most useful in the design of slender compression members to resist buckling. It is central to the column buckling theory developed in Chapter 11 but can also be useful in other ways because it has simple units (length) and can replace two section properties.

---

**EXAMPLE 3–10**

Determine the radius of gyration about the $xx$ axis for the T shape of Figure 3–34.

**Solution**

$$I_{xx} = 122 \text{ in}^4$$

$$r_x = \sqrt{\frac{I_{xx}}{A}}$$

$$= \sqrt{\frac{122 \text{ in}^4}{24 \text{ in}^2}}$$

$$= 2.25 \text{ in}$$

---

**EXAMPLE 3–11M**

Determine the radius of gyration about the $xx$ axis for the plywood box beam of Figure 3–30M.

### Solution

$$r_x = \sqrt{\frac{I_{xx}}{A}}$$

$$I_{xx} = 2170(10)^6 \text{ mm}^4$$

$$A = 6(5280) + 2(12)(600)$$

$$= 46\ 100 \text{ mm}^2$$

$$r_x = \sqrt{\frac{2170(10)^6 \text{ mm}^4}{46\ 100 \text{ mm}^2}}$$

$$r_x = 217 \text{ mm}$$

Both moment of inertia and radius of gyration are properties of areas (usually cross-sectional areas) that depend upon the shape of the area. They both measure the distribution of the area relative to the centroidal axes. A cross section with most of its area located far from its centroid will have a large $I$ value and a correspondingly large $r$ value. One fortunate characteristic of the radius of gyration (as opposed to the moment of inertia) is that one can visually estimate the value of $r$ with considerable accuracy. Estimating $I$ values by inspection requires considerable experience and even then is not easy.

For rectangular cross sections, represented by most timber beams and columns, the radius of gyration imaginary lines are located about 60% of the distance from the centroidal axes out to the edge. For a wide-flange beam section, $r_x$ will locate lines quite a bit closer to the edge, because for this shape most of the moment of inertia is generated by its flanges. These relationships are illustrated in Figure 3–44.

Making such sketches using calculated values will help pinpoint most calculation errors. Some analysts check moment of inertia calculations by finding the corresponding $r$ value and making sure it has a rational value.

## PROBLEMS

**3–27.** Determine $r_x$ and $r_y$ for the wide-flange shape of Figure 3–24M.

**3–28.** Using the book-provided answers for Problem 3–10, determine $r_x$ and $r_y$ for the hollow rectangle.

**3–28M.** Using the book-provided answers for Problem 3–10M, determine $r_x$ and $r_y$ for the slotted rectangle.

**3–29M.** Use the book-provided answers for Problem 3–12M, determine $r_x$ for the tapered section.

**3–30.** Determine $r_x$ for the rotated square of Figure 3–25.

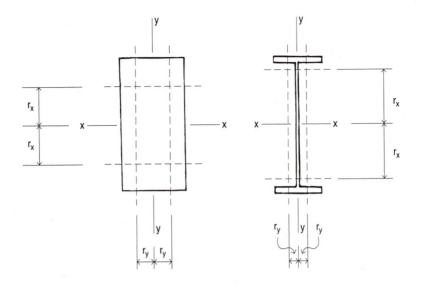

**Figure 3–44**   Radius of gyration relative magnitudes.

**3–31.** Using the book-provided answers for Problem 3–15, determine $r_x$ and $r_y$ for the chevron shape.

**3–31M.** Using the book-provided answers for Problem 3–15M, determine $r_x$ and $r_y$ for the inverted chevron shape.

**3–32.** Figure 3–45 shows two nominal 2 × 6 pieces that are to be fastened together to make a column. What should be the center-to-center spacing $s$ so that $r_y$ will equal $r_x$ for the column cross section?

**3–32M.** Figure 3–45 shows two nominal 38 × 139 pieces that are to be fastened together to make a column. What should be the center-to-center spacing $s$ so that $r_y$ will equal $r_x$ for the column cross section?

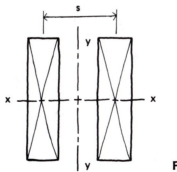

**Figure 3–45**

# 4

# Stress and Strain

## 4-1 TYPES OF STRESS

In simple terms, *stress* is merely the intensity of force. It is force per unit area. The levels and types of stress at different points throughout a building structure are of prime concern to the structural designer. Forces are generated in response to applied loads, dead and live, and those forces always result in a stressed body or an accelerating body. Since most building elements are not meant to accelerate, they develop internal stresses. These stresses, if they become too large, can cause rupture or excess deformation.

Figure 4–1(a) shows a simple bar in tension. The free-body cut in Figure 4–1(b) exposes the stress inside the bar. Even though the loads $P$ are point loads applied along the central axis of the bar, the stress is constant over the cross section except in a zone right next to the end of the bar. The intensity of the stress is simply

$$f_a = \frac{P}{A} \tag{4-1}$$

where $f_a$ = axial stress (psi or ksi) (kPa or MPa)
  $P$ = axial force (lb or kips) (kN)
  $A$ = cross-sectional area (in.$^2$) (m$^2$)

If the stress is too great, it can be reduced by increasing the area of the member. However, it is also possible to reduce the force. This can sometimes be done through a change in geometry or the manner in which the loads develop the force (i.e., a change in the statics of the design).

There are really only two kinds of stress, normal and tangential. *Normal stresses* act at right angles to the surface of the stressed area, while *tangential stresses* act parallel to that surface. Shearing stresses are tangential, while the axial stress shown in Figure 4–1 is normal. Bending results in both normal and tangential stresses, and this will be discussed in detail in Chapters 7 and 8.

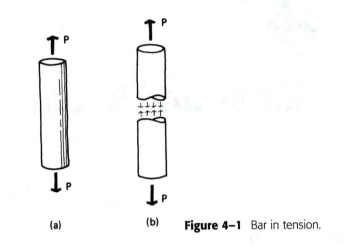

**Figure 4–1**  Bar in tension.

(a)                              (b)

## PROBLEMS

**4–1.** Determine the average axial tensile stress in the bar of Figure 4–1 if $P = 50$ kips and the cross-sectional area of the bar is 4 in.$^2$.

**4–1M.** Determine the average axial tensile stress in the bar of Figure 4–1 if $P = 225$ kN and the cross-sectional area of the bar is 2500 mm$^2$.

**4–2.** A nominal $6 \times 6$ timber is used as a short foundation post. It must support a load of 10,000 lb. Determine the average axial stress in the post. (*Hint*: Actual dimensions for the post may be found in Appendix I.)

**4–2M.** A $139 \times 139$ timber is used as a short foundation post. It must support a load of 45 kN. Determine the average axial stress in the post.

**4–3.** A steel cable used in a suspended roof system has an allowable tensile stress of 150 ksi. It must safely carry a load of 1060 kips. Determine its required diameter.

**4–3M.** A steel cable used in a suspended roof system has an allowable tensile stress of 1050 MPa. It must safely carry a load of 5000 kN. Determine its required diameter.

**4–4.** The bolts in Figure 4–2 are being subjected to shear by a vertical reaction. The shear will act tangentially on a transverse section through each bolt. Determine the average shearing stress in the bolts if the reaction is 15 kips and the diameter of each bolt is 1 in.

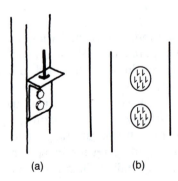

(a)                              (b)

**Figure 4–2**  Beam seat causing shearing stresses across the bolt cross sections.

**4-4M.** The bolts in Figure 4–2 are being subjected to shear by a vertical reaction. The shear will act tangentially on a transverse section through each bolt. Determine the average shearing stress in the bolts if the reaction is 75 kN and the diameter of each bolt is 25 mm.

**4-5.** The bolt in Figure 4–3 is 1 in. in diameter. If the load $P$ is 10 kips, determine the average shearing stress in the bolt.

**4-5M.** The bolt in Figure 4–3 is 25 mm in diameter. If the load $P$ is 45 kN, determine the average shearing stress in the bolt.

**4-6.** The plates in Figure 4–3 are each 3 in. by 0.5 in. in cross section. If the hole in each plate is $\frac{13}{16}$ in. in diameter and the load $P$ is 20 kips, determine the average tensile stress in the plate. [*Hint*: Figure 4–4(c) may be useful.]

**4-6M.** The plates in Figure 4–3 are each 75 mm by 12 mm in cross section. If the hole in each plate is 20 mm in diameter and the load $P$ is 90 kN, determine the average tensile stress in the plate. [*Hint*: Figure 4–4(c) may be useful.]

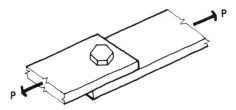

**Figure 4–3**  Bolted connection.

## 4–2  BASIC CONNECTION STRESSES

A simple bolted connection can serve to illustrate several different examples of normal and tangential stresses. The bolted connection of Figure 4–3 could fail in any one of four different ways if subjected to overload. Aside from the obvious bolt shear failure shown in Figure 4–4(a), the hole in the plate could elongate by a compression bearing failure as in Figure 4–4(b). In this case the crushing area to be used for design purposes is the projected rectangle, dimensioned by the bolt diameter for one side and the plate thickness for the other. Bearing stress is a normal stress.

The plate might also fail by excess normal tensile stresses on the area of the plate left after the hole was drilled, as in Figure 4–4(c). This type of stress is called "tension on the net section." Similarly, if the hole was located too close to the end of one of the plates, the tangential failure of Figure 4–4(d) might result. Here the stressed area is actually two parallel sides of a "plug" of material pushed out of the plate by the bolt.

## 4–3  STRAIN

Stresses are usually accompanied by a *strain*, which is a physical change in the size or shape of the stressed body. *Normal stresses* result in a shortening or lengthening of the fibers of that body, while *tangential* or *shearing strain* indicates an angle

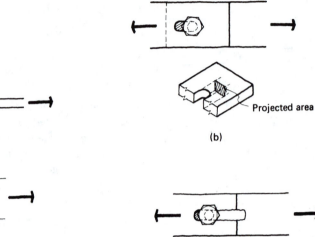

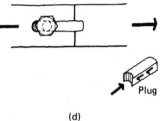

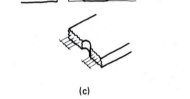

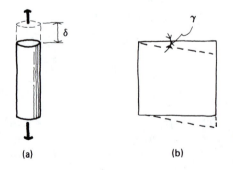

**Figure 4–4** Stresses in a bolted connection: (a) shear through the bolt; (b) excess bearing (crushing) of the plate at the hole; (c) tension on the net section of the plate; (d) end shear-out of the plate.

change. It is interesting to note that stress can never be seen, whereas strain can be seen and precisely measured.

Figure 4–5 shows two types of strain. The normal strain in Figure 4–5(a), called $\delta$, is the *total strain*. Total strain can be either elongation (tensile) or shortening (compressive), and in this case it is elongation. This total strain is the sum of the smaller strains occurring in each individual unit length of the bar. The average unit strain is designated as $\epsilon$.

**Figure 4–5** Normal strain (a) and shearing strain (b).

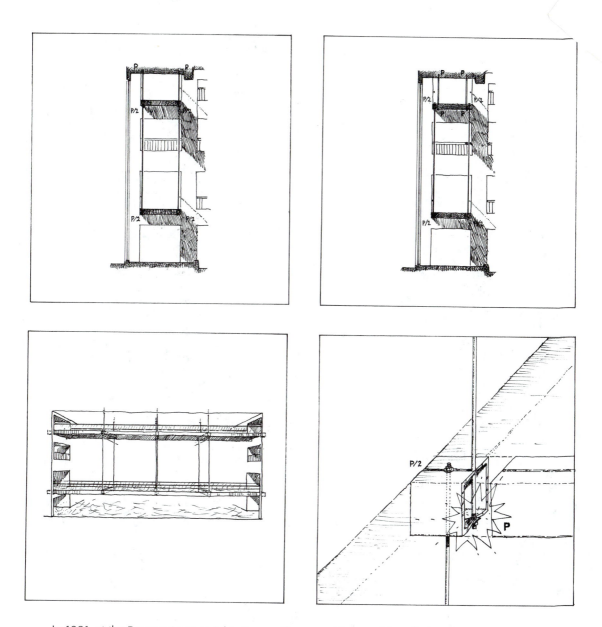

In 1981, at the Regency Hyatt Hotel in Kansas City, two cable-hung "skywalk" bridges collapsed, killing more than a hundred people. The disaster was caused by a simple, fatal error in elementary structural understanding. As designed (cross-section top left; side view lower left), the carrying cables, a single piece bottom to top, received the intended loads *directly* from transverse beams. As built, the cable was in, not one, but two sections, with the load from the lower transverse beams transferred *indirectly*, via the upper transverse beams, to the upper cables (top right illustration, lower right detail). The stress at the threaded-bolt-and-washer connections between upper beams and upper cables thus became **twice** what had been designed for; the resulting strain on these joints allowed the upper cables to pull through the bottom of the welded tubular transverse beams. The upper bridge crashed down onto the lower one, which in turn broke and collapsed to the ground.

Intuitive understanding of the development of axial stresses and the corresponding strains—and their dependencies upon cross-sectional area and material stiffness—can be used by the architect to promote the reading of a certain kind of character into a building. Such manipulations, though they make use of our visual knowledge of structural behavior, may or may not have anything to do with the actual stability of the structure. Thus the builders of Cologne Cathedral chose to make the front faces of the massive nave piers look like almost-independent columns, the only things visibly connecting the vaults above with the ground below. The "columns" are so slender as to seem incapable of taking any appreciable load from the vaults: they hint, even, at being in tension, not columns at all but cables tying down the upward-ballooning roof. At the other extreme, Frank Furness used visual expectations of normal column proportions, and of behavior under axial compression, to suggest that his facade possesses an almost intimidating weight and density.

**EXAMPLE 4–1** Figure 4–6 illustrates a bar 400 in. long with a total strain (shown greatly exaggerated) of 2 in. Each of the 400 inches of the original bar got a tiny bit longer, so that the total effect over the aggregate length summed to 2 in. $\epsilon$ is the amount of strain experienced by each unit length. $\delta$ is the total strain and equal to $\epsilon$ times the number of units in the original length:

$$\delta = \epsilon L \tag{4–2}$$

Solving this for the unit strain, we get

$$\epsilon = \frac{\delta}{L}$$

and for the case in question,

$$\epsilon = \frac{2 \text{ in}}{400 \text{ in}} = 0.005$$

Notice that the average unit strain will always be a pure number, as $\delta$ and $L$ must have like units.

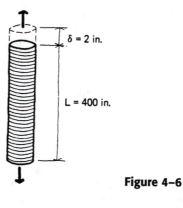

$\delta = 2$ in.

$L = 400$ in.

**Figure 4–6**

**EXAMPLE 4–2M** Figure 4–7M shows a concrete cylinder 300 mm long with a total strain (shown greatly exaggerated) of 0.5 mm. Each of the 300 mm of length got a little shorter, so that the total shortening summed to 0.5 mm. $\epsilon$ is the amount of strain experienced by each unit length. $\delta$ is the total strain and is equal to $\epsilon$ times the number of units in the original length:

$$\delta = \epsilon L$$

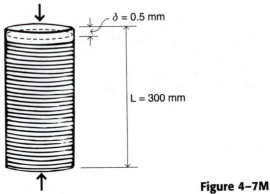

**Figure 4–7M**

Solving this for the unit strain, we get

$$\epsilon = \frac{\delta}{L}$$

and for the case in question,

$$\epsilon = \frac{0.5 \text{ mm}}{300 \text{ mm}} = 0.001\ 67$$

Notice that the average unit strain will always be a pure number, as $\delta$ and $L$ must have like units.

## PROBLEMS

**4–7.** A reinforced concrete column is 15 ft long and under load it shortens ⅛ in. Determine its average unit strain.

**4–7M.** A reinforced concrete column is 5 m long and under load it shortens 3 mm. Determine its average unit strain.

**4–8.** A wood post 20 ft long shortens ¼ in. under load. Determine the average unit strain.

**4–8M.** A wood post 6 m long shortens 6 mm under load. Determine the average unit strain.

**4–9.** A 300-ft-long steel cable is loaded in tension until the average unit strain is 0.004. Determine the total elongation under this load.

**4–9M.** A 100-m-long steel cable is loaded in tension until the average unit strain is 0.004. Determine the total elongation under this load.

**4–10.** In explaining the behavior of concrete under load, one popular theory indicates that concrete will crush when the compressive unit strain reaches 0.003. If a 12-in.-tall concrete test cylinder is failed in a testing machine, what will be its height when it fails?

**4–10M.** In explaining the behavior of concrete under load, one popular theory indicates that concrete will crush when the compressive unit strain reaches 0.003. If a 300-mm-tall concrete test cylinder is failed in a testing machine, what will be its height when it fails?

# 4–4 STRESS VERSUS STRAIN

In 1678, Robert Hooke, an Englishman, observed that most materials were essentially *elastic*, that is, the deformations in a stressed body would disappear upon removal of the load. Furthermore, the relationship between stress and strain (or between load and deformation) was a linear one. Many materials are "springlike," in that if we place a 1-kip (4.5 kN) load on a member, it will strain a certain amount, and if we add an additional 1-kip (4.5 kN) load, an additional strain of that same amount will take place. If we continue this loading process, of course, the material will eventually rupture or, more usually, *yield* (permanently deform), and proportionality between stress and strain will no longer exist. Some materials have definite yield points or stress levels beyond which additional load will cause a great increase in strain. Most steels have very definite yield points, and a graph plotting stress versus strain will show a straight line up to the yield stress and then show an abrupt departure from linearity as failure begins. Other materials, such as concrete (in compression), have no definite yield point, and stress is not proportional to strain except at low levels of stress.

Figure 4–8 illustrates the sharp yield point of mild steel. Figure 4–8(b) is merely an enlarged picture of the elastic region in Figure 4–8(a). The ultimate strength of steel is designated as $F_u$, and the yield strength is called $F_y$. This particular curve is interesting because it shows a large plastic region where strain contin-

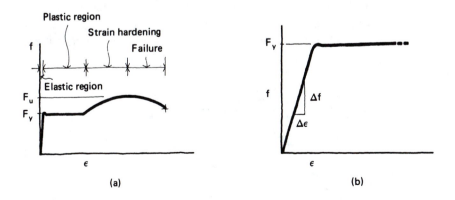

**Figure 4–8** Stress–strain curve for mild steel in tension.

ues with no increase in load, demonstrating the "taffylike" nature of the material. Also indicated is the "strain-hardening" property of mild steel, which causes an increase in strength just before failure.

The curve for structural concrete in compression is shown in Figure 4–9. Notice that the curve for concrete has no real straight-line portion and that stress is only approximately linear with strain and then only for low loads. The ultimate strength of concrete in compression is designated as $f'_c$.

Notice that the curves drop downward, indicating a loss of strength near the material failure point. This happens by virtue of the manner in which the materials are tested and merely illustrates the inability of a testing machine to keep load upon a rapidly disintegrating material.

By definition, once a material has experienced any appreciable yield, it will not return to its original shape when the load is removed but will show some permanent deformation. For all practical purposes, such a material has failed. Therefore, we strive to keep stress levels well below such values when designing structural elements.

The curves of four different materials have been drawn at the same scale in Figure 4–10 so that the reader can sense the relative strengths. The curves for concrete and wood are in compression, while those for steel were taken from tensile tests. (Steel tested in compression behaves much as it does in tension, but in practical applications, buckling failures preclude a true compressive yield. Buckling is discussed in Chapter 11.)

## 4–5    STIFFNESS

The stiffness of a material is a measure of how much that material strains for a given amount of stress. If we place equal loads upon like pieces of wood and steel, the wood will undergo a much larger strain than will the steel. This happens because steel is much stiffer than wood. Because stronger materials are usually stiffer, we tend to confuse stiffness with strength, when in reality the two properties are quite different. The strength of a material can be ascertained as the highest stress achieved during a test of the material (i.e., the largest ordinate at any point on the stress–strain curve). The stiffness, on the other hand, is the slope of the stress–strain curve, given as $\Delta f/\Delta \epsilon$ in Figures 4–8(b) and 4–9. The magnitude of this slope is called

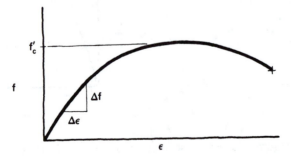

**Figure 4–9**  Stress–strain curve for concrete in compression.

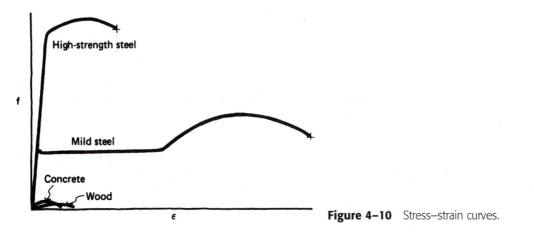

**Figure 4–10** Stress–strain curves.

the *modulus of elasticity* [or *Young's modulus*, after Thomas Young (1773–1829), an English scientist[1]] and is designated by the symbol $E$. It has the same units as stress because it is defined as incremental stress divided by incremental strain, and of course, strain has no units. Values of $E$ for several common materials are tabulated in Appendix E.

The $E$ value or stiffness of concrete is about one-tenth that of steel, while structural wood is about one-half as stiff as concrete. This important property helps to tell us how much a given material will deform or deflect under load. How much will a certain cable stretch? How much will a steel girder deflect? How "bouncy" will a wood joist floor be? The answers to these questions are independent of the material strength but are directly related to material stiffness.

The following examples illustrate how $E$ can be quantified.

**EXAMPLE 4–3** A sample of steel is being stressed in a tensile testing machine. Stress is found to be linear with strain in the elastic region, and when $f = 7400$ psi, $\epsilon = 0.000\ 25$. At a greater load, $f = 22\ 200$ psi and $\epsilon = 0.000\ 75$. Determine the $E$ value for steel.

**Solution**

$$E = \frac{\Delta f}{\Delta \epsilon}$$

$$= \frac{22\ 200 \text{ psi} - 7400 \text{ psi}}{0.000\ 75 - 0.000\ 25}$$

[1] Actually, it was Claude L. M. H. Navier (1785–1836), a French engineer, who first stated the relationship the way we use it today.

$$= \frac{14\,800 \text{ psi}}{0.000\,50}$$

$$E = 29.6(10)^6 \text{ psi}$$

[*Note: E* **is a constant for all hot-rolled steel shapes and is usually taken as** $29(10)^6$ **psi.**]

**EXAMPLE**
**4–4M**

A sample of steel is being stressed in a tensile testing machine. Stress is found to be linear with strain in the elastic region, and when $f = 50$ MPa, $\epsilon = 0.000\,25$. At a greater load, $f = 150$ MPa and $\epsilon = 0.000\,75$. Determine the $E$ value for steel.

*Solution*

$$E = \frac{\Delta f}{\Delta \epsilon}$$

$$= \frac{150 \text{ MPa} - 50 \text{ MPa}}{0.000\,75 - 0.000\,25}$$

$$= \frac{100 \text{ MPa}}{0.000\,50}$$

$$E = 200\,000 \text{ MPa or } 200 \text{ GPa}$$

[*Note: E* **is a constant for all hot-rolled steel shapes and is usually taken as 200 GPa.**]

## PROBLEMS

**4–11.** A concrete cylinder with a cross-sectional area of 28.3 in.$^2$ is to be tested in axial compression. Before loading, two marks are scribed on the cylinder precisely 8.000 in. apart, as in Figure 4–11. When the load is 40 000 lb, the marks are measured and found to be 7.997 in. apart. Determine the $E$ value for this concrete. Assume an approximately linear relationship between stress and strain.

**4–11M.** A concrete cylinder with a cross-sectional area of 18 200 mm$^2$ is to be tested in axial compression. Before loading, two marks are scribed on the cylinder precisely 200.0 mm apart, as in Figure 4–11M. When the load is 182 kN, the marks are measured and found to be 199.9 mm apart. Determine the $E$ value for this concrete. Assume an approximately linear relationship between stress and strain.

**4–12.** The modulus of elasticity for the steel cable of Problem 4–9 is $25(10)^6$ psi. Determine the average stress in the cable under the applied load. Assume a linear stress–strain curve.

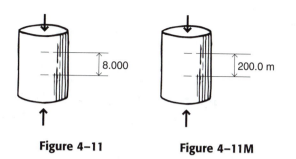

**Figure 4–11**          **Figure 4–11M**

**4–12M.** The modulus of elasticity for the steel cable in Problem 4–9M is 160 GPa. Determine the average stress in the cable under the applied load. Assume a linear stress–strain curve.

**4–13.** A 2-in.-diameter steel hanger rod carries an axial load of 62.8 kips. Under this load its average unit strain is 0.000 69. Determine the modulus of elasticity for this steel.

**4–13M.** A 50-mm-diameter steel hanger rod carries an axial load of 271 kN. Under this load its average unit strain is 0.000 69. Determine the modulus of elasticity for this steel.

**4–14.** The wood post of Problem 4–8 has an $E$ value of $1.3(10)^6$ psi. Determine the average unit stress in the post. Assume a linear stress–strain relationship.

**4–14M.** The wood post of Problem 4–8M has an $E$ value of 9000 MPa. Determine the average unit stress in the post. Assume a linear stress–strain relationship.

**4–15.** A concrete pier 12 in. by 12 in. in section is 6 ft long and shortens a total of 0.05 in. when it is loaded by a 300-kip axial load. Assuming a stress–strain relationship that is approximately linear, determine the $E$ value for this concrete.

**4–15M.** A concrete pier 300 mm by 300 mm in section is 2 m long and shortens a total of 1 mm when it is loaded by a 1300 kN axial load. Assuming a stress–strain relationship that is approximately linear, determine the $E$ value for this concrete.

**4–16.** Carefully describe how stress is different from strain. Can you think of a situation where stress might exist without strain or strain without stress?

# 4–6  TOTAL AXIAL DEFORMATION

A unique relationship can be drawn between load and deformation by combining some of the relationships developed in previous sections. The total deformation of an axially loaded member is given by

$$\delta = \epsilon L \tag{4–2}$$

The average unit strain $\epsilon$ can be expressed in terms of the average unit stress and the modulus, because $E = f/\epsilon$ or $\epsilon = f/E$. Substituting, we get

$$\delta = \frac{fL}{E} \tag{4–3}$$

and since $f = P/A$,

$$\delta = \frac{PL}{AE} \qquad\qquad (4\text{--}4)$$

Equations 4–3 and 4–4 are only valid, of course, if the relationship between stress and strain is essentially linear. For this reason, they are more useful for materials such as steel and less useful for concrete.

**EXAMPLE 4–5** The first-story column of a tall steel building is 20 ft long and must carry an axial load of 5000 kips. If $E = 29(10)^6$ psi and the cross-sectional area of the column is 215 in.$^2$, determine its total shortening.

**Solution**

$$\delta = \frac{PL}{AE}$$

$$= \frac{5\,000\,000\ \text{lb}(20\ \text{ft})(12\ \text{in/ft})}{29(10)^6\ \text{psi}(215\ \text{in}^2)}$$

$$= 0.192\ \text{in}$$

or

$$\delta \approx 0.2\ \text{in}$$

**EXAMPLE 4–6M** The first-story column of a tall steel building is 6 m long and must carry an axial load of 22 000 kN. If $E = 200$ GPa and the cross-sectional area of the column is 140 000 mm$^2$, determine its total shortening.

**Solution**

$$\delta = \frac{PL}{AE}$$

$$= \frac{22\,000\ \text{kN}(6\ \text{m})}{(0.140\ \text{m}^2)[200(10)^6\ \text{kN/m}^2]}$$

$$= 4.71(10)^{-3}\ \text{m}$$

or

$$\delta = 5\ \text{mm}$$

## PROBLEMS

**4–17.** A 1-in.-diameter elevator cable is 1000 ft long and carries a load of 20 kips. If it has a modulus of elasticity of $25(10)^6$ psi, determine its total elongation.

**4–17M.** A 25-mm-diameter elevator cable is 300 m long and carries a load of 100 kN. If it has a modulus of elasticity of 170 GPa, determine its total elongation.

**4–18.** A 25-ft-long wood column is stressed in compression to 700 psi. If $E = 1.4(10)^6$ psi, determine the total shortening of the column.

**4–18M.** An 8-m-long wood column is stressed in compression to 5000 kPa. If $E = 9600$ MPa, determine the total shortening of the column.

**4–19.** With reference to Problem 4–17, suppose that the total elongation of the cable had to be limited to 3 in., what cable diameter would be required?

**4–19M.** With reference to Problem 4–17M, suppose that the total elongation of the cable had to be limited to 75 mm, what cable diameter would be required?

**4–20.** A 500-ft-long roof cable cannot be permitted to stretch more than 30 in. or the roof geometry will change too greatly. If $E = 26(10)^6$ psi and the force in the cable is 1700 kips, determine the required cable diameter needed to avoid excessive elongation.

**4–20M.** A 150-m-long roof cable cannot be permitted to stretch more than 800 mm or the roof geometry will change too greatly. If $E = 155$ GPa and the load is 8000 kN, determine the required cable diameter needed to avoid excessive elongation.

**4–21.** A C6 × 10.5 steel channel is used as a hanger for a skywalk crossing a hotel lobby atrium. It has an $E$ value of $29(10)^6$ psi and must carry a load of 60 kips. How long can it be and still not have its elongation exceed ¼ in.? See Appendix J for the area of the channel.

**4–21M.** A C150 × 16 steel channel is used as a hanger for a skywalk crossing a hotel lobby atrium. It has an $E$ value of 200 GPa and must carry a load of 270 kN. How long can it be and still not have its elongation exceed 6 mm? See Appendix J for the area of the channel.

## 4–7 THERMAL STRESSES AND STRAINS

Whenever a structure is heated or cooled, it changes shape. Most materials expand when heated and contract when cooled. Long building structures must have joints in them to allow these changes to take place. Exposed exterior structural elements often undergo great temperature variations compared to interior ones, and large differential movements can result. If these movements are restrained, stresses can build up in the members themselves and connecting elements.

In many building structures, the small range of ambient temperatures and/or the absence of long uninterrupted structural elements serve to minimize thermal effects and they can be safely ignored. It is helpful to study the magnitudes of thermal movements and stresses to be able to judge when they deserve design attention.

The total change in length of a body due to a temperature change is

$$\delta = \alpha \, \Delta t(L) \tag{4–5}$$

where $\delta$ = total change in length (in.) (m)
$\quad\alpha$ = thermal coefficient for the material $(°F^{-1})$ $(°C^{-1})$
$\quad\Delta t$ = temperature change $(°F)$ $(°C)$
$\quad L$ = original length (in.) (m)

The coefficient $\alpha$ is usually expressed in terms of strain per degree of temperature change. With $\Delta t$ expressed in degrees, it then becomes convenient to think of $\alpha\Delta t$ as equivalent to $\epsilon$, the average unit of strain.

---

**EXAMPLE 4–7** A 125-ft-long masonry wall undergoes a temperature rise from 32°F to 120°F. Determine the total change in length if $\alpha$ for the material is $3.4(10)^{-6}$ in./in. per °F.

**Solution**

$$\delta = \alpha \, \Delta t(L)$$

$$= [3.4(10)^{-6} \, °F^{-1}](88°F)(125 \text{ ft})(12 \text{ in/ft})$$

$$= 0.45 \text{ in}$$

Without intermediate joints to allow movement in the plane of the wall, this motion will accumulate, possibly causing damage to attached walls or other constructions. The real difficulties, however, will develop during the contraction or cooling cycle of the same wall. Masonry and concrete lack the requisite tensile strength needed for large thermal contractions, and vertical cracks will develop in a long unjointed wall.

---

**EXAMPLE 4–8M** A 40-m-long masonry wall undergoes a temperature rise from 0°C to 40°C. Determine the total change in length if $\alpha$ for the material is $6.1(10)^{-6}$ m/m per °C.

**Solution**

$$\delta = \alpha \, \Delta t(L)$$

$$= [6.1(10)^{-6} \, °C^{-1}](40°C)(40 \text{ m})$$

$$= 9.76(10)^{-3} \text{ m}$$

$$= 10 \text{ mm}$$

(See note at the end of Example 4–7.)

**EXAMPLE 4–9** A high-rise building 700 ft tall has an exposed steel frame. On a sunny day in winter, the columns on the south side reach 120°F while those on the north side remain at 10°F. Under these extreme conditions, what will be the overall difference in length of the columns? Assume that $\alpha = 6.5(10)^{-6} \, °F^{-1}$.

*Solution*

$$\delta = \alpha \, \Delta t(L)$$

$$= [6.5(10)^{-6} \, °F^{-1}](110°F)(700 \text{ ft})(12 \text{ in/ft})$$

$$= 6.0 \text{ in}$$

**EXAMPLE 4–10M** A high-rise building 220 m tall has an exposed steel frame. On a sunny day in winter, the columns on the south side reach 50°C while those on the north side remain at $-10°C$. Under these extreme conditions, what will be the overall difference in length of the columns? Assume that $\alpha = 11.7(10)^{-6} \, °C^{-1}$.

*Solution*

$$\delta = \alpha \, \Delta t(L)$$

$$= [11.7(10)^{-6} \, °C^{-1}](60°C)(220 \text{ m})$$

$$= 0.154 \text{ m}$$

$$= 154 \text{ mm}$$

Unlike the contraction of a material such as masonry, unrestrained thermal elongation seldom results in a structural failure of the piece itself. However, as soon as that element is attached to others or restrained in any way, stresses will develop. A fully restrained bar fixed at the ends will build up compressive stress while attempting to elongate during a temperature increase. The resulting stress will have the same magnitude as if the bar had been allowed to expand and then axially loaded until it was "squeezed" back to its initial length. In other words, the change in length under free thermal expansion would be

$$\delta_{case \, 1} = \alpha \, \Delta t(L)$$

The change in length due to an applied compressive load will be

$$\delta_{case \, 2} = \frac{PL}{AE}$$

or

$$\delta_{case\ 2} = \frac{fL}{E}$$

Since the actual elongation for a "fixed" end bar is zero,

$$\delta_{case\ 1} - \delta_{case\ 2} = 0$$

or

$$\delta_{case\ 1} = \delta_{case\ 2}$$

$$\alpha\ \Delta t(L) = \frac{fL}{E}$$

Solving for the axial stress, we get

$$f = E\alpha\ \Delta t \tag{4-6}$$

This stress would, of course, be tensile if we cooled the bar instead of heating it. Equation 4–6 is interesting in that it illustrates that axial stresses (assuming a constant cross section), which develop in a restrained body due to a temperature change, are independent of any dimension of the body.

---

**EXAMPLE 4–11**

A straight concrete bridge is restrained by two canyon walls. If $E = 3.5(10)^6$ psi and $\alpha = 5.5(10)^{-6}$ °F$^{-1}$, determine the compressive stress developed during a temperature increase of 60°F.

*Solution*

$$f = E\alpha\ \Delta t$$

$$= 3.5(10)^6\ psi[5.5(10)^{-6}\ °F^{-1}](60°F)$$

$$= 1155\ psi$$

---

**EXAMPLE 4–12M**

A straight concrete bridge is restrained by two canyon walls. If $E = 25$ GPa and $\alpha = 10(10)^{-6}$ °C$^{-1}$, determine the compressive stress developed during a temperature increase of 35°C.

*Solution*

$$f = E\alpha\ \Delta t$$

$$= 25\ GPa[10(10)^{-6}\ °C^{-1}](35°C)$$

$$= 8.75(10)^{-3} \text{ GPa}$$

$$= 8750 \text{ kPa}$$

---

**EXAMPLE**
**4–13M**

A steel cable 25 mm in diameter is stretched tightly between two massive supports. If $E = 175$ GPa and $\alpha = 11(10)^{-6} \text{ °C}^{-1}$, determine the force in the cable developed due to a temperature drop of 40°C.

**Solution**

$$P = fA$$

$$= E\alpha \, \Delta t A$$

$$A = \pi r^2 = \pi (0.0125 \text{ m})^2$$

$$= 491(10)^{-6} \text{ m}^2$$

$$= 175 \text{ GPa}[11(10)^{-6} \text{ °C}^{-1}](40\text{°C})[491(10)^{-6} \text{ m}^2]$$

$$= 37.8(10)^{-6} \text{ GN}$$

$$= 37.8 \text{ kN}$$

---

# PROBLEMS

---

**4–22.** An unloaded steel roof cable is 400 ft long at 75°F. Determine its length at 10°F and at 140°F. Assume that $\alpha = 6.0(10)^{-6} \text{ °F}^{-1}$.

**4–22M.** An unloaded steel roof cable is 125.00 m long at 25°C. Determine its length at $-10$°C and at 70°C. Assume that $\alpha = 11(10)^{-6} \text{ °C}^{-1}$.

**4–23.** Steel roof trusses in an uninsulated attic space have members that are 32 ft long. The temperature differential from summer to winter is 120°F. If $\alpha = 6.0(10)^{-6} \text{ °F}^{-1}$, determine the total change in length of one of the members.

**4–23M.** Steel roof trusses in an uninsulated attic space have members that are 10 m long. The temperature differential from summer to winter is 70°C. If $\alpha = 11(10)^{-6} \text{ °C}^{-1}$, determine the total change in length of one of the members.

**4–24.** A long concrete bearing wall has vertical expansion joints placed every 70 ft. Determine the required width of the gap in a joint if it is wide open at 20°F and just barely closed at 110°F. Assume that $\alpha = 5.5(10)^{-6} \text{ °F}^{-1}$.

**4–24M.** A long concrete bearing wall has vertical expansion joints placed every 22 m. Determine the required width of the gap in a joint if it is wide open at $-5$°C and just barely closed at 40°C. Assume that $\alpha = 10(10)^{-6} \text{ °C}^{-1}$.

**4–25.** A 60-ft-long masonry wall has a parapet extending above the roof elevation. No expansion joints have been provided. The parapet portion heats up to an average temperature of 130°F in the summer sun. The portion of the wall below the parapet is exposed to the outside air from one side only; its average temperature is only 80°F. If $\alpha = 3.4(10)^{-6}$ °F$^{-1}$, how much longer will the parapet be than the wall beneath it?

**4–25M.** A 20-m-long masonry wall has a parapet extending above the roof elevation. No expansion joints have been provided. The parapet portion heats up to an average temperature of 60°C in the summer sun. The portion of the wall below the parapet is exposed to the outside air from one side only; its average temperature is only 30°C. If $\alpha = 6(10)^{-6}$ °C$^{-1}$, how much longer will the parapet be than the wall beneath it?

**4–26.** A large steam pipe is built with no provision for thermal expansion. If the ends are fixed, what is the level of compressive stress developed as the pipe goes from 60°F to 240°F? Assume that buckling does not occur. Assume that $E = 29(10)^6$ psi and $\alpha = 6.5(10)^{-6}$ °F$^{-1}$.

**4–26M.** A large steam pipe is built with no provision for thermal expansion. If the ends are fixed, what is the level of compressive stress developed as the pipe goes from 20°C to 120°C? Assume that buckling does not occur. Assume that $E = 200$ GPa and $\alpha = 11.7(10)^{-6}$ °C$^{-1}$.

**4–27.** An expansion loop is placed in the pipe of Problem 4–26 as shown in Figure 4–12. The loop is 6 ft long at 60°F. How long is it at 240°F?

**4–27M.** An expansion loop is placed in the pipe of Problem 4–26M as shown in Figure 4–12M. The loop is 2.0 m long at 20°C. How long is it at 120°C?

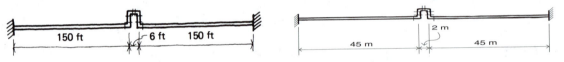

**Figure 4–12** Pipe with expansion loop.                                    **Figure 4–12M**

**4–28M.** An aluminum curtain wall panel is attached to large concrete columns when the temperature is 18°C. No provision is made for differential thermal movement. Because of insulation between the two, the sun heats the wall to 55°C but the column only gets to 33°C. Determine the consequent compressive stress in the curtain wall. (Ignore the effect of small tensile stresses in the concrete columns.) Assume that $E$ of the aluminum is 70 GPa, $\alpha$ for the concrete is $10(10)^{-6}$ °C$^{-1}$, and $\alpha$ for the aluminum is $22(10)^{-6}$ °C$^{-1}$.

# 4-8  ALLOWABLE STRESS DESIGN AND LIMIT STATES DESIGN

Structural problems are basically of two types: analysis or design. *Analysis* (sometimes called review) is the process of determining the types and magnitudes of stresses and deformations in a *given* structure when it is subjected to known or assumed loads. Most of the problems in this chapter have been of the analysis type. The process of design has quite a different goal. In a *design* situation we are trying to proportion the size and shape of a structure so that it can carry the known or assumed loads in a safe manner.

These two attitudes are not as distinct and separate as they might appear at first glance, because often the design process becomes an analytical trial-and-error procedure in order to determine the "best" size or configuration for a structure. (The phrase "trial and error," while much used and generally understood, can be misleading, as there really is no "error" involved. The writer prefers "trial-and-check" or "select-and-try" as being more descriptive terminology.)

What determines the "best" structure for a given architectural and construction endeavor is quite an impossible question to answer definitively. In pure structural design terms, the word *best* can sometimes be interpreted as "efficiency," that is, efficient use of the material(s) involved through (a) optimum manipulation of the geometry or statics present and (b) loading the materials with types of stress which they can most easily "take" or resist. Here the word *efficiency* is used, in a narrow sense, to mean just enough material to do the job without waste. Too much material would be "overdesign," and not enough for proper safety would be "underdesign."

As with any type of design, many orders and sequences of compromises are involved, and it is a rare (or nonexistent) effort in which the structural, constructional, and functional considerations become coincident. The structural design process can never be accomplished in isolation and becomes a mixture of efficiency and compatibility determinations.

There are two different approaches to structural design currently in use. One is called the *allowable stress design (ASD)* method, in which the structure is shaped and the elements are proportioned so that certain "allowable" stresses are not exceeded. (This approach is also referred to as the *working stress method* or *service load method*.) These allowable stresses are determined as percentages of the failure strengths of materials under various kinds of stress. For example, the allowable stress in shear for a certain species and grade of wood might be 70 psi (500 kPa), whereas its failure stress in shear might be 280 psi (2000 kPa).

The allowable bending stress in structural steel might be determined as two-thirds its yield value [i.e., steel that has a yield strength of 50 ksi (345 MPa) can be safely stressed to 33 ksi (230 MPa)]. The difference between the two values in each case constitutes a "margin of safety" or factor of safety obtained by dividing the failure stress by the allowable stress. Currently, this design method is used for wood structures and for some steel structures. Elements are proportioned so that the computed stresses present under the expected loads (both dead and live) will be less than the allowable stresses. The behavior of the structure under overload or failure load is not considered. Representative allowable stresses for several types of wood are given in Appendix H. Allowable stresses for different steels are given as required in various chapters.

The other approach to design is called *limit states design*. It is based upon the ultimate strength or load capacity of structural elements and systems. It is called *strength design* by the reinforced concrete industry and *load and resistance factor design* (LRFD) by the steel industry. Here, the factor of safety is applied in a manner quite different from the allowable stress method. In this method, various factors of safety are applied directly to the loads which are known or assumed to be acting on the structure. Thus, we conceptually increase the loads that the structure must be proportioned to take by multiplying those loads by specific factors. Such increased or factored loads are called *ultimate loads* or *design loads*, and we then design the member or structure to *fail* under the application of these increased loads.

Limit states design is quite distinct from allowable stress design in that the approach proportions a structure to fail under a specified overload and the actual stresses present in the structure under normal loads are not computed. The margin of safety is present in the degree of overload specified. Strength design is widely used for reinforced concrete structures, and overload load factors of 1.4 for dead loads and 1.7 for live loads have been specified by the American Concrete Institute. For LRFD design in steel, the comparable values are 1.2 and 1.6, respectively.

(Various sections of this chapter have focused on the fact that structures might become unusable because of excess deformation or deflection as well as by actual material rupture. The two different approaches to structural design presented here concern only the stresses in a loaded structure and are not involved with "failure" by excess deformation. This is guarded against by specifying certain permissible values of deformation or deflection which may not be exceeded when the expected loads act on the structure. There is no easily quantifiable factor of safety involved.)

It is important to note that no matter which design approach is adopted, the stresses and deformations due to the actual loads will be reasonably close in magnitude. The limit states approach can take advantage of the redundancy of indeterminate structures and is therefore preferred by many analysts in such cases.

The factors of safety in each case are not up to the designer but are usually specified by code bodies and industry associations and are based upon research and experience with the particular materials. These factors can vary in magnitude and are determined by considering the answers to such questions as:

How statistically consistent are the properties of the material?

How homogeneous is the material?

What is the statistical frequency of flaws in the material?

What level of supervision and inspection will be present?

How easy is it to make errors when fabricating the material?

How disastrous is the failure that might occur?

How important is this piece to the integrity of the whole structure?

How accurately can the probable loads be predicted?

How accurate are the assumptions that have been made in the structural design process?

How good or proven is the theory used in the structural design?

One can see from the last three questions that a factor of ignorance is needed within the factor of safety. Many structures are standing today because the effects of errors were safely negated by providing for an overload that never occurred.

# 5

# Properties of Structural Materials

## 5–1 INTRODUCTION

It is important for the structural designer to realize that different engineering materials have different characteristics and will exhibit different behaviors under load. A knowledge of such characteristics or properties will help to ensure proper use of these materials, both architecturally and structurally.

It is assumed that the reader will have already been exposed to the study of materials through courses in building construction or materials science. This chapter will only highlight a few selected structural materials in the interest of emphasizing the range of structural characteristics and their diversity. Tables of properties of selected structural materials are given in Appendix E.

## 5–2 NATURE OF WOOD

Wood is a natural material and has a broad range of physical properties because of the different characteristics of its many species. Softwoods such as fir, pine, and hemlock are most often used for structural applications because they are more plentiful (grow fast and tall) and are easier to fabricate. These woods are generally strong in tension and compression in a direction parallel to the grain and weak when stressed perpendicularly to the grain. Wood is also weak in shear because of its tendency to split along the natural grain laminations. The allowable stresses for three selected species are given in Appendix H.

Wood is light and soft compared to most other structural materials and is easily shaped and fastened together. A minimum of materials-handling equipment is needed to erect wood structures because of their weight. Wood is also very versatile in terms of its adaptability to the making of geometric shapes and even nonlinear forms.

Most softwoods are fairly ductile and will not fail suddenly when overloaded. Because of their lack of homogeneity or uniformity, the allowable stresses are quite

In considering the properties of structural materials, it is important for the architect to realize that although an understanding is necessary to make a successful building, the materials themselves do not usually *dictate* architectural form. This may be illustrated by two contrasting sets of comparisons. Here are plans and interior views of four churches with generally the same architectural intent: the Ste. Chapelle in Paris, constructed in stone, and Notre-Dame du Raincy, outside Paris, built in concrete by Auguste Perret (on this page); and Otto Bartning's Steel Church in Germany and Richard Munday's wooden Trinity Church in Newport, Rhode Island (on the following page).

Disregarding the variation in size of the four buildings, they are very similar in plan. In fact, on the basis of plan alone, it would be hard to tell which church is of which material. The reason is this: Despite the fact that all four seek to make the lightest possible impression on the interior, none of the designs strains the structural capacities of its material. Formal desires, rather than the structural properties of materials, by and large determine the nature of architectural mass and space relations.

In bridges, on the other hand, conditions are generally more extreme, formal preconceptions fewer, and concern for economy of materials high, and so bridge forms and the characteristics of the materials making the forms are often highly correlated. Stone works best in compression and is conveniently transported in pieces, so arched bridges of stone are a natural result. Wooden bridges take into account the facts that their material will come in sticks of greater or lesser, but certainly of limited size, and will take compression or tension with equal capacity. Suspension bridges are unthinkable without some homogeneous material that is quite good in tension and available in extreme lengths like steel, cable or rope. Spans of reinforced concrete, at least the best of them, take advantage of its compressive strength and its ability to make large continuous elements and smooth connections.

low compared to failure stresses. Consequently, when wood structures are properly engineered, a statistically high margin of safety is present. Wood is often known as the "forgiving" material because of its apparent ability to sustain loads not accounted for when the structure was designed.

Wood, on the other hand, is not very stiff. It is subject to excessive deflection and creep deformation if not designed with these characteristics in mind (see Section 5–6). It is prone to damage by fire and to deterioration by moisture and insects. It expands and contracts with variations in humidity, markedly so in the direction perpendicular to the grain. Timber structures that are to be exposed to the elements must be carefully treated or highly maintained to preserve their integrity.

The American Forest and Paper Association publishes the "National Design Specification," which is the primary reference guide for timber designers. The Association also publishes a number of useful bulletins and manuals, as do other groups, such as the Western Wood Products Association and the Southern Forest Products Association.

## 5–3  CONCRETE AND REINFORCED CONCRETE

Concrete is a man-made conglomerate stone composed of essentially four ingredients: portland cement, water, sand, and coarse aggregate. The cement and water combine to make a paste that binds the sand and stones together. Ideally, the aggregates are graded so that the volume of paste is at a minimum, merely surrounding every piece with a thin layer. Most structural concrete is stone concrete, but structural lightweight concrete (roughly two-thirds the density of stone concrete) is becoming increasingly popular.

Concrete is essentially a compressive material having almost no tensile strength. As explained in Chapter 8, shearing stresses are always accompanied by tension, so concrete's weakness in tension also causes it to be weak in shear. These deficiencies are overcome by using steel bars for reinforcement at the places where tensile and shearing stresses are generated. Under load, reinforced concrete beams actually have numerous minute cracks which run at right angles to the direction of major tensile stresses. The tensile forces at such locations are being taken completely by the steel "re-bars."

The compressive strength of a given concrete is a function of the quality and proportions of its constituents and the manner in which the fresh concrete is cured. [*Curing* is the provision of an appropriate environment surrounding freshly-placed concrete while it gains its initial strength. During this time (7 to 14 days in duration), the concrete should be kept at a reasonable temperature and must be prevented from "drying out," as the presence of water is necessary for the chemical action to progress.]

Coarse aggregate that is hard and well graded is particularly essential for quality concrete.

The most important factor governing the strength, however, is the percentage of water used in the mix. A minimum amount of water is needed for proper hydration of the cement. Additional water is needed for handling and placing the concrete, but excess amounts cause the strength to drop markedly.

These and other topics are fully covered in the booklet, "Design and Control of Concrete Mixtures," published by the Portland Cement Association. This is an excellent reference, treating both concrete mix design and proper construction practices. The American Concrete Institute publishes a widely adopted code specifying the structural requirements for reinforced concrete.

Concrete is known as the "formable" or "moldable" structural material. Compared to other materials, it is easy to make curvilinear members and surfaces with concrete. It has no inherent texture but adopts the texture of the forming material, so it can range widely in surface appearance. It is relatively inexpensive to make, both in terms of raw materials and labor, and the basic ingredients of portland cement are available the world over. (It should be noted, however, that the necessary reinforcing bars for concrete may not be readily available in less-developed countries.)

The best structural use of reinforced concrete, in terms of the characteristics of the material, is in those structures requiring continuity and/or rigidity. It has a monolithic quality which automatically makes fixed or continuous connections. These moment-resistant joints are such that many low-rise concrete buildings do not require a secondary bracing system for lateral loads. In essence, a concrete beam joins a concrete column very differently from the way steel and wood pieces join, and the sensitive designer will not ignore this difference. (These remarks do not apply to precast structural elements, which are usually not joined in a continuous manner.)

Concrete is naturally fireproof and needs no separate protection system. Because of its mass, it can also serve as an effective barrier to sound transmission.

In viewing the negative aspects, concrete is unfortunately quite heavy, and it is often noted that a concrete structure expends a large portion of its capacity merely carrying itself. Attempts to make concrete less dense, while maintaining high quality levels, have generally resulted in increased costs. Nevertheless, use of lightweight concrete can sometimes result in overall economies.

Concrete requires more quality control than most other building materials. Modern transit-mixed concrete suppliers are available to all U.S. urban areas and the mix is usually of a uniformly high quality. Field- or job-mixed concrete requires knowledgable supervision, however. In any type of concrete work, missing or mislocated reinforcing bars can result in elements with reduced load capacities. Poor handling and/or curing conditions can seriously weaken any concrete. For these and other reasons, most building codes require independent field inspections at various stages of construction.

Proper concrete placement is also somewhat dependent upon the ambient weather conditions. Extremely high temperatures and, more important, those below (or near) freezing can make concrete work very difficult.

## 5–4 STRUCTURAL STEEL

Steel is the strongest and stiffest building material in common use today. Relative to wood and concrete, it is a high-technology material made by highly refined and controlled processes. Structural steel has a uniformly high strength in tension and compression and is also very good in shear. It comes in a range of yield strengths made by adjusting the chemistry of the material in its molten state. It is the most consistent of all structural materials and is, for all practical purposes, homogeneous and *isotropic,* meaning it has like characteristics in all directions. (By contrast, wood is *anisotropic.*)

The greatest asset of steel is its strength and "plastic reserve," as shown in Figure 4–7. It is highly ductile and deforms greatly before failing if overloaded. Because of steel's strength, the individual members of a frame are usually small in cross section and have very little visual mass. Steel is a linear material and can be economically made into a visual curve only by using a segmented geometry. It is most appropriately used in rectilinear structures where bolted or welded connections are easy to make. The structural shapes (i.e., pipes, tubes, channels, angles, and wide-flange sections) are manufactured to uniform dimensions having low tolerances. They are fully prepared (cut, trimmed or milled, drilled or punched, etc.) in a fabrication shop, remote from the site, and then delivered ready for erection. Such structures go up rapidly with a minimum of on-site labor. The most popular form of construction used today is referred to as "shop-welded, field-bolted." In this method the various clip angles, beam seats, and so on are welded to the members in a shop and then the members are bolted together in the field.

A major disadvantage of structural steel is its need to be fire-protected in most applications. It loses its strength at around 1100°F (600°C) and will then yield rapidly under low loads. A few municipalities require that all structural steel be fire-protected, and most codes will not permit any exposed elements to be within approximately 12 ft (4 m) of a combustible fire source.

The making of steel requires large physical plants and a high capital outlay, and therefore relatively few countries of the world have extensive mill facilities. The cost of manufacturing, coupled with the cost of transportation, can make steel a relatively expensive material. Just the same, in most urban areas, concrete and steel are quite competitive with one another in terms of in-place construction costs.

Continuity in the connections is much harder to achieve in steel than in concrete, and most buildings are constructed with simple connections or ones that are only partially moment-resistant. Some type of lateral load bracing system is almost always required in a steel-framed building and must be considered early in the design process.

Rolled steel is manufactured in a wide range of strengths. The standard low-carbon mild steel in use today has a yield strength of 36 ksi (250 MPa). However, recent changes in industry practices have made 50-ksi (345-MPa) steel as economical to produce as the 36 (250) grade steel, and the stronger material is increasing in popularity. Steel plate can be obtained with an $F_y$ value of 100 ksi (700 MPa), and most standard shapes can be rolled in steel as strong as 65 ksi (450 MPa), although this can be expensive. Examples and problems in this text are limited to shapes of $F_y = 36$ ksi (250 MPa) and $F_y = 50$ ksi (345 MPa).

Information about the various kinds of steel available can be obtained directly from manufacturers and fabricators. The reader is also advised to purchase the latest edition of the *Manual of Steel Construction,* published by the American Institute of Steel Construction. (With reference to Section 4–8, it is published in two versions: ASD and LRFD.) It is an indispensable reference work for the design professional.

## 5–5   MASONRY AND REINFORCED MASONRY

Like concrete, brick and concrete masonry units are strong in compression and weak in tension. These materials have traditionally been used in walls, both bearing and nonbearing. Usually, wall thicknesses required by code specifications to prevent lateral instability are such that the actual compressive stresses are low. Crushing is seldom an important design constraint.

Masonry walls are more permanent than wood walls and provide effective barriers to both fire and noise. They are less expensive and often more attractive than formed concrete walls. Brick generally has more variation of pattern and texture than does concrete block, but is also more expensive.

It is becoming increasingly common to use reinforced concrete block for retaining walls and structural pilasters. In this construction, individual reinforcing bars are grouted in some of the vertically aligned cells of the concrete units and serve as tensile reinforcement. This greatly increases the lateral load capacity of the block. Reinforcing can also be placed in special channel-shaped blocks to serve as lintels and tie beams. Brick can be reinforced by using two wythes to create a cavity for grout and reinforcing bars. The brick not only serves as formwork but also carries compressive forces under load.

## 5–6   CREEP

Section 4–3 explained how structural elements change their size and shape upon application of load. This is called *elastic strain,* and, provided that we do not stress the material too greatly, such deformation will disappear upon removal of the load. Most materials, if left under load for a long time, will exhibit an additional strain referred to as *creep.* In most cases these strains will remain after removal of the load.

The amount of creep, which takes place under long-term load, seems to vary directly with the stress level present and the ambient temperature and inversely with the material stiffness. Many plastics creep considerably in just a short period of time. Steel exhibits very little creep except at elevated temperatures. Concrete and wood both creep appreciably if stressed highly for long periods of time.

Members that must support constantly applied loads such as dead weight should be "overdesigned" so that the stresses will be low. For example, the increased deflection (over a couple of years) of a reinforced concrete beam carrying a heavy masonry wall can be double the initial elastic deflection. Many cantilevered portions of wood structures develop an unsightly sag with time which could have been prevented or minimized through the proper consideration of creep.

# 6

# Shear and Moment

## 6–1 DEFINITIONS AND SIGN CONVENTIONS

A transverse load on a linear element such as a beam or column will generate essentially two kinds of stress, shearing and flexural. *Shearing stresses* are the result of internal shearing forces, and *flexural stresses* result from internal resisting couples (moments). Both of these effects are responses to the externally applied forces and will vary along the length of a member. Their magnitudes and senses at any section will be dependent upon the loads, span, and support conditions of the member.

Usually, the structural designer is interested in the maximum values of these shears and moments, which exist in the many elements or parts of a structure. These maximum values will help to make judgments as to the soundness of the overall scheme and in the planning of the geometry of the various structural elements. Member sizes are also determined and spans sometimes modified on the basis of maximum shear and moment values.

In order to study these forces, we use free-body diagrams, and the techniques of statics. By cutting a transverse section through a beam or column, we expose the internal forces and make them external via a free-body diagram (see Figure 2–45). Statics can then be applied to solve for the magnitudes and senses of the unknown values. Graphs or plots are made to show how forces change from section to section. A plot illustrating how the internal shearing force $V$ varies over the length of a member is called a *shear diagram* or *V diagram*. A *moment diagram* or *M diagram* is a similar plot showing the variation of the internal moment throughout the member. Shears and moments are plotted according to the following sign conventions.

Shearing forces that tend to cause the slippage failures shown in Figure 6–1 will be denoted by the sign accompanying them. Notice in Figure 6–2 that for positive shear, forces exist (for equilibrium) which are *up* on the left-hand face and *down* on the right-hand face of an element. The opposite forces act on the cut faces when they are subjected to negative shear. (*Note:* The sign convention for shear is somewhat arbitrary, and the opposite plus and minus associations are preferred by some writers.)

The sign convention for internal bending moment is considerably more straightforward, as Figure 6–3 shows. Positive moment generates concave upward

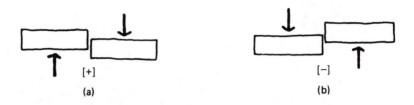

**Figure 6–1**   Sign convention for shearing forces.

curvature, causing compression in the top fibers and tension in the bottom fibers. Negative moment causes concave downward curvature and, of course, the opposite types of fiber strain. This convention is the standard one for curvature in mathematics and is universally accepted. Since the convention is related to strain, it is possible to look at the probable deflected shape of a beam under load, for example, and determine what portions of the span have positive or negative internal moments. The uniformly loaded beam in Figure 6–4 has an overhang, and from the deflected shape we can see that negative internal moments exist over part of the beam's length and positive internal moments in another portion. The implication here is

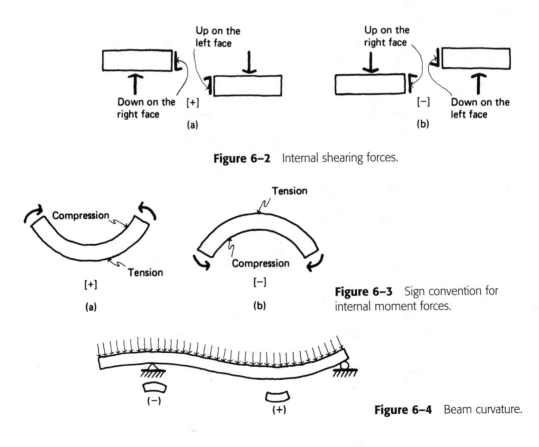

**Figure 6–2**   Internal shearing forces.

**Figure 6–3**   Sign convention for internal moment forces.

**Figure 6–4**   Beam curvature.

*and ∴ curvature = 0*

that there is likely a transverse section or portion of the span where the bending moment is <u>zero</u> to accommodate the required sign change. Such a section, termed a *point of inflection,* is almost always present in overhanging beams.

The most important feature of these sign conventions is that they are different from the conventions used in statics. When using the three equations of equilibrium, forces up and to the right are plus and counterclockwise moments are plus. The new sign conventions are used only for plotting the shear and moment diagrams. It is important that the two conventions not become confused.

## 6-2  SHEAR AND MOMENT EQUATIONS

The most basic way to obtain $V$ and $M$ diagrams is to graph specific values from statics equations that have been written so they are valid for appropriate portions of the member. (In these explanations we shall assume that the member is a beam, acted upon by downward loads, but actually the member could be turned at any angle. In the general case, we are determining shears and moments due to transverse loads.) The following examples will illustrate how we can write and plot $V$ and $M$ equations. In each case, the uniform load of the beam's own weight has been neglected.

---

**EXAMPLE 6–1**  Construct the $V$ and $M$ diagrams for the beam in Figure 6–5.

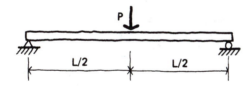

**Figure 6–5**  Simple beam with a concentrated load at midspan.

### Solution

If we let $x$ designate any point along the length of a beam and assume the origin to be at the left end ($x = 0$ there), we can write expressions for $V$ and $M$ in terms of constants (the external loads and reactions) and the distance $x$. To examine $V_x$ and $M_x$ in the left-hand half of the beam in Figure 6–6, we cut through that portion. This will make the unknowns external and we can use statics on the resulting free body of Figure 6–7. Notice that the unknowns have been assumed positive by the new

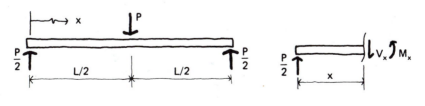

**Figure 6–6**                          **Figure 6–7**

convention. This will mean that answers yielded by statics which come out plus (as assumed) will be plotted as positive ordinates, and those that turn out minus (not as assumed) must be plotted as negative values. Now, applying the equations of equilibrium to find the unknowns, we get

$$\Sigma F_y = 0$$

$$\frac{P}{2} - V_x = 0$$

$$\left(0 \le x \le \frac{L}{2}\right) V_x = \frac{P}{2}$$

Taking moments at the cut face (thereby eliminating $V_x$), we get

$$\Sigma M_x = 0$$

$$-\frac{P}{2}(x) + M_x = 0$$

$$\left(0 \le x \le \frac{L}{2}\right) M_x = \frac{P}{2}x$$

For the right-hand half and using the free body in Figure 6–8, we get

$$\Sigma F_y = 0$$

$$\frac{P}{2} - P - V_x = 0$$

$$\left(\frac{L}{2} \le x \le L\right) V_x = \frac{P}{2} - P$$

$$\Sigma M_x = 0$$

$$-\frac{P}{2}(x) + P\left(x - \frac{L}{2}\right) + M_x = 0$$

$$\left(\frac{L}{2} \le x \le L\right) M_x = \frac{P}{2}(x) - P\left(x - \frac{L}{2}\right)$$

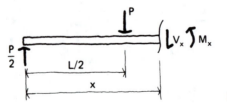

**Figure 6–8**

Substituting finite values of $x$ (e.g., $x = 0$, $L/4$, $L/2$, $3L/4$, and $L$), we can plot the equations as shown in Figure 6–9. In this case the $V$ diagram does not vary with $x$ except in sign. The ordinates on the moment diagram are all positive, as might be verified by the deflected shape of this beam.

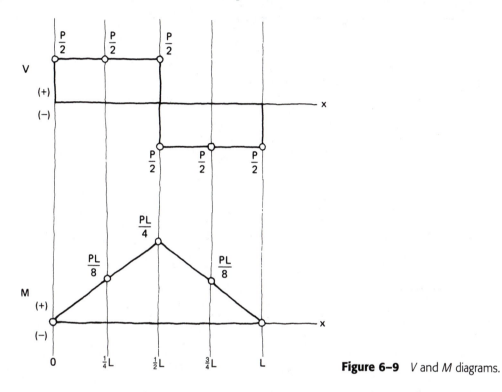

**Figure 6–9** $V$ and $M$ diagrams.

From the diagrams we can also see the necessity for writing an equation for each half of this beam. Any type of load change or application (including reactions) will cause discontinuities in the diagrams so that each set of $V$ and $M$ equations is only valid for a specific part of the beam length. However, equations with appropriate interval limits can always be written for each beam portion.

**EXAMPLE 6–2** Construct the $V$ and $M$ diagrams for the beam in Figure 6–10.

**Solution**

Since the uniform load is constant over the entire span, only one set of equations will be necessary. Using Figure 6–11, we get

$$\Sigma F_y = 0$$

$$\frac{wL}{2} - wx - V_x = 0$$

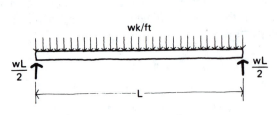

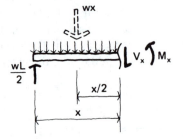

**Figure 6–10**   Simple beam with a uniform load.

**Figure 6–11**   Free-body diagram.

$$(0 \leqslant x \leqslant L) \; V_x = \frac{wL}{2} - wx$$

$$\Sigma M_x = 0$$

$$\frac{-wL}{2}(x) + wx\left(\frac{x}{2}\right) + M_x = 0$$

$$(0 \leqslant x \leqslant L) \; M_x = \frac{wLx}{2} - \frac{wx^2}{2}$$

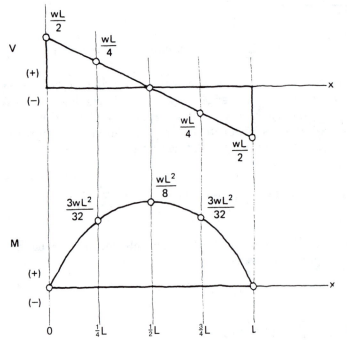

**Figure 6–12**   *V* and *M* diagrams.

Substituting the values of $x$ equal to the quarter points of the span, we get the diagrams shown in Figure 6–12. Ordinates that lie above the reference line are taken as positive and those below, as negative.

The shear diagram reflects the linear variation with $x$ as required by the shear equation. The moment diagram is a parabolic curve, which follows the second-power function of $x$ present in the moment equation. This can be confusing, as one is led to think that bending is a second-power function of the span alone. Moment is always linear with span; it is just that with a uniform load, as opposed to a concentrated one, a change in length means a change in load. For this reason the plot becomes parabolic. To understand this better, compare the maximum moment for the concentrated load of Example 6–1 (i.e., $M = PL/4$) to the maximum moment for the uniform load, which is $M = wL^2/8$. Since $P$ is the total load in the first case, let $wL = W$ to get a comparable load for the uniform case. Then $M$ will be $WL/8$, indicating that moment varies linearly with span and linearly with load. **(This also shows clearly that concentrated loads will generate double the moment caused by uniform loads.)**

---

**EXAMPLE 6–3** Construct the $V$ and $M$ diagrams for the beam in Figure 6–13. The free-body diagrams are shown in Figures 6–14 through 6–16.

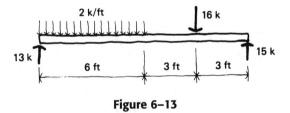

**Figure 6–13**

### Solution

Three separate sets of $V$ and $M$ equations will be used.

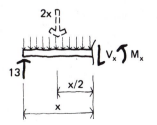

**Figure 6–14**  FBD for interval $(0 \leqslant x \leqslant 6)$.

$$\Sigma F_y = 0$$

$$13 - 2x - V_x = 0$$

$$(0 \leqslant x \leqslant 6)\ \ V_x = 13 - 2x$$

$$\Sigma M_x = 0$$

$$-13x + 2x\left(\frac{x}{2}\right) + M_x = 0$$

$$(0 \leqslant x \leqslant 6)\ \ M_x = 13x - x^2$$

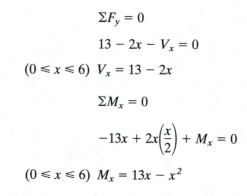

**Figure 6–15**  FBD for interval $(6 \leqslant x \leqslant 9)$.

$$\Sigma F_y = 0$$

$$13 - 12 - V_x = 0$$

$$(6 \leqslant x \leqslant 9)\ \ V_x = 1$$

$$\Sigma M_x = 0$$

$$-13x + 12(x - 3) + M_x = 0$$

$$(6 \leqslant x \leqslant 9)\ \ M_x = 13x - 12x + 36$$

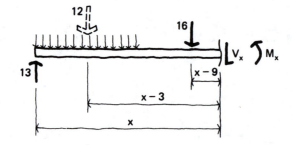

**Figure 6–16**  FBD for interval $(9 \leqslant x \leqslant 12)$.

$$\Sigma F_y = 0$$

$$13 - 12 - 16 - V_x = 0$$

$$(9 \leqslant x \leqslant 12) \ V_x = -15$$

$$\Sigma M_x = 0$$

$$-13x + 12(x-3) + 16(x - 9) + M_x = 0$$

$$(9 \leqslant x \leqslant 12) \ M_x = 13x - 28x + 180$$

Using these equations and values of $x$ as needed, the diagrams of Figure 6–17 can be drawn. Notice that equations that contain the variable $x$ plot as straight horizontal lines, that equations containing $x$ raised only to the first power plot as straight sloping lines, and that equations involving $x^2$ plot as parabolic curves. Note also that the points of curve change on the diagrams are common to two equations, and either equation may be used to find the ordinate value. Downward-acting uniform loads will result in moment curves that are concave downward, as verified by the value of 30 at $x = 3$ on the $M$ diagram. Concentrated loads will always cause a sudden change in ordinate on the shear diagram and require two values of $V$ at that section. (Actually, these two values of $V$, differing by the magnitude of the concentrated load, are a small distance apart because the load does occupy a short length of beam. In theory, however, we assume a true point load and the distance becomes infinitesimal.)

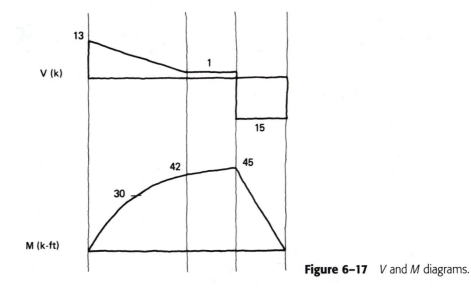

**Figure 6–17** $V$ and $M$ diagrams.

Using the guidelines given above, it is usually possible to construct shear and moment diagrams using only the values of $x$ at the points of load change (i.e., the interval limits).

Cantilevers are the stuff of architects' and engineers' dreams of near-flight, of ecstatic visionary breadth. Leon Krier's 1974 House for a Painter in Tuscany evokes this dream; the romantic artist can stand, windswept, out on a long cantilevered metal balcony, yet still retreat to a traditional timber-trussed, solid stone tower. No such retreat was offered in the 1922–23 Suspended Restaurants project by the young architect Simbirchev, in the then-young Soviet Union. Cantilever dreams do not come cheap, and neither project was built. The 1958–59 United Air Lines Hangar in San Francisco, with roofs cantilevering nearly 150 feet out from a central support, used both the dream of flight and, with its tapered steel girders, the literal structural principle of airplane wings; Myron Goldsmith of Skidmore, Owings & Merrill designed it. At a much smaller scale but also with evocative intentions, Paul Rudolph clearly distinguished between sheltering, tree-like, branching-cantilever roof supports and the protected enclosure within and below them, at the Greeley Laboratories, Yale School of Forestry, in 1957–59.

**EXAMPLE
6–4M**

Construct the $V$ and $M$ diagrams for the beam in Figure 6–18M(a).

### Solution

The external reactions are found in Figure 6–18M(b), and the FBD needed to write the equations appears in Figure 6–18M(c).

$$\Sigma F_y = 0$$

$$36 - 12x - V_x = 0$$

$$(0 \leqslant x \leqslant 3) \quad V_x = 36 - 12x$$

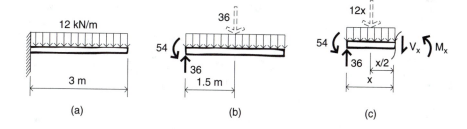

**(a)**                    **(b)**                    **(c)**

**Figure 6–18M**    Cantilever beam with a uniform load.

$$\Sigma M_x = 0$$

$$54 - 36x + 12x\left(\frac{x}{2}\right) + M_x = 0$$

$$(0 \leqslant x \leqslant 3) \quad M_x = -54 + 36x - 6x^2$$

This example serves well to illustrate the difference in the two sign conventions. In the FBD of Figure 6–18M(c), the moment reaction is counterclockwise or plus in the $\Sigma M_x = 0$ statics equation. It causes tension in the top fiber, however (as the deflected shape of the beam would be concave downward), and is plotted as a *negative* ordinate on the moment diagram of Figure 6–19M. Also notice that, just as a point load (or reaction) causes a sudden jump in the shear diagram, an externally applied moment will cause a sudden jump in the moment diagram. (**Actually, this is the only way that such a change can occur in a moment diagram, which means that the moment will always be zero at the end of a beam unless there is an applied moment load or a wall reaction at that point.**)

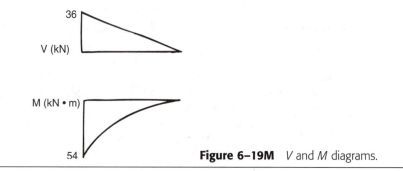

**Figure 6–19M**  *V* and *M* diagrams.

As an exercise the student should turn the cantilever beam of Example 6–4M end for end and again construct the *V* and *M* diagrams. (Go ahead, do it!) Since the origin is always kept at the left end, *x* will then equal zero at the free end and the two equations will change. The moment diagram remains all negative and is merely turned right about, as would be expected, but the shear diagram changes sign, illustrating that the shear sense is not related to the beam's physical behavior.

## PROBLEMS

**6–1.** Construct the *V* and *M* diagrams for the simple beam in Figure 6–20.

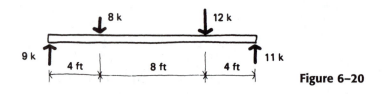

**Figure 6–20**

**6–1M.** Construct the *V* and *M* diagrams for the simple beam in Figure 6–20M.

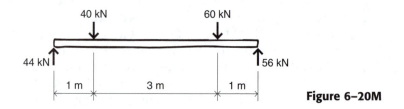

**Figure 6–20M**

**6–2.** Construct the *V* and *M* diagrams for the overhanging beam in Figure 6–21.

**6–2M.** Construct the *V* and *M* diagrams for the overhanging beam in Figure 6–21M.

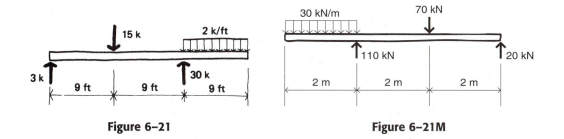

**Figure 6–21**          **Figure 6–21M**

**6–3.** Determine the reactions and construct the $V$ and $M$ diagrams for the beam in Figure 6–22.

**6–3M.** Determine the reactions and construct the $V$ and $M$ diagrams for the beam in Figure 6–22M.

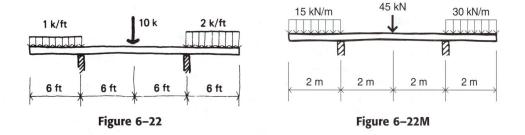

**Figure 6–22**          **Figure 6–22M**

**6–4.** Determine the reactions and construct the $V$ and $M$ diagrams for the cantilever beam in Figure 6–23.

**6–4M.** Determine the reactions and construct the $V$ and $M$ diagrams for the cantilever beam in Figure 6–23M.

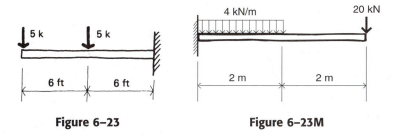

**Figure 6–23**          **Figure 6–23M**

**6–5.** The beam in Figure 6–24 is partially restrained by its supports, resulting in applied couple loads at its ends. Determine the reactions and construct the $V$ and $M$ diagrams.

**6–5M.** The beam in Figure 6–24M is partially restrained by its supports, resulting in applied couple loads at its ends. Determine the reactions and construct the $V$ and $M$ diagrams.

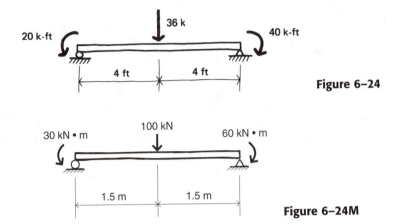

**Figure 6–24**

**Figure 6–24M**

## 6–3    SIGNIFICANCE OF ZERO SHEAR

The examples and problems of Section 6–2 had shear and moment diagrams that could be drawn by connecting the ordinates at the various points of curve change with either straight or curved lines. With the exception of Example 6–2, the maximum value of moment in each case occurred at one of these points. (For Example 6–2, the location of the maximum bending was obviously at midspan by symmetry.) For many loading situations, the point of maximum moment cannot be found so conveniently. However, it must be located before the magnitude of the maximum moment can be determined. Example 6–5 illustrates this situation.

**EXAMPLE 6–5**  Construct the $V$ and $M$ diagrams for the partially loaded beam of Figure 6–25.

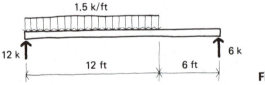

**Figure 6–25**

### Solution
The $V$ and $M$ equations for the two intervals will be

$$(0 \leqslant x \leqslant 12) \quad V_x = 12 - 1.5x$$

$$(0 \leqslant x \leqslant 12) \quad M_x = 12x - 0.75x^2$$

$$(12 \leqslant x \leqslant 18) \quad V_x = -6$$

$$(12 \leqslant x \leqslant 18) \quad M_x = 108 - 6x$$

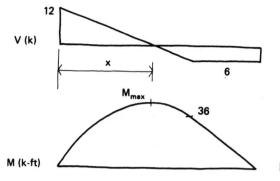

**Figure 6–26** *V* and *M* diagrams.

The location of the point of maximum moment is not obvious from Figure 6–26. The distance $x$ can be found rather easily, however, by remembering that the first derivative of an equation represents the slope. Borrowing the maxima–minima concept from calculus, we can find where the slope levels off to zero by taking $dM_x/dx$ and setting it equal to zero. For the interval $0 \le x \le 12$, we have

$$M_x = 12x - 0.75x^2$$

and

$$\frac{dM_x}{dx} = 12 - 1.5x$$

Setting this expression to zero, we get

$$12 - 1.5x = 0$$

$$x = 8$$

By substitution, the value of the maximum moment is then

$$Mx_{max} = 12(8) - 0.75(8)^2$$

$$= 48 \text{ kip-ft}$$

It is most important to notice that the first derivative of the moment equation is equal to the *V* equation. Hence, when the moment becomes a maximum (or minimum) between points of curve change, it will do so where the shear is zero. This means that the value of $x$ could have been determined from the slope of the *V* diagram (Figure 6–27). The slope is known as 1.5 kips per foot of length; therefore, to reduce the left-hand shear ordinate from 12 to 0 will require that

$$\frac{12 \text{ kips}}{1.5 \text{ kips/ft}} = 8 \text{ ft}$$

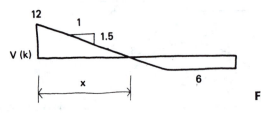

**Figure 6–27**  Similar triangles.

The relationships between the two diagrams are explored more fully in the next section.

## PROBLEMS

**6–6.** Construct the $V$ and $M$ diagrams for the beam in Figure 6–28.

**6–6M.** Construct the $V$ and $M$ diagrams for the beam in Figure 6–28M.

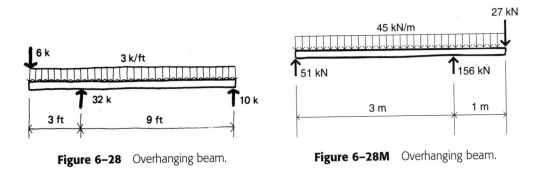

**Figure 6–28**  Overhanging beam.

**Figure 6–28M**  Overhanging beam.

**6–7.** What is the value of maximum moment in the beam of Figure 6–29? It carries a uniform load and is subjected to an applied moment at its left end.

**6–7M.** What is the value of maximum moment in the beam of Figure 6–29M? It carries a point load and is subjected to an applied moment at its left end.

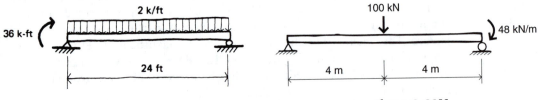

**Figure 6–29**

**Figure 6–29M**

## 6–4  LOAD, SHEAR, AND MOMENT RELATIONSHIPS

Let us look further at what goes on inside a beam by studying the forces acting on a small length of the uniformly loaded simple span of Example 6–2. In Figure 6–30 we have added a load diagram which is nothing more than a plot of the transverse loads (including reactions) that act on a beam. Up loads are taken as positive and down loads as negative.

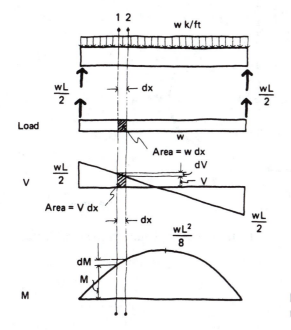

**Figure 6–30**  Load, shear, and moment diagrams.

Figure 6–31 shows a small elemental length of beam taken from between sections 1 and 2. This element must be in equilibrium under the forces shown: therefore,

$$\Sigma F_y = 0$$

$$V - w\,dx - (V - dV) = 0 \qquad (6\text{–}1a)$$

$$dV = w\,dx$$

or

$$w = \frac{dV}{dx} \qquad (6\text{–}2a)$$

Similarly, for rotational equilibrium, we can take moments about any point. Point $c$ conveniently eliminates a force, and thus

$$\Sigma M_c = 0$$

$$M + dM + w\,dx\left(\frac{dx}{2}\right) - V\,dx - M = 0$$

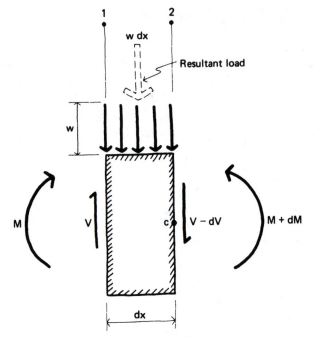

**Figure 6–31**  Free-body diagram.

Considering the third term to be small enough to be neglected gives us

$$dM = V \, dx \qquad\qquad\qquad (6\text{--}1b)$$

or

$$V = \frac{dM}{dx} \qquad\qquad\qquad (6\text{--}2b)$$

The four equations, 6–1a, 6–1b, 6–2a, and 6–2b, establish some very useful relationships among the load, shear, and moment diagrams. They enable us to construct the diagrams rapidly, without writing the shear and moment equations.

   Look first at Equations 6–2a and 6–2b and let the distance $dx$ in Figure 6–30 approach zero.

| | |
|---|---|
| $w = \dfrac{dV}{dx}$ | At any point along the length of the beam, the *ordinate* on the load diagram is equal to the *slope* of the shear diagram. |
| $V = \dfrac{dM}{dx}$ | At any point along the length of the beam, the *ordinate* on the shear diagram is equal to the *slope* of the moment diagram. |

   This slope–ordinate relationship holds for both magnitude and sign. (A positive slope is one that is up and to the right; a negative slope is down and to the right.)

Looking at the diagrams, we see that in the *left* half of the beam, the ordinate of the $V$ diagram and the slope of the $M$ diagram are both positive and decreasing as we move from left to right. At midspan, the ordinate of the $V$ diagram and the slope of the $M$ diagram are both zero. Moving from left to right for the *right* half of the beam, we find that the ordinate of the $V$ diagram and the slope of the $M$ diagram are both negative and increasing.

Now look at Equations 6–1a and 6–1b and let the distance $dx$ in Figure 6–30 be a small but finite interval.

| | |
|---|---|
| $dV = w \, dx$ | Over any beam length interval, the *net area* under the load curve is equal to the *change in ordinate* on the shear diagram. |
| $dM = V \, dx$ | Over any beam length interval, the *net area* under the shear curve is equal to the *change in ordinate* on the moment diagram. |

For the left half of the beam, the area under the shear curve is

$$A = \frac{1}{2}\left(\frac{wL}{2}\right)\left(\frac{L}{2}\right) = \frac{wL^2}{8}$$

This is the change in ordinate on the $M$ diagram from $x = 0$ to $x = L/2$, that same interval. For the entire beam length, the *net area* under the shear curve is zero because the positive area equals the negative area. The change in ordinate on the $M$ diagram from $x = 0$ to $x = L$ is zero, both points having a value of $M = 0$.

These relationships support and verify the point made previously concerning maximum moment and zero shear. We can also now state that for a beam portion having no load ($w \, dx = 0$), the shear must be constant ($dV = 0$). Furthermore, if the moment is constant over a beam length ($dM = 0$), no shear can exist ($V \, dx = 0$).

The reader should study the illustrations in Figures 6–32 or 6–32M to become familiar with the diagram relationships. It is recommended that Problems 6–1 through 6–7 or Problems 6–1M through 6–7M be reworked for practice using the new techniques before attempting any new problems. *Always remember to sketch the probable deflected shape and make sure that it can be rationalized with the moment diagram in each case.* Many careless errors can be found or prevented this way. *Also, remember that the moment will always be zero at the end of a beam unless that end has a fixed (wall) reaction.*

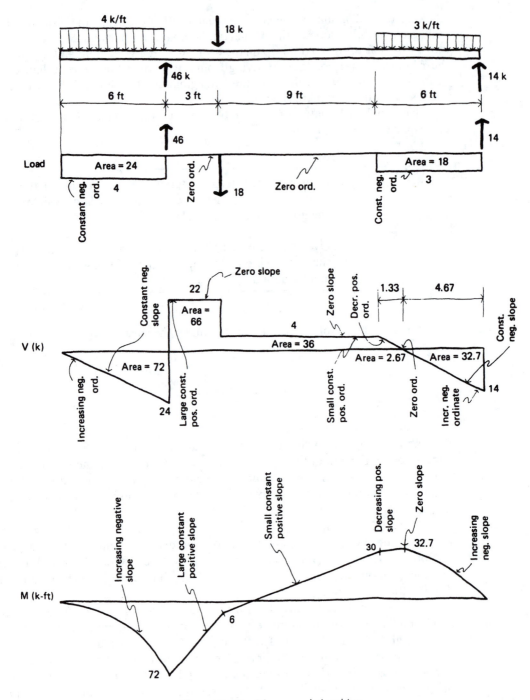

**Figure 6–32** Diagram relationships.

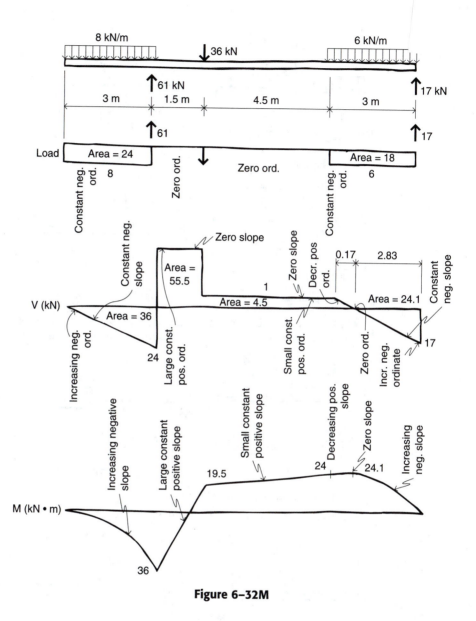

**Figure 6–32M**

## PROBLEMS

**6–8.** Construct the V and M diagrams for the beam in Figure 6–33.

**6–8M.** Construct the V and M diagrams for the beam in Figure 6–33M.

**6–9.** Construct the V and M diagrams for the beam in Figure 6–34.

**6–9M.** Construct the V and M diagrams for the beam in Figure 6–34M.

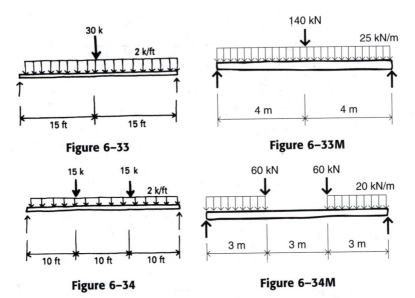

**Figure 6–33**

**Figure 6–33M**

**Figure 6–34**

**Figure 6–34M**

**6–10.** Construct the $V$ and $M$ diagrams for the beam in Figure 6–35.

**6–10M.** Construct the $V$ and $M$ diagrams for the beam in Figure 6–35M.

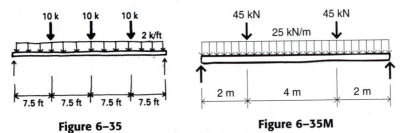

**Figure 6–35**

**Figure 6–35M**

**6–11.** Construct the $V$ and $M$ diagrams for the long-span girder of Figure 6–36.

**6–11M.** Construct the $V$ and $M$ diagrams for the long-span girder of Figure 6–36M.

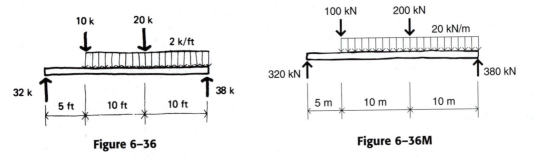

**Figure 6–36**

**Figure 6–36M**

**6–12.** Construct the $V$ and $M$ diagrams for the cantilever beam of Figure 6–37.

**6–13M.** Construct the $V$ and $M$ diagrams for the beam in Figure 6–38M.

**6–14.** Construct the $V$ and $M$ diagrams for the beam with two overhanging ends in Figure 6–39.

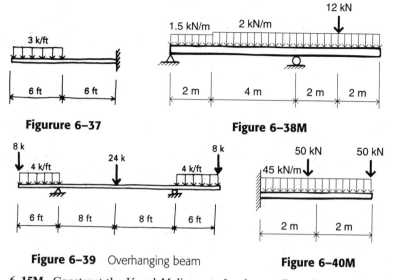

**Figurure 6–37**

**Figure 6–38M**

**Figure 6–39** Overhanging beam

**Figure 6–40M**

**6–15M.** Construct the $V$ and $M$ diagrams for the cantilever beam of Figure 6–40M.

**6–16.** Construct the $V$ and $M$ diagrams for the cantilever beam of Figure 6–41.

**6–16M.** Construct the $V$ and $M$ diagrams for the cantilever beam of Figure 6–41M.

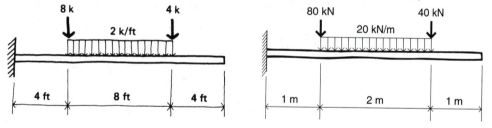

**Figure 6–41**

**Figure 6–41M**

**6–17.** Construct the $V$ and $M$ diagrams for the beam in Figure 6–42. Pinned connections may be assumed.

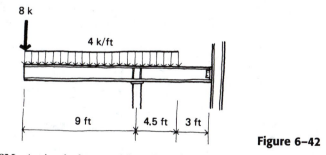

**Figure 6–42**

**6–18M.** A simple beam of length $L$ (m) supporting a uniform load of $w$ (kN/m) has a midspan moment of $wL^2/8$ (kN·m). How much moment, $M$ (in terms of $w$ and $L$), should be applied to the ends of the beam in Figure 6–43M to reduce that midspan moment by a factor of 3?

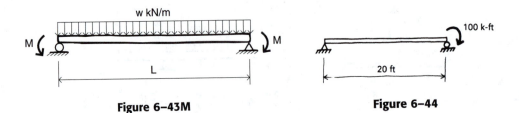

**Figure 6–43M**                    **Figure 6–44**

**6–19.** Construct the $V$ and $M$ diagrams for the simple beam with an applied moment load in Figure 6–44.

**6–20M.** Construct the $V$ and $M$ diagrams for the hinged beam of Figure 6–45M. (*Hint:* Make sure that the moment diagram goes to zero at the hinge.)

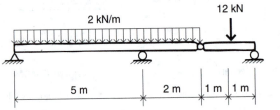

**Figure 6–45M**   Hinged beam.

**6–21.** Construct the $V$ and $M$ diagrams for the beam system in Figure 6–46.

**6–22.** Construct the $V$ and $M$ diagrams for the overhanging beam of Figure 6–47.

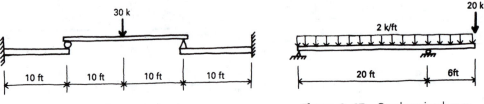

**Figure 6–46**                    **Figure 6–47**   Overhanging beam.

**6–23.** Construct the $V$ and $M$ diagrams for the concrete beam in Figure 6–48. The columns may be assumed to provide fixed ends and the connections to act as hinges.

**6–23M.** Construct the $V$ and $M$ diagrams for the concrete beam in Figure 6–48M. The columns may be assumed to provide fixed ends and the connections to act as hinges.

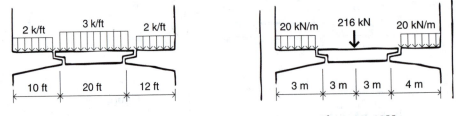

**Figure 6–48**                    **Figure 6–48M**

**6–24M.** Construct the $V$ and $M$ diagrams for the cantilevered-suspended beam system in Figure 6–49M.

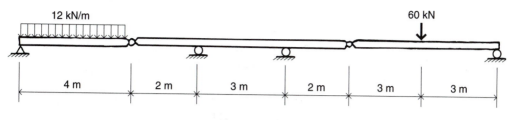

**Figure 6–49M**

**6–25M.** Construct the $V$ and $M$ diagrams for the simple beam with applied moment loads in Figure 6–50M.

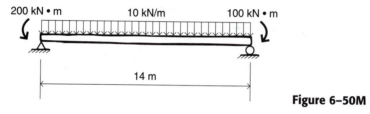

**Figure 6–50M**

# 6–5 UNIFORMLY VARYING LOADS

The uniformly varying load of Figure 6–51 may be thought of as the type of load provided by a pile of sand in a storage bin or the horizontal pressure of water against a dam. More frequently, however, such loads in building structures result from structural framing layouts, which employ diagonal elements or openings. Chapter 10 treats tributary areas for different framing patterns, and this will not be discussed here. At this point we are interested only in the shears and moments that result from such loads. Basically, the uniformly varying load causes each diagram to move up one order so that the $V$ diagram has parabolic curves and the $M$ diagram, cubic curves. This means that in some instances it is necessary to evaluate the areas under parabolic curves to utilize the area–ordinate relationship. Appendix F provides the areas for two parabolic cases with the stipulation that the curve pass through the apex of the parabola.

**EXAMPLE 6–6** Construct the shear and moment diagrams for the cantilever in Figure 6–51.

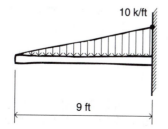

**Figure 6–51** Cantilever with uniformly varying load

*Solution*

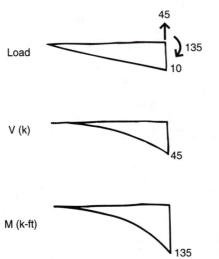

**Figure 6–52**  Load, shear, and moment diagrams.

---

**EXAMPLE 6–7M**

Construct the shear and moment diagrams for the beam in Figure 6–53M.

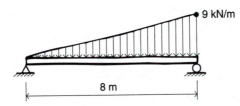

**Figure 6–53M**  Simple beam with uniformly varying load.

*Solution*

To find $M_{\text{max}}$ we need to determine $x$, the location of the point of zero shear. We can use the FBD of Figure 6–54M(b) and let $V_x$ take a value of zero.

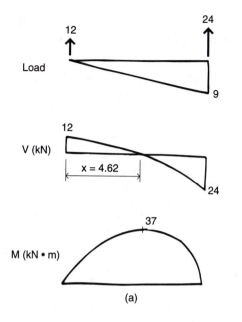

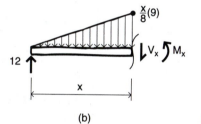

(b)

**Figure 6–54M**

$$\Sigma F_y = 0$$

$$12 - \frac{1}{2}\left(\frac{x}{8}\right)(9)x = 0$$

$$x = 4.62 \text{ m}$$

The half-parabola in the $V$ diagram (altitude = 12 kN, base = 4.62 m) *does* include an apex at its left end, and according to Appendix F, its area will be

$$A = \tfrac{2}{3}bh$$

$$= \tfrac{2}{3}(12)(4.62)$$

$$= 37.0 \text{ kN·m}$$

In this case, the maximum moment will have the same value. This can be verified from Case 6 in Appendix K, where

$$M_{max} = \frac{wL^2}{9\sqrt{3}}$$

$$= \frac{9(8)^2}{9\sqrt{3}}$$

$$= 37.0 \text{ kN·m}$$

## PROBLEMS

**6–26.** Construct the $V$ and $M$ diagrams for the cantilever beam in Figure 6–55.

**6–26M.** Construct the $V$ and $M$ diagrams for the cantilever beam in Figure 6–55M.

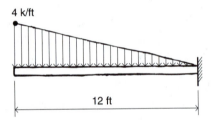

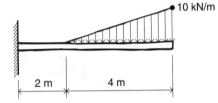

**Figure 6–55** Cantilever beam with uniformly varying load.

**Figure 6–55M** Cantilever beam with uniformly varying load.

**6–27.** Construct the $V$ and $M$ diagrams for the timber beam in Figure 6–56.

**6–27M.** Construct the $V$ and $M$ diagrams for the beam in Figure 6–56M.

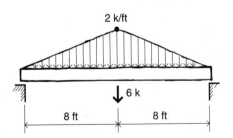

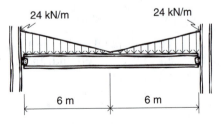

**Figure 6–56** Timber beam with uniformly varying load and a point load.

**Figure 6–56M** Steel beam with pin connections.

**6–28.** Construct the $V$ and $M$ diagrams for the beam in Figure 6–57.

**6–28M.** Construct the $V$ and $M$ diagrams for the beam in Figure 6–57M.

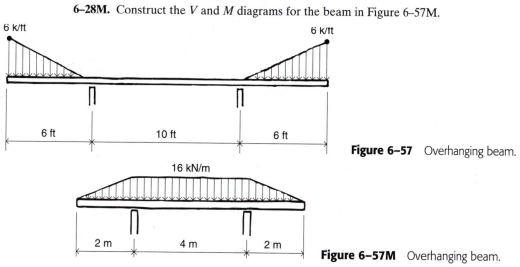

**Figure 6–57** Overhanging beam.

**Figure 6–57M** Overhanging beam.

# 7

# Flexural Stresses

## 7-1  INTRODUCTION

Chapter 1 makes reference to the structural inefficiency of bending as a means to carry load. Compared to axial tension and compression, bending generates much higher stresses in members, because the fibers of material are not stressed uniformly. We know from experience that if we want to break something in half, it is much easier to break it in bending than to try to pull it apart in tension or buckle it in compression.

Although it might be nice (for the sake of structural efficiency) to eliminate bending stresses in our structures, clearly this is not possible. To attempt to carry all loads in direct tension or compression would lead to some very awkward configurations and/or very small spans. The fact is that true structural efficiency, in many cases, would result in a false economy, because of the resulting spatial or architectural inefficiencies. [It is also true that bending action takes place in most structural elements anyway, even those designed and configured to carry loads by other means. As soon as we build stiffness into a structure, by thickness or shape (moment of inertia), it will have a tendency to carry load by bending.]

Many functions require clear spans in the short-to-moderate range of 15 to 60 ft (5 to 18 m), and most often some form of beam construction is preferable to arches, cables, vaults, folded plates, or other "more efficient" structures. A post-and-beam system will usually provide more usable building volume than the more form-resistant structures. For anything but the longer spans, beams will give the largest span/depth ratio, enabling the designer to reduce the space between the ceiling and the finish floor above. This can be essential in highrise buildings.

Beams are also relatively insensitive to the type and location of loads they can receive. For example, trusses will only take concentrated loads at the panel points, and vaults and domes are quite unsuitable for either concentrated or line loads. Although it is true that bending action results in high stresses and deflections, it is equally true that strong, stiff structural materials are both readily available and inexpensive, particularly when compared to some nonstructural building materials and labor.

A beam can be loosely defined as any structural element that carries transverse loads and has two of its dimensions much smaller than the third. Most beams are straight and of constant cross section, but some are curved or angled and some have varying cross sections. All beams develop two kinds of stress: flexural (sometimes called bending), which is a normal stress, and shearing, which is tangential. All beams also deflect under load, and these three items—flexural stress, shearing stress, and deflection—are the parameters by which we determine the required size and/or shape of the cross section. Of these, flexural stress governs most frequently. The remainder of this chapter and Chapter 8 will be devoted to analyzing and understanding beam stresses. Beam deflection is covered in Chapter 9. These are combined in Chapter 10, where beam design is treated.

## 7-2  FLEXURAL STRAIN

Beams bend under load such that transverse sections remain plane as represented by the lines on the beam in Figure 7–1(b). There is a compression zone and a tension zone, which are separated by a horizontal neutral plane. The neutral plane (called the *neutral axis*) is located at the centroid of the cross section, and the beam fibers are squeezed or stretched in direct proportion to their distance from this neutral axis. The top and bottom fibers, or extreme fibers, undergo the most strain and do the most work, while those close to the neutral axis have strain levels near zero and are least effective. A fiber located halfway between the neutral axis and the top or bottom edge of a beam will have half the strain of a fiber located at that edge. (Verify this by making a small beam of an easy-to-bend material, such as polyurethane foam, and drawing parallel lines on one of its sides.) It is important to understand the linearity of bending strain and the corresponding stresses that develop, and this subject is fully examined in Appendix A.

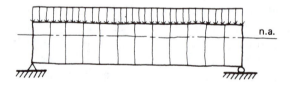

(a)

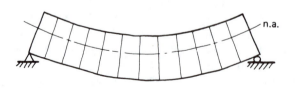

(b)

**Figure 7–1**  Flexural strain.

The correct location of the neutral axis, critical to the proper understanding of flexural stresses, was theorized in 1713 by Antoine Parent (1666–1716), a French scientist. However, it remained largely unknown until the extensive work by Claude L. M. H. Navier, also French, in 1826.

## 7–3  FLEXURAL STRESS

External bending loads are resisted by internal stresses that build up in the beam fibers. These stresses are directly related to flexural strain by the stiffness of the material. Their direction is always normal to the transverse section. Assuming that the beam is made of a reasonably homogenous material, the stresses will vary as the strains do (i.e., linear with the distance from the neutral axis). Bending stress distributions for two different beam shapes are illustrated in Figure 7–2. Because of symmetry, the top and bottom fiber stresses in a rectangular section will be equal in magnitude. This is not true for the T shape, where higher stresses exist in the lower fibers of the stem than in those of the flange because of their greater distance from the neutral axis. In all cases, the stresses above and below the neutral axis will be opposite in sense.

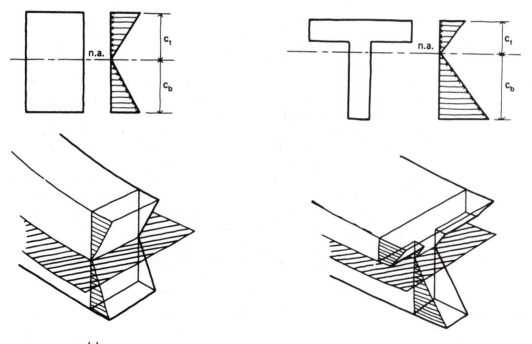

(a)                                                (b)

**Figure 7–2**  Bending stress distribution.

Flexural stresses are a direct response to bending moments and therefore will vary over the length of a beam as well as with the distance from the neutral axis. The general flexure formula is

$$f_y = \frac{My}{I} \tag{7-1}$$

where $f_y$ = flexural stress at fiber level $y$ (psi or ksi) (kPa or MPa)
  $M$ = bending moment at the transverse section being examined (lb-in. or kip-in.) (kN · m)
  $y$ = vertical distance from the neutral axis to level $y$ (in.) (m)
  $I$ = moment of inertia of the cross section with respect to the neutral (centroidal) axis (in.$^4$) (m$^4$)

(*Note:* The examples and problems in this book assume that bending occurs about the strong axis, and therefore, $I$ takes the value of $I_x$.)

When we are interested only in the maximum bending stresses at a given transverse section, we can use

$$f_b = \frac{Mc}{I} \tag{7-1a}$$

where $f_b$ = extreme fiber bending stress (psi or ksi) (kPa or MPa)
  $c$ = distance to extreme fiber (in.) (m)

(In Equation 7–1a, $c$ will have the value illustrated as distance $c_t$ or $c_b$ in Figure 7–2, and $f_b$ will then be a top fiber stress or a bottom fiber stress, respectively.)

The derivation of these formulas and certain restrictions concerning their use are given in Appendix A. The reader will notice that the sense of the bending stress at a given point in a beam depends on the sense of the bending moment and whether the point is above or below the neutral axis.

(In some of the examples and problems that follow, the effects of member self-weight as part of the dead load have been ignored. Although this is not recommended as sound engineering practice, the writer's own experience indicates that this component of the load seldom controls the member size in short-span building structures of timber or steel. Larger spans in these materials and all reinforced concrete beams, however, have significant self-weights, which cannot be ignored.)

---

**EXAMPLE 7–1** The wood joist in Figure 7–3 is nominal 2 × 10. (The actual dimensions may be found in Appendix I.)
  (a) Determine the maximum bending stress.
  (b) Determine the stress due to bending at a point 4.5 ft in from one of the ends and 3 in. below the top edge.

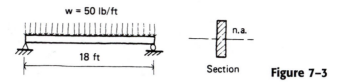

**Figure 7–3**

### Solution

(a) By symmetry the neutral axis (n.a.) is located at middepth. The maximum bending stress will occur where the moment is a maximum and will be compressive in the top fiber and tensile in the bottom fiber (Figure 7–4). From Appendix I, $I = 98.9$ in.[4].

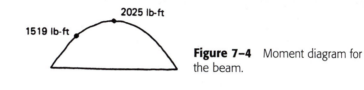

**Figure 7–4**   Moment diagram for the beam.

$$f_b = \frac{Mc}{I}$$

$$= \frac{2025 \text{ lb-ft } (12 \text{ in/ft}) (4.625 \text{ in})}{98.9 \text{ in}^4}$$

$$f_{b_{top}} = 1140 \text{ psi compression}$$

$$f_{b_{bottom}} = 1140 \text{ psi tension}$$

(b) The point is above the neutral axis, and in the presence of positive moment the stress will be compressive.

$$f_y = \frac{My}{I}$$

$$= \frac{1519 \text{ lb-ft } (12 \text{ in/ft}) (1.625 \text{ in})}{98.9 \text{ in}^4}$$

$$= 300 \text{ psi compression}$$

**EXAMPLE 7–2M**   The wood joist in Figure 7–5M is 38 × 235 mm in section.

(a) Determine the maximum bending stress.
(b) Determine the stress due to bending at a point 1.5 m in from one of the ends and 100 mm below the top edge.

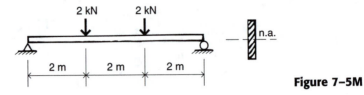

**Figure 7–5M**

### Solution

(a) By symmetry the neutral axis (n.a.) is located at mid-depth. The maximum bending stress will occur where the moment is a maximum and will be compressive in the top fiber and tensile in the bottom fiber (Figure 7–6M). From Appendix I, $I = 41.1(10)^6$ mm$^4$.

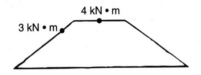

**Figure 7–6M**   Moment diagram for the beam.

$$f_b = \frac{Mc}{I}$$

$$= \frac{4.0 \text{ kN} \cdot \text{m}(0.118 \text{ m})}{41.1(10)^{-6} \text{ m}^4}$$

$$f_{b_{\text{top}}} = 11\ 500 \text{ kPa compression}$$

$$f_{b_{\text{bottom}}} = 11\ 500 \text{ kPa tension}$$

(b) The point is above the neutral axis, and in the presence of positive moment the stress will be compressive.

$$f_y = \frac{My}{I}$$

$$= \frac{3.0 \text{ kN} \cdot \text{m}(0.018 \text{ m})}{41.1(10)^{-6} \text{ m}^4}$$

$$= 1310 \text{ kPa compression}$$

**EXAMPLE 7–3** Determine the maximum bending stress in the beam of Figure 7–7.

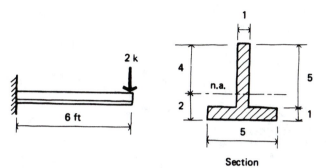

**Section**

**Figure 7–7** Inverted T-beam used as a cantilever.

### Solution

The maximum bending stress will be at the cross section where the moment maximizes and at the fiber level farthest from the n.a. (Figure 7–8). In this case, these conditions are met at the top fiber, where the beam enters the wall. The stress will be tensile. Using the parallel axis techniques described in Section 3–4, we can find the moment of inertia with respect to the neutral axis as

$$I = 33.3 \text{ in}^4$$

(Students who need to review the parallel axis theorem should verify this $I$ value. See Section 3–4.)

The maximum moment is 12 kip-ft.

$$f_{b_{top}} = \frac{Mc}{I}$$

$$= \frac{12 \text{ kip-ft } (12 \text{ in/ft}) \ (4 \text{ in})}{33.3 \text{ in}^4}$$

$$= 17.3 \text{ ksi}$$

**12 k-ft**

**Figure 7–8** Moment diagram for the beam.

**EXAMPLE 7–4M**

Determine the maximum bending stress in the beam of Figure 7–9M.

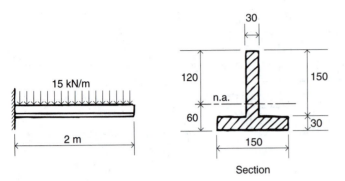

Section

**Figure 7–9M**  Inverted T-beam used as a cantilever.

### Solution

The maximum bending stress will be at the cross section where the moment maximizes and at the fiber level farthest from the n.a. (Figure 7–10M). In this case, these conditions are met at the top fiber, where the beam enters the wall. The stress will be tensile. Using the parallel axis theorem described in Section 3–4, we get

$$I = 27(10)^6 \text{ mm}^4$$

The maximum moment is 30 kN · m.

$$f_b = \frac{Mc}{I}$$

$$= \frac{30 \text{ kN} \cdot \text{m}(0.120 \text{ m})}{27(10)^{-6} \text{ m}^4}$$

$$= 133(10)^3 \text{ kPa}$$

$$= 133 \text{ MPa}$$

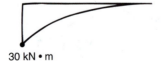

30 kN • m

**Figure 7–10M**  Moment diagram for the beam.

**EXAMPLE 7–5**  A steel W12 × 16 is used for the beam in Figure 7–11. Determine the maximum tensile and compressive stresses due to bending. (The designation W12 × 16 indicates a wide-flange beam with a *nominal* depth of 12 in. and a self-weight of 16 lb per foot of length.)

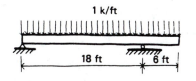

**Figure 7–11**  Overhanging beam with uniform load.

### Solution
As with the rectangle, the symmetrical W shape has its *c* distance to the top fiber equal to the *c* distance to the bottom fiber, so that (for any given section) the top and bottom stresses are equal in magnitude. This means that the maximum tensile and compressive stresses must both occur where the moment is a maximum, in this case, 32 kip-ft (Figure 7–12). From Appendix J, $I = 103$ in.[4] and $d = 11.99$ in.

$$f_{b_{top}} = \frac{Mc}{I}$$

$$= \frac{32 \text{ kip-ft } (12 \text{ in/ft}) \, (11.99 \text{ in/2})}{103 \text{ in}^4}$$

$$= 22.4 \text{ ksi compression}$$

and

$$f_{b_{bottom}} = \frac{Mc}{I}$$

$$= \frac{32 \text{ kip-ft } (12 \text{ in/ft}) \, (6 \text{ in})}{103 \text{ in}^4}$$

$$= 22.4 \text{ ksi tension}$$

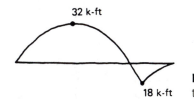

**Figure 7–12**  Moment diagram for the beam.

**EXAMPLE 7–6M**     A steel W310 × 21 is used for the beam in Figure 7–13M. Determine the maximum tensile and compressive stresses due to bending. (The designation W310 × 21 indicates a wide-flange beam with a *nominal* depth of 310 mm and a self-mass of 21 kg per meter of length.)

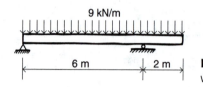

9 kN/m

6 m          2 m

**Figure 7–13M**   Overhanging beam with uniform load.

### Solution

As with the rectangle, the symmetrical W shape has its $c$ distance to the top fiber equal to the $c$ distance to the bottom fiber, so that (for any given section) the top and bottom stresses are equal in magnitude. This means that the maximum tensile and compressive stresses must both occur where the moment is a maximum, in this case, 32 kN·m (Figure 7–14M). From Appendix J, $I = 36.9(10)^6$ mm⁴ and $d = 303$ mm.

$$f_{b_{top}} = \frac{Mc}{I}$$

$$= \frac{32 \text{ kN} \cdot \text{m}(0.152 \text{ m})}{36.9(10)^{-6} \text{ m}^4}$$

$$= 132 \text{ MPa compression}$$

and

$$f_{b_{bottom}} = \frac{Mc}{I}$$

$$= \frac{32 \text{ kN} \cdot \text{m}(0.152 \text{ m})}{36.9(10)^{-6} \text{ m}^4}$$

$$= 132 \text{ MPa tension}$$

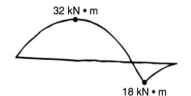

32 kN • m

18 kN • m

**Figure 7–14M**   Moment diagram for the beam.

**EXAMPLE 7–7** To save construction depth with a precast plank system, a structural T is used in lieu of the W shape in Example 7–5. Its properties are given in Figure 7–15. Determine the maximum tensile and compressive stresses due to bending.

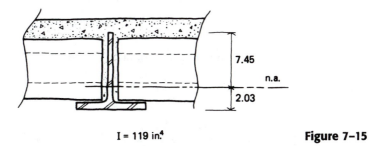

7.45

n.a.

2.03

$I = 119$ in.$^4$

**Figure 7–15**

**Solution**

The structural T is not symmetrical with respect to its extreme fiber distances, and the possibility exists that one of the maximum stresses (tensile or compressive) may occur where the moment is not at its maximum absolute value. The compression stress will maximize at fibers 1 and 2 in Figure 7–16, and the tension stress will maximize at fibers 3 and 4. For illustrative purposes we shall examine all four points in this example. (The moment diagram from Example 7–5 is, of course, still valid.)

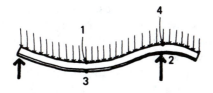

**Figure 7–16** Deflected shape of structural T-beam.

For the maximum positive moment section,

$$f_{b_1} = \frac{Mc}{I}$$

$$= \frac{32 \text{ kip-ft } (12 \text{ in/ft}) (7.45 \text{ in})}{119 \text{ in}^4}$$

$$= 24.0 \text{ ksi compression}$$

and

$$f_{b_3} = \frac{Mc}{I}$$

$$= \frac{32 \text{ kip-ft } (12 \text{ in/ft}) (2.03 \text{ in})}{119 \text{ in}^4}$$

$$= 6.55 \text{ ksi tension}$$

For the maximum negative moment section,

$$f_{b_4} = \frac{Mc}{I}$$

$$= \frac{18 \text{ kip-ft } (12 \text{ in/ft}) (7.45 \text{ in})}{119 \text{ in}^4}$$

$$= 13.5 \text{ ksi tension}$$

and

$$f_{b_2} = \frac{Mc}{I}$$

$$= \frac{18 \text{ kip-ft } (12 \text{ in/ft}) (2.03 \text{ in})}{119 \text{ in}^4}$$

$$= 3.69 \text{ ksi compression}$$

A comparison of the four values just determined shows that the maximum compression occurs where the moment is maximum (i.e., at point 1). The tensile stress, however, maximizes at point 4, where the moment is only 18 kip-ft. This happens, of course, because of the different $c$ distances involved in the asymmetrical T shape. Accordingly, when sections having dissimilar $c$ distances are used for beams that have both positive and negative curvature, the largest bending stress of a specified sense will not necessarily occur where the moment is maximum. In order to determine the maximum bending stresses in such cases, one either has to check the stress levels at four points (as we did) or compare the ratio of $M$ values to the ratio of $c$ distances to ascertain where the equation $Mc/I$ will maximize. What remains true, however, is that the *absolute* maximum value of bending stress (24.0 ksi in our example) will always occur at the section having the *absolute* maximum moment.

---

**EXAMPLE 7–8M**

To save construction depth with a precast plank system, a structural T is used in lieu of the W shape in Example 7–6M. Its properties are given in Figure 7–17M. Determine the maximum tensile and compressive stresses due to bending.

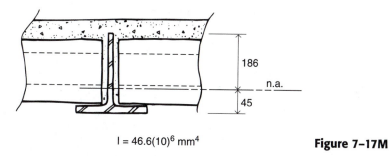

$I = 46.6(10)^6$ mm$^4$

**Figure 7–17M**

## Solution

The structural T is not symmetrical with respect to its extreme fiber distances, and the possibility exists that one of the maximum stresses (tensile or compressive) may occur where the moment is not at its maximum absolute value. The compression stress will maximize at fibers 1 and 2 in Figure 7–18M, and the tension stress will maximize at fibers 3 and 4. For illustrative purposes we shall examine all four points in this example. (The moment diagram from Example 7–6M is, of course, still valid.)

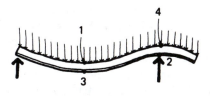

**Figure 7–18M** Deflected shape of structural T beam.

For the maximum positive moment section,

$$f_{b_1} = \frac{Mc}{I}$$

$$= \frac{32 \text{ kN} \cdot \text{m}(0.186 \text{ m})}{46.6(10)^{-6} \text{ m}^4}$$

$$= 128 \text{ MPa compression}$$

and

$$f_{b_3} = \frac{Mc}{I}$$

$$= \frac{32 \text{ kN} \cdot \text{m}(0.045 \text{ m})}{46.6(10)^{-6} \text{ m}^4}$$

$$= 30.9 \text{ MPa tension}$$

For the maximum negative moment section,

$$f_{b_4} = \frac{Mc}{I}$$

$$= \frac{18 \text{ kN} \cdot \text{m}(0.186 \text{ m})}{46.6(10)^{-6} \text{ m}^4}$$

$$= 71.8 \text{ MPa tension}$$

and

$$f_{b_2} = \frac{Mc}{I}$$

$$= \frac{18 \text{ kN} \cdot \text{m}(0.045 \text{ m})}{46.6(10)^{-6} \text{ m}^4}$$

$$= 17.4 \text{ MPa compression}$$

A comparison of the four values just determined shows that the maximum compression occurs where the moment is maximum (i.e., at point 1). The tensile stress, however, maximizes at point 4, where the moment is only 18 kN · m. This happens, of course, because of the different $c$ distances involved in the asymmetrical T shape. Accordingly, when sections having dissimilar $c$ distances are used for beams that have both positive and negative curvature, the largest bending stress of a specified sense will not necessarily occur where the moment is maximum. In order to determine the maximum bending stresses in such cases, one either has to check the stress levels at four points (as we did) or compare the ratio of $M$ values to the ratio of $c$ distances to ascertain where the equation $Mc/I$ will maximize. What remains true, however, is that the *absolute* maximum value of bending stress (128 MPa in our example) will always occur at the section having the *absolute* maximum moment.

---

**EXAMPLE 7–9** Nominal 2 × 12 joists are spaced 2 ft apart (2 ft on center) to make a flat roof. They must support a total load of 50 psf over a simple span of 17 ft. Determine the maximum flexural stress.

**Solution**
This example introduces the concept of tributary area, which is discussed in Chapter 10. Each joist must support an amount of floor half the distance to its neighboring parallel joist on each side. In this case this "half-distance" is 1 ft. Therefore, each linear foot of each joist must support 2 ft² of floor area. Each foot of joist is responsible for an area of floor that is 1 ft by 2 ft. Therefore,

$$w = 50 \text{ psf}(2 \text{ ft}) = 100 \text{ lb/ft}$$

The maximum moment exists at midspan and can be obtained from the moment diagram or from Appendix K as

$$M = \frac{wL^2}{8}$$

$$= \frac{(100 \text{ lb/ft}) (17 \text{ ft})^2}{8}$$

$$= 3610 \text{ lb-ft}$$

Using Appendix I, we find that

$$f_b = \frac{Mc}{I}$$

$$= \frac{3610 \text{ lb-ft} (12 \text{ in/ft}) (5.625 \text{ in})}{178 \text{ in}^4}$$

$$= 1370 \text{ psi}$$

This will occur as compression in the top fiber and tension in the bottom fiber.

---

**EXAMPLE 7–10M**

Nominal 38 × 285 joists are spaced 600 mm apart (0.6 m on center) to make a flat roof. They must support a total load of 2.5 kN/m² over a simple span of 5 m. Determine the maximum flexural stress.

### Solution

This example introduces the concept of tributary area, which is discussed in Chapter 10. Each joist must support an amount of floor half the distance to its neighboring parallel joist on each side. In this case this "half-distance" is 0.3 m. Therefore, each linear meter of joist must support 0.6 m of floor area. Each meter of joist is responsible for an area of floor that is 1 m by 0.6 m. Therefore,

$$w = (2.5 \text{ kN/m}^2) (0.6 \text{ m}) = 1.5 \text{ kN/m}$$

The maximum moment exists at midspan and can be obtained from the moment diagram or from Appendix K as

$$M = \frac{wL^2}{8}$$

$$= \frac{(1.5 \text{ kN/m}) (5 \text{ m})^2}{8}$$

$$= 4.7 \text{ kN} \cdot \text{m}$$

Using Appendix I, we find that

$$f_b = \frac{Mc}{I}$$

$$= \frac{4.7 \text{ kN} \cdot \text{m} \ (0.143 \text{ m})}{73.3(10)^{-6} \text{ m}^4}$$

$$= 9170 \text{ kPa}$$

This will occur as compression in the top fiber and tension in the bottom fiber.

## PROBLEMS

**7–1.** A nominal $2 \times 10$ is used as a simple beam to support a uniform load of 67 lb/ft over a span of 16 ft. Determine the top and bottom extreme fiber stresses at:
(a) midspan.
(b) 4 ft from the left end.

**7–1M.** A nominal $38 \times 235$ is used as a simple beam to support a uniform load of 1 kN/m over a span of 5 m. Determine the top and bottom extreme fiber stresses at:
(a) midspan.
(b) 1 m from the left end.

**7–2.** Two nominal $2 \times 8$s are nailed side by side to make a 6-ft-long cantilever beam. Determine the maximum bending stress if its only load is a concentrated 400-lb load located at the free end.

**7–2M.** Two nominal $38 \times 185$s are nailed side by side to make a 2-m-long cantilever beam. Determine the maximum bending stress if its only load is a concentrated 1.8-kN load located at the free end.

**7–3.** A $W24 \times 68$ is used as a simple beam spanning 30 ft. It must carry a uniform load of 1 kip/ft and two concentrated loads at the third points of 10 kips each. Determine the maximum bending stress.

**7–3M.** A $W610 \times 101$ is used as a simple beam spanning 10 m. It must carry a uniform load of 15 kN/m and two concentrated loads at the third points of 45 kN each. Determine the maximum bending stress.

**7–4.** A $W21 \times 50$ is used for the overhanging beam of Figure 7–19. Assuming pinned connections, determine the magnitude and sense of the extreme fiber stresses
(a) at the section where the moment is maximum.
(b) under the left-hand concentrated load.

**7–4M.** A $W530 \times 74$ is used for the overhanging beam of Figure 7–19M. Assuming pinned connections, determine the magnitude and sense of the extreme fiber stresses
(a) at the section where the moment is maximum.
(b) under the left-hand concentrated load.

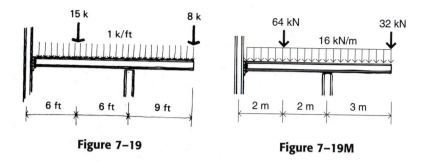

**Figure 7–19**                    **Figure 7–19M**

**7–5.** Douglas fir 2 × 12 joists span 17 ft between two bearing walls. If the maximum allowable bending stress is 1300 psi, determine the permissible uniform load in lb/ft.

**7–5M.** Douglas fir 38 × 285 joists span 5 m between two bearing walls. If the maximum allowable bending stress is 9000 kPa, determine the permissible uniform load in kN/m.

**7–6.** Assuming that an allowable bending stress of 1500 psi controls the design of the beam in Problem 7–1, determine the maximum permissible span.

**7–6M.** Assuming that an allowable bending stress of 10 500 kPa controls the design of the beam in Problem 7–1M, determine the maximum permissible span.

**7–7.** If the allowable bending stress is 33 ksi, will a W30 × 99 be satisfactory for the beam of Figure 7–20? Assume that bending will control. (*Hint:* The span is large, so watch the effect of the weight of the beam itself, which in this case is 0.099 kips per foot.)

**7–7M.** If the allowable bending stress is 230 MPa, will a W760 × 147 be satisfactory for the beam of Figure 7–20M? Assume that bending will control. (*Hint:* The span is large, so watch the effect of the weight of the beam itself, which in this case is 147 kg per meter. Note that kg can be converted to kN by multiplying by 0.0098 kN/kg. This is further explained in Example 7–14M in the next section.)

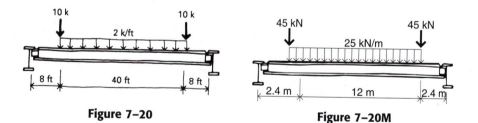

**Figure 7–20**                    **Figure 7–20M**

**7–8.** Simply supported nominal 4 × 14s span 18 ft and are spaced 8 ft on center (o.c.). If the total load is 40 psf, determine the maximum bending stress. (*Hint:* Each linear foot of beam must support 8 ft² of floor area.)

**7–8M.** Simply supported nominal 89 × 335s span 6 m and are spaced 2.4 m on center (o.c.). If the total load is 2 kN/m², determine the maximum bending stress. (*Hint:* Each linear meter of beam must support 2.4 m² of floor area.)

**7–9.** Steel W18 × 76 beams are located 10 ft o.c. and make simple spans of 30 ft to girders that support them. They must support a lightweight concrete-on-steel-deck floor system. The total load is 100 psf. Each beam also carries a point load of 20 kips at midspan. Determine the maximum bending stress.

**7–9M.** Steel W460 × 113 beams are located 3 m o.c. and make simple spans of 10 m to girders that support them. They must support a lightweight concrete-on-steel-deck floor system. The total load is 5 kN/m². Each beam also carries a point load of 90 kN at midspan. Determine the maximum bending stress.

**7–10.** The built-up timber beam of Figure 7–21 is made of two 2 × 10 members enclosing a 2 × 6. The beams are spaced 8 ft apart in a direction normal to the page and must carry a floor load of 30 psf. Determine the maximum tensile and compressive stresses due to bending. Compare your answers to the values given in Appendix H and comment.

**7–10M.** The built-up timber beam of Figure 7–21M is made of two 38 × 235 members enclosing a 38 × 139. The beams are spaced 2.4 m apart in a direction normal to the page and must carry a floor load of 1.5 kN/m². Determine the maximum tensile and compressive stresses due to bending. Compare your answers to the values given in Appendix H and comment.

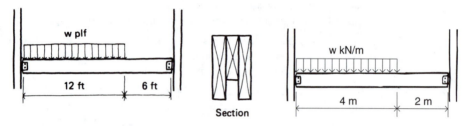

**Figure 7–21**　Balcony beam simply supported by two columns.　　　　　　　　　　　　　　　**Figure 7–21M**

**7–11.** The 24-ft-long timber 4 × 16 beams of Figure 7–22 are spaced 4 ft o.c. They must carry a total floor load of 40 psf and a wall load (from the roof) of 1 kip per running foot of wall. Determine the maximum bending stress.

**7–11M.** The 8-m-long timber 89 × 385 beams of Figure 7–22M are spaced 1.2 m o.c. They must carry a total floor load of 2 kN/m² and a wall load (from the roof) of 14 kN per running meter of wall. Determine the maximum bending stress.

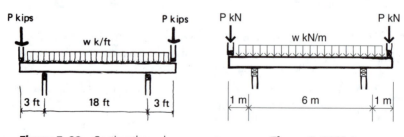

**Figure 7–22**　Section through a residential floor.　　　　　　　　　　　　　　　**Figure 7–22M**

**7–12.** The steel beam of Figure 7–23 is composed of a W section and a channel welded together and has the $I$ value given. Determine the maximum compressive and tensile stresses due to bending.

**7–12M.** The steel beam of Figure 7–23M is composed of a W section and a channel welded together and has the $I$ value given. Determine the maximum compressive and tensile stresses due to bending.

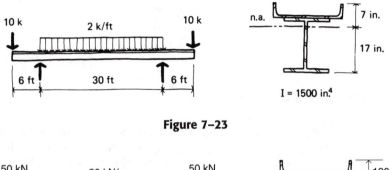

**Figure 7–23**

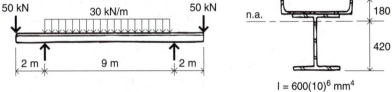

**Figure 7–23M**

**7–13.** The cross section of the beam in Figure 7–24 is built up of a $2 \times 6$, a $2 \times 12$, and a $2 \times 4$ as illustrated. Determine the maximum tensile and compressive stresses due to bending.

**7–13M.** The cross section of the beam in Figure 7–24M is built up of a $38 \times 139$, a $38 \times 285$, and a $38 \times 89$ as illustrated. Determine the maximum tensile and compressive stresses due to bending.

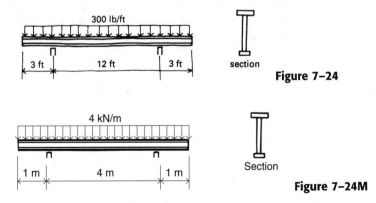

**Figure 7–24**

**Figure 7–24M**

**7–14M.** Assume that the joist of Example 7–2M has a hole bored through it at midspan. If the hole is located as in Figure 7–25M, determine the extreme fiber bending stresses.

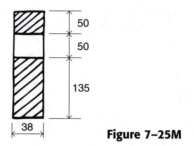

**Figure 7–25M**

# 7–4   SECTION MODULUS

The general formula for flexural stress, $f_b = Mc/I$, can be simplified slightly if we restrict our analyses to problems involving only extreme fiber stresses. (Since these are the stresses that usually control, this restriction is of little consequence.) Notice that the maximum flexural stress is really a function of only two items, the bending moment and the dimensions of the cross section. If we can combine the two cross-sectional factors $c$ and $I$ into one term, the general equation will be easier to use, particularly when applied in a design situation. The quantity $I/c$ has been given the special name of *section modulus* and the symbol $S$. It is a measure of bending resistance which includes both the moment of inertia and the depth. Its units are length cubed. The formula for maximum flexural stress will then be

$$f_b = \frac{M}{S} \tag{7–2}$$

In a section that is symmetrical about the neutral axis, the $c$ distances to the tensile and compressive fibers will be equal and the section modulus will have only one value. For a T shape or other unsymmetrical section, where the neutral axis is not at middepth, the larger $c$ dimension should be used in $S$ so that Equation 7–2 will compute the larger of the two extreme fiber stresses. (It is probably just as easy to use the straight $Mc/I$ formula for unsymmetrical shapes.)

For a rectangular section, $S$ can be stated in terms of the width and depth, by-passing the $I$ computation.

$$S = \frac{I}{c}$$

$$= \frac{bd^3/12}{d/2} \tag{7–3}$$

$$= \frac{bd^2}{6}$$

Section modulus values for the strong axes ($S_x$) of some common timber rectangles have been computed and are given in Appendix I. For selected steel shapes, $S_x$ values are listed in Appendix J. The examples and problems in this book all assume strong-axis bending.

---

**EXAMPLE 7–11**

Determine the maximum bending stress in a 4 × 10 timber which is used as a uniformly loaded cantilever 6 ft long. The total load is 300 plf.

**Solution**

From Appendix I, $S$ = 49.9 in.³, and the moment diagram is given in Figure 7–26.

$$f_b = \frac{M}{S}$$

$$= \frac{5400 \text{ lb-ft}(12 \text{ in/ft})}{49.9 \text{ in}^3}$$

$$= 1300 \text{ psi}$$

**Figure 7–26**  Moment diagram for the cantilever beam.

5400 lb-ft

---

**EXAMPLE 7–12M**

Determine the maximum bending stress in an 89 × 235 mm timber which is used as a uniformly loaded cantilever 2 m long. The total load is 4 kN/m.

**Solution**

From Appendix I, $S$ = 819(10)³ mm³, and the moment diagram is given in Figure 7–27M.

$$f_b = \frac{M}{S}$$

$$= \frac{8 \text{ kN} \cdot \text{m}}{819(10)^{-6} \text{ m}^3}$$

$$= 9770 \text{ kPa}$$

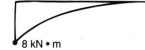

**Figure 7–27M**  Moment diagram for the cantilever beam.

8 kN • m

**EXAMPLE
7–13**

The W36 × 260 beam in Figure 7–28 must carry three column loads of 75 kips each. Determine the maximum bending stress.

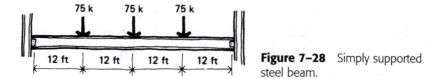

**Figure 7–28**  Simply supported steel beam.

**Solution**
From Appendix J, $S = 953$ in.$^3$, and the moment diagram is shown in Figure 7–29.

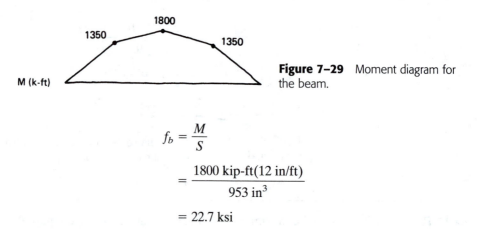

**Figure 7–29**  Moment diagram for the beam.

$$f_b = \frac{M}{S}$$

$$= \frac{1800 \text{ kip-ft}(12 \text{ in/ft})}{953 \text{ in}^3}$$

$$= 22.7 \text{ ksi}$$

This is very close to the allowable bending stress of 24 ksi for mild steel, and the size of the span indicates that the bending stress due to the dead weight of the beam itself should not be ignored. The self-weight is a uniform load that (in this case) causes a maximum moment at midspan which should be added to the applied load moment at that point. The self-weight moment is

$$M_{\text{s.w.}} = \frac{wL^2}{8}$$

where $w$ is the self-weight of the beam (plf or klf). In this case $w = 260$ plf or 0.260 klf. Therefore,

$$M_{\text{s.w.}} = \frac{0.260 \text{ kip/ft}(48 \text{ ft})^2}{8}$$

$$= 75 \text{ kip-ft}$$

The actual bending stress after inclusion of the self-weight moment will be, by ratio,

$$f_{b_{new}} = \frac{M + M_{s.w.}}{M}(f_b)$$

$$= \frac{1800 + 75}{1800}(22.7 \text{ ksi})$$

$$= 23.6 \text{ ksi}$$

This is still within the allowable of 24 ksi, so a larger beam will not be required.

**EXAMPLE 7–14M**

The W920 × 365 beam in Figure 7–30M must carry three column loads of 300 kN each. Determine the maximum bending stress.

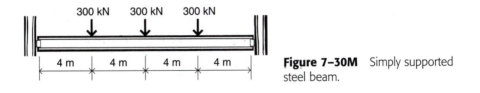

**Figure 7–30M**  Simply supported steel beam.

### Solution
From Appendix J, $S = 14\,700(10)^3 \text{ mm}^3$, and the moment diagram is given in Figure 7–31M.

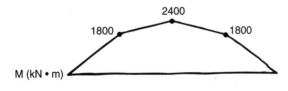

**Figure 7–31M**  Moment diagram for the beam in Figure 7–30M.

$$f_b = \frac{M}{S}$$

$$= \frac{2400 \text{ kN} \cdot \text{m}}{14\,700(10)^{-6} \text{ m}^3}$$

$$= 163(10)^3 \text{ kPa}$$

$$= 163 \text{ MPa}$$

This is very close to the allowable bending stress of 165 MPa for mild steel, and the size of the span indicates that the bending stress due to the dead weight of the beam itself should not be ignored. The self-weight is a uniform load that (in this case) causes a maximum moment at midspan which should be added to the applied load moment at that point. The self-weight moment is

$$M_{s.w.} = \frac{wL^2}{8}$$

where $w$ is the self-weight of the beam (kN/m). To obtain this weight, we must convert from the mass value given by the shape designation. Because of the earth's gravity, a mass of 1 kg will exert a force of about 9.8 N. Therefore, the weight of the steel beam can be obtained as

$$w = (365 \text{ kg/m})(9.8 \text{ N/kg})$$

$$= 3580 \text{ N/m}$$

$$= 3.58 \text{ kN/m}$$

$$M_{s.w.} = \frac{(3.58 \text{ kN/m})(16 \text{ m})^2}{8}$$

$$= 115 \text{ kN} \cdot \text{m}$$

The actual bending stress after inclusion of the self-weight moment will be, by ratio,

$$f_{b_{new}} = \frac{M + M_{s.w.}}{M} (f_b)$$

$$= \frac{2400 + 115}{2400} (163 \text{ MPa})$$

$$= 171 \text{ MPa}$$

There is an overstress of about 4%, and a larger beam should be used.

The following examples introduce the subject of *structural design*. When we are finding the stresses in members due to loads acting upon them, we are engaged in *analysis* or *review*. When we select members or determine cross-sectional sizes and shapes such that certain allowable stresses and deformations are not exceeded, we are doing *design* rather than analysis. Structural design is treated in more detail in Chapter 10.

**EXAMPLE 7–15**

A timber joist must span 15 ft and carry a total uniform load of 67 plf (lb/ft). Assuming that flexure controls the design, select the smallest adequate 2 × ? section. Use an allowable bending stress of 1150 psi.

**Solution**

$$M = \frac{wL^2}{8}$$

$$= \frac{(67 \text{ lb/ft})(15 \text{ ft})^2}{8}$$

$$= 1880 \text{ lb-ft}$$

$$S_{required} = S_r = \frac{M}{F_b}$$

$$S_r = \frac{1880 \text{ lb-ft}(12 \text{ in/ft})}{1150 \text{ psi}}$$

$$= 19.6 \text{ in}^3$$

From Appendix I, a 2 × 10 with a section modulus of 21.4 in.$^3$ is the smallest adequate size.

---

**EXAMPLE
7–16M**

A timber joist must span 4.8 m and carry a total uniform load of 1 kN/m. Assuming that flexure controls the design, select the smallest adequate 38 × ? mm section. Use an allowable bending stress of 8500 kPa.

**Solution**

$$M = \frac{wL^2}{8}$$

$$= \frac{(1 \text{ kN/m})(4.8 \text{ m})^2}{8}$$

$$= 2.88 \text{ kN} \cdot \text{m}$$

$$S_{required} = S_r = \frac{M}{F_b}$$

$$S_r = \frac{2.88 \text{ kN} \cdot \text{m}}{8500 \text{ kN/m}^2}$$

$$= 0.000\ 339 \text{ m}^3$$

$$S_r = 339(10)^3 \text{ mm}^3$$

From Appendix I, a 38 × 235 mm joist with an $S$ value of 350(10)$^3$ mm$^3$ is the smallest adequate size.

## PROBLEMS

**7–15.** Determine the maximum bending stress in a W18 × 40 steel beam that carries a midspan concentrated load of 30 kips on a simple span of 24 ft.

**7–15M.** Determine the maximum bending stress in a W460 × 60 steel beam that carries a midspan concentrated load of 125 kN on a simple span of 8 m.

**7–16.** A large beam simply spans 74 ft and carries an applied load of 2 klf. Assuming that an allowable bending stress of 33 ksi will control the beam size, select the lightest adequate W shape from those listed below. Include the effect of member self-weight.
(a) W36 × 150, $S = 504$ in$^3$
(b) W36 × 160, $S = 542$ in$^3$
(c) W36 × 170, $S = 580$ in$^3$
(d) W36 × 182, $S = 623$ in$^3$

**7–16M.** A large beam simply spans 22 m and carries an applied load of 30 kN/m. Assuming that an allowable bending stress of 230 MPa will control the beam size, select the lightest adequate W shape from those listed below. Include the effect of member self-weight.
(a) W920 × 223, $S = 8\ 260(10)^3$ mm$^3$
(b) W920 × 238, $S = 8\ 880(10)^3$ mm$^3$
(c) W920 × 253, $S = 9\ 500(10)^3$ mm$^3$
(d) W920 × 271, $S = 10\ 200(10)^3$ mm$^3$

**7–17.** Select southern pine 2 × ? floor joists for each of the following conditions. Assume that bending will control the selection, and use the given allowable stresses.
(a) $w = 67$ plf, $L = 18$ ft, $F_b = 1121$ psi
(b) $w = 67$ plf, $L = 15$ ft, $F_b = 1207$ psi
(c) $w = 53$ plf, $L = 18$ ft, $F_b = 1207$ psi
(d) $w = 53$ plf, $L = 12$ ft, $F_b = 1380$ psi

**7–17M.** Select southern pine 38 × ? mm floor joists for each of the following conditions. Assume that bending will control the selection, and use the given allowable stresses.
(a) $w = 0.8$ kN/m, $L = 6$ m, $F_b = 7730$ kPa
(b) $w = 1$ kN/m, $L = 4$ m, $F_b = 8324$ kPa
(c) $w = 0.8$ kN/m, $L = 5$ m, $F_b = 8324$ kPa
(d) $w = 0.8$ kN/m, $L = 4$ m, $F_b = 9515$ kPa

**7–18.** Ignoring any deflection limitation and other controlling factors, how far can a W36 × 300 span before the flexural stress from its own weight will reach an allowable value of 33 ksi?

**7–18M.** Ignoring any deflection limitation and other controlling factors, how far can a W920 × 446 span before the flexural stress from its own weight will reach an allowable value of 230 MPa?

**7–19.** The beam in Figure 7–32 is a hemlock member having an actual cross section of 4 × 10 in. Determine the flexural stress
(a) at the right-hand support.
(b) under the point load.
(*Hint*: Use $S = bd^2/6$.)

**7–19M.** The beam in Figure 7–32M is a hemlock member having an actual cross section of 100 × 250 mm. Determine the flexural stress

(a) at the right hand support.

(b) under the point load.

(*Hint*: Use $S = bd^2/6$.)

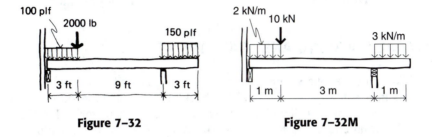

**Figure 7–32**          **Figure 7–32M**

**7–20.** The beam in Figure 7–33 is built up of three nominal 2 × 12s as shown.

(a) Determine the maximum flexural stress in the upright stems.

(b) Determine the maximum flexural stress in the crossweb.

(*Hint*: Use $f = My/I$ because the individual $S$ values are not additive.)

**7–20M.** The beam in Figure 7–33M is built up of three 38 × 285 mm pieces as shown.

(a) Determine the maximum flexural stress in the upright stems.

(b) Determine the maximum flexural stress in the crossweb.

(*Hint*: Use $f = My/I$ because the individual $S$ values are not additive.)

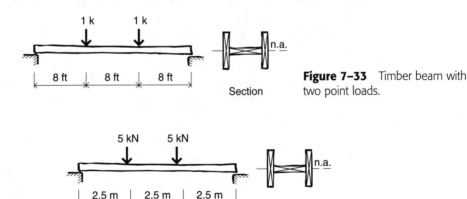

**Figure 7–33**  Timber beam with two point loads.

Section

**Figure 7–33M**

**7–21.** A triangular opening in a building floor causes the total load on a W18 × 35 to be as illustrated in Figure 7–34. Determine the maximum bending stress.

**7–21M.** A triangular opening in a building floor causes the total load on a W460 × 52 to be as illustrated in Figure 7–34M. Determine the maximum bending stress.

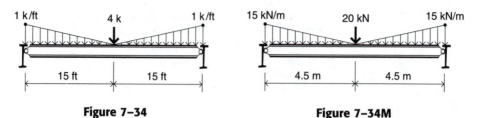

**Figure 7–34**                    **Figure 7–34M**

## 7–5  LATERAL BUCKLING AND STABILITY

Whenever a long, slender column is loaded in compression along its axis, it tends to deflect sideways, or *buckle*. This buckling phenomenon occurs even though the stresses remain well within the elastic range of the material. It occurs rapidly once a certain critical load is reached and is a function of the modulus of elasticity and cross-sectional shape rather than of material strength. (Elastic column buckling is discussed in Chapter 11.) This same behavior occurs in the compression zones of long slender beams.

Whenever compressive stresses exist over a length of beam, such as in the top of a simple beam or along the bottom of a cantilever, there exists a tendency for the compressive fibers to buckle laterally or "get out of the way of the compressive forces" (Figure 7–35). It makes no difference whether the loads are applied from above or below. The buckling is caused by the horizontal force resultant of the internal moment couple, not by the fact that loads push downward from above. Even though the tension fibers tend to remain straight, the section undergoes a rotation or twisting action, which reduces both the effective depth and the moment of inertia.

The examples and problems presented earlier all assumed that lateral buckling was not a factor or was prevented from happening in some manner. Certain beams are inherently stable against any lateral buckling tendency by virtue of cross-sectional shape. For example, a rectangle with a width greater than its depth and loaded vertically in a plane of symmetry will have no lateral stability problem. A

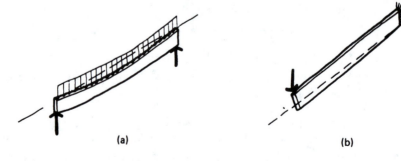

(a)                                                    (b)

**Figure 7–35**  Lateral buckling.

wide-flange beam having a compression flange that is both wide and thick, so as to provide a resistance to bending in a horizontal plane, will have considerable resistance to lateral buckling.

A beam that is not laterally stiff in cross section must be braced every so often along its compressive side in order to develop its full moment capacity. Sections not so braced or laterally supported by secondary members will fail prematurely (or at best be unsafe in terms of maintaining a proper factor of safety). Sometimes such lateral bracing occurs naturally because of other design considerations. The plywood subfloor nailed (and frequently glued) to the tops of wood joists of simple residential spans provides excellent lateral support. Open web bar joists, with their ends welded to the top flanges of the beams that carry them, provide lateral bracing for those flanges. Other situations, such as the overhanging beams of Figure 7–36, require specific bracing elements. In this case, the four beams are tied together by the spandrel channels at their ends and one bay has X-bracing (of rods or angles) connecting the critical compression flanges.

Reinforced concrete beams usually have cross-sectional dimensions such that lateral buckling is not a consideration. As mentioned previously, small timber joists and beams almost always have adequate lateral support for their top edges provided by the attached floor deck and required bridging. Similarly, the overhangs involved in small-scale timber construction are usually short so that the lateral stability of the underside is not a critical concern. Larger solid-sawn timber sections and glued-laminated beams, however, can easily have lateral support problems. The American Institute of Timber Construction (AITC) provides equations so that designers can compute reduced allowable stresses when adequate lateral support cannot be provided.

The issue of lateral stability occurs more frequently when designing with steel than with other materials. Its inherent strength means smaller sections, and because of relatively high material costs, such sections tend to be efficiently configured for bending (i.e., deep and narrow). The American Institute of Steel Construction (AISC) has developed equations to determine the reduced bending capacities for members with inadequate lateral support and provides a series of graphs as design aids.

In a few design situations, it may become desirable not to use the full moment capacity of a section but rather to use reduced allowable stresses (calling for lower

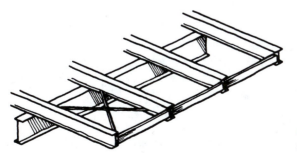

**Figure 7–36** Lateral stability for overhanging steel beams.

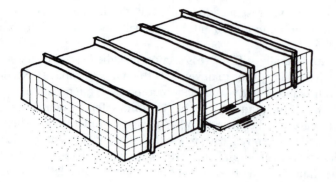

**Figure 7–37**  Concept sketch of Crown Hall.

loads or larger members) in order to maintain the same margin of safety. An excellent illustration of this approach is found in Crown Hall (on the Illinois Institute of Technology campus) by Mies van der Rohe (see Figure 7–37). Here the large clear-span steel plate girders that frame the roof are exposed, with the roof deck attached to the bottom or tension flange.

Clearly, the architect desired a strong statement of the horizontal structure, achieved by exposing these girders, and was similarly not concerned by the increase in their size required by the absence of lateral support elements.

The proper analysis of beams that lack lateral support has not been included in this basic text and is more properly treated in a context involving applied analysis and design procedures for specific materials.

# 8

# Shearing Stresses

## 8–1  NATURE OF SHEARING STRESSES

As introduced in Section 4–1, shearing stresses are tangential stresses that act parallel to the planes which they stress. Figure 8–1 shows how the shearing force in a beam provides shearing stresses on both vertical and horizontal planes within the beam. The two vertical stresses $f_v$ must be equal in magnitude and opposite in sense to ensure vertical equilibrium. However, under the action of those two stresses alone, the element would rotate in a clockwise manner. Clearly, this couple must be negated by the action of another couple, shown as the dashed arrows. If the small element is taken as a differential one, the magnitude of the horizontal stresses must also have the value $f_v$. This principle is sometimes phrased as "cross-shears are equal." In other words, a shearing stress cannot exist on an element without a like stress located 90° around the corner.

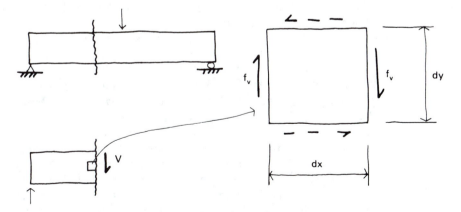

**Figure 8–1**  Development of shearing stresses in a beam.

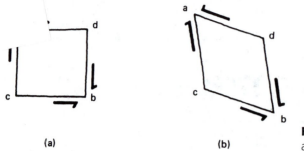

**(a)**                    **(b)**

**Figure 8–2**   Shear causes tension and compression.

## 8–2   DIAGONAL TENSION AND COMPRESSION

Shearing stresses also create tensile and compressive stresses. The square element of Figure 8–2(a) is being acted upon by four shearing stresses as just explained. The stressed element will appear as in Figure 8–2(b) as it deforms, developing a tensile stress along a line from *a* to *b* and compressive stress along a line joining *c* and *d*. In the absence of any other stresses acting on the element, the lines of tension and compression will be oriented at 45° to the original shear planes.

If the thickness of the element is designated as *dz*, an equation of equilibrium in the *ab* direction can be used to solve for the magnitude of $f_t$, the tensile stress. Referring to Figure 8–3,

$$\Sigma F_{ab} = 0$$

$$f_t\!\left(\sqrt{2}\right)\!dL(dz) - 2\!\left[\frac{\sqrt{2}}{2}(f_v)\right]\!dL(dz) = 0$$

$$f_t = f_v$$

which illustrates that the diagonal tension developed by shearing stress is equal to the shearing stress itself. A similar proof could be made for the diagonal compression stress in the *cd* direction.

It is important to note that a material which is weak in either tension or compression will also be effectively weak in shear. Thus, it was explained in Chapter 5 that concrete is weak in shear because of its lack of strength in tension. Concrete beams are strengthened by specially placed reinforcing bars (called *stirrups*) to prevent diagonal tension cracking. Figure 8–4 shows potential diagonal tension cracks being crossed by several stirrups. (Stirrups are placed vertically rather than normal to the potential cracks for reasons of construction ease.)

Some deep steel girders have relatively thin webs, which tend to buckle in compression along 45° lines, as shown happening on the upper beam of Figure 8–5. Vertical plate stiffeners shown welded in place on the lower sketch constitute one way to prevent such failure. The reader can "see" diagonal buckling as it occurs in a thin member by applying shearing forces along the opposite edges of a piece of paper.

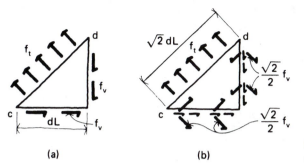

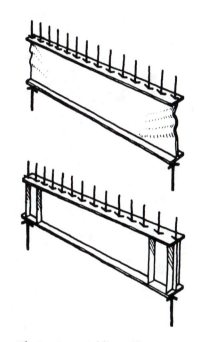

**Figure 8–3** Free-body cut through *cd*, showing stresses.

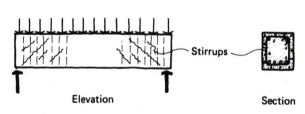

**Figure 8–4** Concrete beam reinforced for shear.

**Figure 8–5** Adding stiffeners to a steel girder for shear.

## 8–3 BASIC HORIZONTAL SHEARING STRESS EQUATION

In some ways it is easier to visualize shearing stresses acting on horizontal planes than upon vertical ones. For example, if you make a beam by laying several planks flatwise on top of one another, there would exist horizontal slippage planes as shown in Figure 8–6. As the top fibers of each plank get shorter in compression, they have to "slip past" the bottom fibers of the plank above. The bottom fibers, in each case, are themselves getting longer because of the bending tensile strain. Now if we glued all the planks together, so as to simulate a solid one-piece cross section, there

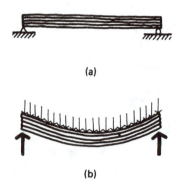

**(a)**

**(b)**

**Figure 8–6**  Beam made of planks.

would be less deflection, and horizontal shearing stresses would develop in the glue planes. These same stresses occur in solid pieces, of course, and are particularly important in the design of wood beams because most softwood shears rather easily parallel to the grain. (Examine the values given in Appendix H and see that this is reflected in the relative magnitudes of the allowable stresses. The horizontal shearing stress value $F_v$ is quite low.)

There is a close relationship between flexural stress and shearing stress. Clearly, the slippage deformations of Figure 8–6 would not take place in the absence of bending. Indeed, the derivation of the general shearing stress formula in Appendix B proves that such stresses are caused by the *change* in moment from one beam section to the next. This is also implied in Chapter 6, where it is stated that the magnitude of the ordinate on the $V$ diagram (the shear force) is equal to the slope of the moment diagram. Zero shear can only exist when the slope of the moment diagram is zero.

Since shearing stresses must exist on all four planes of an element in order to exist at all, it follows that shearing stresses will be zero at the top and bottom edges of a beam where there is no material present to provide one of the four stresses. Unlike flexural stresses, which maximize at the extreme top and bottom fibers of a section, shearing stresses tend to maximize near the center of a beam cross section and go to zero at the extreme fibers. The general equation for horizontal (or vertical) shearing stress in beams is

$$f_v = \frac{VQ}{Ib} \tag{8–1}$$

where  $f_v$ = shearing stress (psi or ksi) (kPa or MPa)

  $V$ = vertical shear force at the transverse section being examined (lb or kips) (kN)

  $Q$ = statical moment of that area of cross section between the horizontal plane under investigation and the near edge of the beam, taken with respect to the neutral axis (in.$^3$) (m$^3$)

$I$ = moment of inertia of the cross section with respect to the neutral axis (in.$^4$) (m$^4$)

$b$ = width of cross section at the horizontal plane under investigation (in.) (m)

The term represented by $Q$ in the formula is not nearly so complicated as its written definition implies. $Q$ is really nothing more than a shape factor that represents how bending forces (which cause the shear) are distributed with respect to the neutral axis. Bending stress is linear, but bending force is a function of the stressed area as well and is not linear. As illustrated in Appendix B, this will cause the shear stress in beams to vary as a square function, or parabolically, over the depth of the cross section. A few examples will illustrate this more clearly.

---

**EXAMPLE 8–1** The 2 × 10 wood joist of Example 7–1 has the shear diagram shown in Figure 8–7. Determine the distribution of horizontal shearing stresses on a transverse section just to the inside of either support.

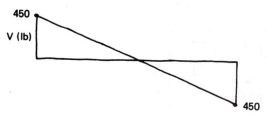

**Figure 8–7** Shear diagram for the wood joist in Example 7–1.

### Solution

The shearing force $V$ has its maximum value of 450 lb at these two locations. For a given transverse section, $V$ and $I$ are constants, and the shearing stress varies with the value of the ratio $Q/b$. For this cross section, $b$ is a constant also, so $f_v$ will vary over the depth directly with the value of $Q$.

At the neutral axis, using Figure 8–8(a),

$$Q = 4.62 \text{ in } (1.5 \text{ in}) (2.31 \text{ in})$$

$$= 16 \text{ in}^3$$

and the shearing stress will be

$$f_{v_{\text{n.a.}}} = \frac{VQ}{Ib}$$

$$= \frac{450 \text{ lb}(16 \text{ in}^3)}{98.9 \text{ in}^4(1.5 \text{ in})}$$

$$= 48 \text{ psi}$$

Halfway to the edge of the section, at the $d/4$ level, $Q$ is obtained using Figure 8–8(b).

$$Q = 2.31 \text{ in } (1.5 \text{ in})(3.47 \text{ in})$$
$$= 12 \text{ in}^3$$

The shearing stress is

$$f_{vd/4} = \frac{VQ}{Ib}$$
$$= \frac{450 \text{ lb}(12 \text{ in}^3)}{98.9 \text{ in}^4(1.5 \text{ in})}$$
$$= 36 \text{ psi}$$

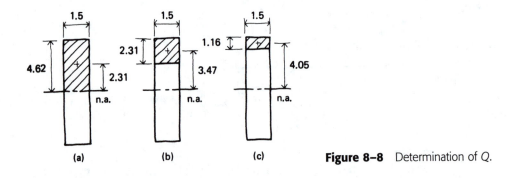

**Figure 8–8**  Determination of Q.

At the $d/8$ level, where $Q$ is obtained using Figure 8–8(c),

$$Q = 1.16(1.5)(4.05)$$
$$= 7 \text{ in}^3$$
$$f_{vd/8} = \frac{VQ}{Ib}$$
$$= \frac{450 \text{ lb}(7 \text{ in}^3)}{98.9 \text{ in}^4(1.5 \text{ in})}$$
$$= 21 \text{ psi}$$

The values will be identical at symmetrical levels below the neutral axis, and a plot of the shearing stress is given in Figure 8–9. The distribution would look the same at the other transverse sections of the beam, but the values would be less because of the decrease in $V$.

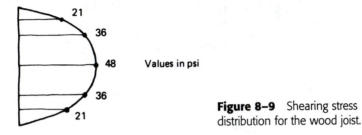

Values in psi

**Figure 8–9**  Shearing stress distribution for the wood joist.

---

**EXAMPLE 8–2M**

Determine the shearing stress distribution for the beam in Figure 8–10M.

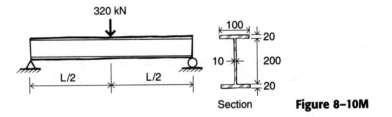

Section          **Figure 8–10M**

**Solution**

$V$ is constant for this beam at 160 kN. $I$ can be determined as $1320(10)^6$ mm$^4$. Using Figure 8–11M, $Q$ at level 1 (the neutral axis) is found to be

$$Q = 100 \text{ mm } (10 \text{ mm})(50 \text{ mm}) + 20 \text{ mm } (100 \text{ mm})(110 \text{ mm})$$

$$= 270\,000 \text{ mm}^3$$

and     $$f_{v_1} = \frac{VQ}{Ib}$$

$$= \frac{160 \text{ kN}[270(10)^{-6} \text{ m}^3]}{[1320(10)^{-6} \text{ m}^4](0.010 \text{ m})}$$

$$= 3270 \text{ kPa}$$

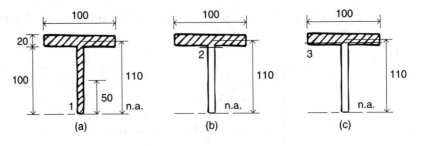

**Figure 8–11M**

At level 2, $b$ remains the same but

$$Q = 20 \text{ mm } (100 \text{ mm})(110 \text{ mm})$$
$$= 220\,000 \text{ mm}^3$$

and
$$f_{v_2} = \frac{VQ}{Ib}$$

$$= \frac{160 \text{ kN}[220(10)^{-6} \text{ m}^3]}{[1320(10)^{-6} \text{ m}^4](0.010 \text{ m})}$$

$$= 2670 \text{ kPa}$$

At level 3, just inside the flange, $b$ takes a sharp increase, which will cause a drop in the shearing stress. $Q$ has the same value as at level 2.

and
$$f_{v_3} = \frac{VQ}{Ib}$$

$$= \frac{160 \text{ kN}[220(10)^{-6} \text{ m}^3]}{[1320(10)^{-6} \text{ m}^4](0.100 \text{ m})}$$

$$= 267 \text{ kPa}$$

The final distribution is shown in Figure 8–12M. It varies according to the ratio $Q/b$.

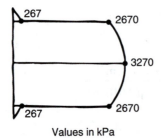

Values in kPa

**Figure 8–12M**   Shearing stress distribution for the wide-flange shape.

**EXAMPLE 8–3** Determine the plane of maximum shear stress for the cross section in Figure 8–13.

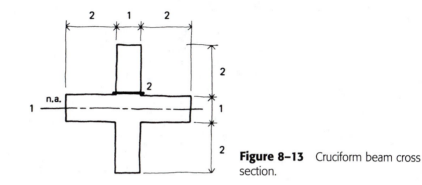

**Figure 8–13** Cruciform beam cross section.

### Solution

The stress will maximize at the plane where the ratio $Q/b$ maximizes. $Q$ will always be a maximum at the neutral axis; therefore,

$$\left(\frac{Q}{b}\right)_1 = \frac{0.5 \text{ in } (5 \text{ in})(0.25 \text{ in}) + 2 \text{ in } (1 \text{ in})(1.5 \text{ in})}{5 \text{ in}}$$

$$= 0.725 \text{ in}^2$$

The other possible place for $Q/b$ to reach its largest value would be at the junction of the two rectangles, plane 2.

$$\left(\frac{Q}{b}\right)_2 = \frac{2 \text{ in } (1 \text{ in})(1.5 \text{ in})}{1 \text{ in}}$$

$$= 3.0 \text{ in}^2$$

Clearly, the shearing stress will maximize at this junction. A distribution plot would appear as in Figure 8–14.

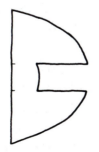

**Figure 8–14** Qualitative shearing stress distribution for the cruciform shape.

**EXAMPLE**
**8–4M**

Assume that a 38 × 235 mm joist is fabricated from two pieces of wood as shown in Figure 8–15M. If the nails are spaced every 100 mm apart, determine the shearing force on each nail. Let $V = 3$ kN.

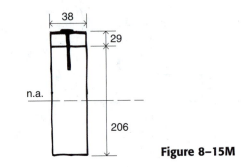

**Figure 8–15M**

**Solution**
It can be shown that $Q$ for that "nailed-together" plane is 114 000 mm³, and the shearing stress (that would exist in a solid piece) is 219 kPa. Each nail must take the shearing force that would be present at that level in a 100-mm length of solid joist, as shown in Figure 8–16M.

$$F = f_v A$$

$$= 219 \text{ kPa } (0.038 \text{ m})(0.100 \text{ m})$$

$$= 0.832 \text{ kN}$$

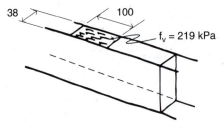

**Figure 8–16M**  Cutaway showing the stressed area for one nail.

With reference to Figure 8–7, notice that the spacing of the nails can be farther apart as we come in from the ends of the joist. Where $V$ is zero, no nails would be needed.

## PROBLEMS

**8–1.** Determine the shearing stress distribution for the beam in Figure 7–7. Give the values in 1-in. increments of depth.

**8–2M.** Determine the shearing stress distribution at the wall for the beam in Figure 7–9M. Give values in 30-mm increments of depth.

**8–3.** Show by taking successive trial planes (or by the calculus) that the shea.. a triangular cross section, such as that of Figure 8–17, maximizes at $h/2$.

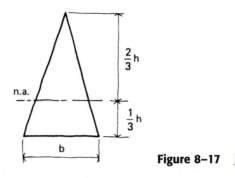

**Figure 8–17**

**8–4.** A rectangular timber 11½ in. deep is used as a simple beam spanning 12 ft. It supports a total uniform load of 300 plf. If the maximum allowable shearing stress for this wood is 95 psi, determine the required safe minimum width.

**8–4M.** A rectangular timber 300 mm deep is used as a simple beam spanning 4 m. It supports a total uniform load of 4.5 kN/m. If the maximum allowable shearing stress for this wood is 650 kPa, determine the required safe minimum width.

**8–5.** Locate the plane of maximum shearing stress for the channel section of Figure 8–18.

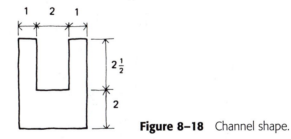

**Figure 8–18**  Channel shape.

**8–6M.** If $V = 10$ kN, plot the shearing stress distribution for the rotated square section of Figure 8–19M. Give values in 30-mm increments of depth.

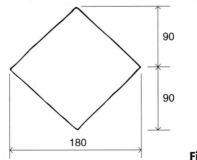

**Figure 8–19M**

**8–7.** The neutral axis for the beam cross section of Problem 7–10 is located 4.2 in. down from its top edge. Its moment of inertia can be computed as 241 in.$^4$. The maximum shearing force $V$ in the beam occurs at the left end and is 1920 lb. Determine the shearing stress at a section located:
(a) at the neutral axis.
(b) just below the bottom edge of the 2 × 6. Sketch the shearing stress distribution for this cross section.

**8–7M.** The neutral axis for the beam cross section of Problem 7–10M is located 106 mm down from its top edge. Its moment of inertia can be computed as $100(10)^6$ mm$^4$. The maximum shearing force $V$ in the beam occurs at the left end and is 8.5 kN. Determine the shearing stress at a section located:
(a) at the neutral axis.
(b) just below the bottom edge of the 38 × 139. Sketch the shearing stress distribution for this cross section.

**8–8.** Determine the value of the maximum shearing stress in the beam of Problem 7–13. Compare your answer to the allowable values in Appendix H and comment.

**8–8M.** Determine the value of the maximum shearing stress in the beam of Problem 7–13M. Compare your answer to the allowable values in Appendix H and comment.

**8–9M.** In each nail in the section shown in Figure 8–20M can take 500 N of force, determine the required spacing of each set of four nails in
(a) the left portion of the beam.
(b) the right portion of the beam.

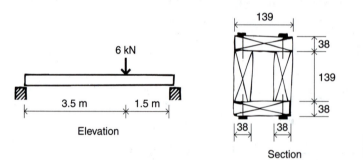

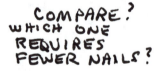

**Figure 8–20M** Wood box-type beam.

# 8–4 HORIZONTAL SHEARING STRESSES IN TIMBER BEAMS

The general expression for shearing stress due to bending is $f_v = VQ/Ib$, as presented in Section 8–3. For a rectangular section, as most timber beams are, the maximum value of shearing stress occurs at the neutral axis. Since $Q$ and $I$ can both be expressed in terms of $b$ and $d$ for a rectangular section, a simpler expression for this maximum $f_v$ can be developed.

$$f_{v_{max}} = \frac{VQ_{max}}{Ib}$$

$$Q_{max} = b\left(\frac{d}{2}\right)\left(\frac{d}{4}\right)$$

$$= \frac{bd^2}{8}$$

$$I = \frac{bd^3}{12}$$

$$f_{v_{max}} = \frac{Vbd^2(12)}{bd^3(8)b} \qquad (8\text{--}2)$$

$$bd = A$$

$$f_{v_{max}} = \frac{3V}{2A}$$

From Equation 8–2, the maximum shearing stress is 50% larger than the average value, which can be represented by a rectangular block. The total shearing force resistance or shear capacity is stress times area, and this is represented as a volume in Figure 8–21. The area under a parabola is ⅔ the base times the height (see Appendix F), so its altitude must be 50% greater to achieve a "stress volume" equal to a rectangular volume.

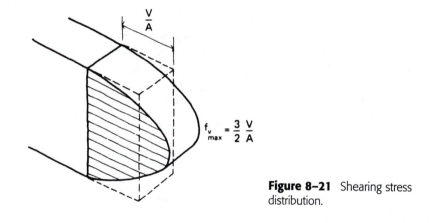

**Figure 8–21** Shearing stress distribution.

**EXAMPLE 8–5** A hem-fir No. 2 4 × 6 beam spans 10 ft and supports a uniform load of 180 plf. Is it adequate in shear?

**Solution**
The maximum shear force will be equal to one of the reactions.

$$V = \frac{wL}{2}$$

$$= \frac{180 \text{ lb/ft}(10 \text{ ft})}{2} = 900 \text{ lb}$$

$$f_v = \frac{3V}{2A}$$

$$A = 19.25 \text{ in}^2$$

$$f_v = \frac{3(900 \text{ lb})}{2(19.25 \text{ in}^2)}$$

$$= 70 \text{ psi}$$

From Appendix H the allowable shearing stress for hem-fir is 75 psi, so the section is adequate in shear.

---

**EXAMPLE 8–6M**     A hem-fir No. 2 89 × 185 mm beam spans 3.5 m and supports a uniform load of 3 kN/m. Is it adequate in shear?

**Solution**
The maximum shear force will be equal to one of the reactions.

$$V = \frac{wL}{2}$$

$$= \frac{3 \text{ kN/m } (3.5 \text{ m})}{2} = 5.25 \text{ kN}$$

$$f_v = \frac{3V}{2A}$$

$$A = 16\,500 \text{ mm}^2$$

$$f_v = \frac{3 (5.25 \text{ kN})}{2(0.0165 \text{ m}^2)}$$

$$= 477 \text{ kPa}$$

From Appendix H the allowable shear stress for hem-fir is 517 kPa, so the section is adequate in shear.

---

# PROBLEMS

**8–10.** Determine the maximum shearing stress in the beam of Problem 7–1.

**8–10M.** Determine the maximum shearing stress in the beam of Problem 7–1M.

**8–11.** Determine the maximum shearing stress in the beam of Problem 7–2.

**8–11M.** Determine the maximum shearing stress in the beam of Problem 7–2M.

**8–12.** A uniformly loaded southern pine No. 2 joist must span 16 ft and carry a total load of 67 plf. Will a 2 × 10 section be adequate in shear?

**8–12M.** A uniformly loaded southern pine No. 2 joist must span 5 m and carry a total load of 0.8 kN/m. Will a 38 × 235 mm section be adequate in shear?

**8–13.** If Douglas fir No. 2 is used for the 4 × 14s of Problem 7–8, will they be adequate in shear?

**8–13M.** If Douglas fir No. 2 is used for the 89 × 335s of Problem 7–8M, will they be adequate in shear?

**8–14.** The beam in Figure 8–22 is a doubled 2 × 8. Compute the maximum shearing stress caused by the five concentrated loads.

**8–14M.** The beam in Figure 8–22M is a doubled 38 × 185 mm section of Douglas fir No. 2. Is it adequate in shear?

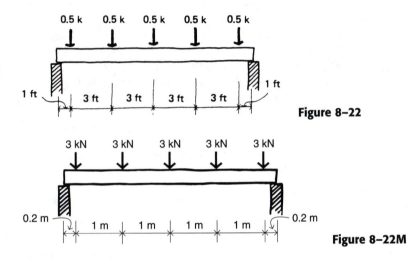

Figure 8–22

Figure 8–22M

**8–15.** Compute the maximum shearing stress in the overhanging 2 × 12 joist in Figure 8–23.

**8–15M.** Compute the maximum shearing stress in the overhanging 38 × 285 joist in Figure 8–23M.

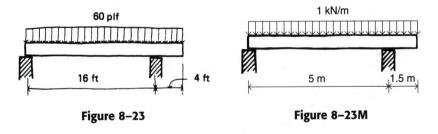

Figure 8–23

Figure 8–23M

**8–16.** Assume that the cross section of the beam in Problem 7–10 is replaced by a nominal 6 × 10. Compute the maximum shearing stress.

**8–16M.** Assume that the cross section of the beam in Problem 7–10M is replaced by a nominal 139 × 240. Compute the maximum shearing stress.

**8–17.** Compute the maximum shearing stress in the 4 × 16s of Problem 7–11. Compare your answer to the values given in Appendix H for allowable shearing stress $F_v$ and comment.

**8–17M.** Compute the maximum shearing stress in the 89 × 385s of Problem 7–11M. Compare your answer to the values given in Appendix H for allowable shearing stress $F_v$ and comment.

## 8–5  HORIZONTAL SHEARING STRESSES IN STEEL BEAMS

As developed in Section 8–3, the distribution of shearing stresses in a W shape is as shown in Figure 8–24. Almost all of the shearing force is resisted by stresses in the web, and very little work is done by the flanges—the opposite, of course, being the case for flexural stresses. The calculation of the exact maximum stress magnitude using $VQ/Ib$ can become difficult because of the presence of fillets where the flanges join the web. A high level of accuracy is even harder to achieve in channels of I shapes which have sloping flange surfaces. Accordingly, the American Institute of Steel Construction recommends the use of a much simpler approximate formula for the common steel shapes:

$$f_v = \frac{V}{th} \tag{8-3}$$

where $t$ = web thickness (in.)(m)
      $h$ = total beam depth (in.)(m)

(*Note*: Values of $h$ are given as $d$ for selected steel sections in Appendix J.) This formula gives the average unit shearing stress for the web over the full beam depth, ignoring any contribution of the flange projections (Figure 8–25). Depending upon the particular steel shape, this formula can be as much as 20% in error in the nonconservative direction. This means that when a shearing stress com-

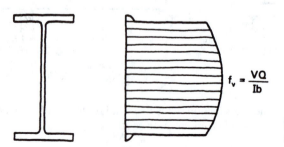

$$f_v = \frac{VQ}{Ib}$$

**Figure 8–24**  Shearing stress distribution in a wide-flange beam.

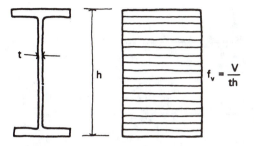

**Figure 8-25** Simplified shearing stress distribution.

puted by Equation 8–3 gets to within 20% of the maximum allowable stress, the actual maximum stress (computable by $VQ/Ib$) might be exceeding the allowable by a small amount.

Fortunately, this low level of accuracy is seldom a problem, for two reasons:

1. Structural steel is very strong in shear.
2. Most beams and girders in buildings, unlike those in some machines, have very low shearing stresses.

A rolled steel beam has to be very short and very heavily loaded, or have a large concentrated load adjacent to a support, in order for shear to control. In determining the size of a steel beam, flexural stresses will usually govern. Excessive deflection will occasionally dictate the use of a larger section, but shear will almost never govern the design.

When shearing stresses do become excessive, steel beams do not fail by ripping along the neutral axis as might happen in wood. Rather, it is the compression buckling of the relatively thin web which constitutes a shear failure. This can be diagonal buckling, as discussed in Section 8–2, or a type of vertical buckling, illustrated in Figure 8–26. The AISC has provided several design formulas for determining when extra bearing area must be provided at concentrated loads or when web stiffeners are needed to prevent such failures (Figure 8–27).

Beams of normal depth seldom present any major problems, and detailed design considerations will not be given here. A word of caution is given with respect to large built-up plate girders, however. Such sections usually have deep, thin webs and are particularly susceptible to buckling action. For these beams,

**Figure 8-26** Web buckling in steel beams.

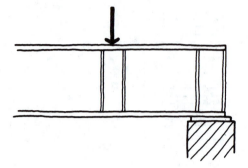

**Figure 8–27**   Bearing plate and web stiffeners.

shear can be a determining factor in the overall structural design. The reader may wish to consult a textbook on structural steel design for further material on web buckling.

**EXAMPLE 8–7**  Determine the average shearing stress for the W16 × 57 in Figure 8–28 if $V =$ 25 kips.

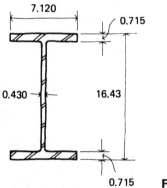

**Figure 8–28**

**Solution**

$$f_v = \frac{V}{th}$$

$$= \frac{25 \text{ kips}}{0.430 \text{ in}(16.43 \text{ in})}$$

$$= 3.5 \text{ ksi}$$

(*Note*: This is much less than the allowable shearing stress of 20.0 ksi for the steel most often used in building construction today.)

---

**EXAMPLE 8–8M**

Determine the average shearing stress for the W640 × 89 beam in Figure 8–29M if $V = 100$ kN.

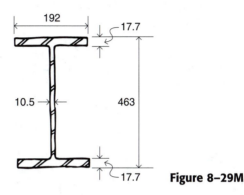

**Figure 8–29M**

*Solution*

$$f_v = \frac{V}{th}$$

$$= \frac{100 \text{ kN}}{(0.0105 \text{ m})(0.463 \text{ m})}$$

$$= 20\ 600 \text{ kPa}$$

or $$f_v = 20.6 \text{ MPa}$$

(*Note:* This is much less than the allowable shearing stress of 140 MPa for the steel most often used in building construction today.)

---

# PROBLEMS

**8–18.** Determine the average shearing stress where $V$ is a maximum for the beam in Problem 7–3. Let $t = 0.415$ in. $h$ may be obtained as the value $d$ in Appendix J.

**8–18M.** Determine the average shearing stress where $V$ is a maximum for the beam in Problem 7–3M. Let $t = 10$ mm. $h$ may be obtained as the value $d$ in Appendix J.

**8–19.** Determine the average shearing stress where $V$ is a maximum for the beam in Problem 7–4. Let $t = 0.380$ in. $h$ may be obtained as the value $d$ in Appendix J.

**8–19M.** Determine the average shearing stress where $V$ is a maximum for the beam in Problem 7–4M. Let $t = 9$ mm. $h$ may be obtained as the value $d$ in Appendix J.

**8–20.** Determine the percentage error that accrues by using the average shearing stress formula instead of the exact one for the beam in Example 8–7.

**8–20M.** Determine the percentage error that accrues by using the average shearing stress formula instead of the exact one for the beam in Example 8–8M. Let $I = 410(10)^6$ mm$^4$.

**8–21.** Regular low-alloy steel has an allowable shearing stress of approximately 20.0 ksi. The beam in Figure 8–30 has an unusual looking pattern that will cause very high shearing forces. If this beam is a W30 $\times$ 173 with $t = 0.655$ in., will it be adequate in shear?

**8–22M.** A W760 $\times$ 257 of $Fy = 345$-MPa steel has been selected for the beam in Figure 8–31M. The loading pattern is such that very high shearing stresses will be present. The allowable in shear is 140 MPa. If $h = 773$ mm and $t = 17$ mm, will it be adequate in shear?

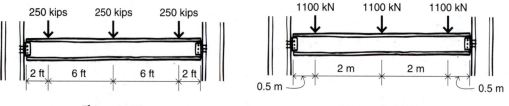

**Figure 8–30**          **Figure 8–31M**

# 9

# Deflection and Indeterminate Beams

## 9–1  INTRODUCTION

The *deflection* of beams is an important topic in structural design. As noted previously, the design of a beam for a particular load generally involves the investigation of bending stresses, shearing stresses, and deflection. Building codes limit the permissible deflection of a beam just as stresses or loads are limited. Some typical values are given in Table 9–1.

Excessive deflection can cause cracking of nonstructural materials that are attached to beams. Many cracks in nonbearing partition walls are due to such deflection. Doors and windows can bind up or become inoperable due to distortion of their openings by structural deflection. More important, flat or nearly flat roof surfaces are subject to "ponding," a continued buildup of water that can result eventually in a dishlike collapse.

Deflection is particularly critical in situations where a large portion of the total load is dead as opposed to live. Timber or concrete beams will both creep if subjected to permanent loading and will eventually sag enough to become unsightly and possibly unsafe. Therefore, when a beam supports a heavy wall or roof, for example, special design consideration needs to be given to deflection control.

Floor beams that are closely designed for bending stresses but not adequately limited in deflection can often be too "springy" or "bouncy" when loaded by impact or vibrated by a machine or vehicle to a natural frequency. This "springiness," although seldom a real structural safety problem, can be most annoying and in some cases make a space unfit for occupancy.

Over the last 100 years we have been able to produce steels of increasing strength; however, the stiffness ($E$) of these steels remains at about 29 million psi (200 GPa) (the stiffness of mild, low-alloy steel) and is independent of strength. This means that as strength increases, a larger percentage of steel beams will be designed with deflection controlling (or governing), as opposed to flexure or shear.

**TABLE 9–1  Typical Deflection Limitations Expressed as a Fraction of the Span**

|                          | Total Load          | Live Load Only      |
|--------------------------|:-------------------:|:-------------------:|
| Roof beams[a]            | $\dfrac{L}{180}$    | $\dfrac{L}{240}$    |
| Floor beams              | $\dfrac{L}{240}$    | $\dfrac{L}{360}$    |

[a] Floor beam values should be used in place of these if a plaster ceiling is attached directly to the structural members.

Beam deflections may be calculated easily by using deflection formulas available in a number of structural handbooks and design aids. A few simple cases are given in Appendix K. These formulas can also be used to approximate deflection magnitudes, for more complicated loading patterns and conditions, by "modifying" the actual conditions in the conservative direction so as to "fit" one of the tabled situations. The results so obtained will be overestimates of actual deflections and will enable the designer to ascertain if further and more accurate computations are necessary. To do this with any accuracy, an exposure to beam deflection theory is helpful, and for this reason further discussion of this technique will be deferred to a later section.

A knowledge of deflection theory will also help the designer to visualize more easily the deflected shapes of beams, frames, and other structures, determinate or indeterminate. Often a reasonably accurate image (or sketch) of how a structure deforms under load will help immeasurably in understanding how the loads are being resisted. Points of high and low stress can be ascertained, and this in turn can help decide whether a given structural choice is rational or irrational for those loads.

Deflection theory involves the study of the slopes and deflections of the neutral axis upon application of the loads. The deflected position taken by the neutral axis is called the *elastic curve*.

# 9–2  MOMENT-AREA METHOD

To introduce deflection theory the writer prefers the *moment-area method* because it emphasizes the relationship between the area under the moment curve and the resulting beam deflections and because it is often useful in providing a background for the future study of indeterminate structures.

The method is a semigraphical one first developed by Barré de Saint-Venant, a French scientist, and makes use of curves called *M/EI diagrams*. Beam deflection is inversely proportional to both *E* and *I*, and an *M/EI* diagram is nothing more than a moment diagram in which every ordinate has been divided by $E$ and $I$. (In this treatment it is assumed that the product $EI$ remains constant over the length of a given beam.)

The entire method can be stated in two theorems with reference to Figure 9–1. In general the method finds slopes and deflections only *indirectly*, and careful attention should be paid to the equivalencies presented in the theorems. Proof of the theorems may be found in Appendix G.

> *First moment-area theorem:* The change in slope between any two points, $A$ and $B$, on the elastic curve is equal to the net area under the $M/EI$ curve between those two points.

(Note that this theorem finds only a *change* in slope and does not directly find a slope.)

> *Second moment-area theorem:* The vertical distance from point $B$ on the elastic curve to a tangent line from point $A$ is equal to the statical moment of the net area under the $M/EI$ curve between points $A$ and $B$ taken about the vertical line through $B$.

(Note that this theorem finds only a vertical distance, often called a *tangential deviation*, and does not directly find a deflection.)

The theorems and the notes will both become clear after a few example problems. (Note that in these examples the symbol $\nabla$, normally a differential operator, is used to mean "triangle side.")

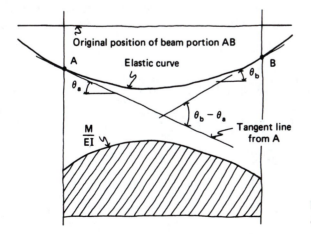

**Figure 9–1** Portion of elastic curve and $M/EI$ diagram.

**EXAMPLE 9–1**  Determine the slope at the left end and the deflection at midspan for the beam in Figure 9–2.

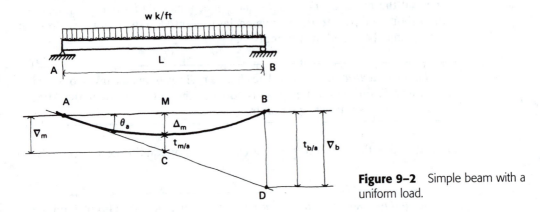

**Figure 9–2**  Simple beam with a uniform load.

### Solution
To find $\theta_a$, it is helpful to note that $\tan \theta_a = \theta_a$ for very small angles, and that

$$\tan \theta_a = \frac{\nabla_b}{L}$$

in Figure 9–2, where $\nabla_b$, the triangle side $BD$ of triangle $ABD$, is geometrically equal to $\nabla_{b/a}$, a tangential deviation. $\nabla_{b/a}$ is a "vertical distance from point $B$ on the elastic curve to a tangent line from point $A$, and we can use the second moment-area theorem to find it.

The $M/EI$ curve for the beam is shown in Figure 9–3, and it is easy to compute its statical moment about a vertical line through $B$. See Appendix F for areas and centroidal distance of parabolic curves.

$$t_{b/a} = \frac{2}{3}\left(\frac{wL^2}{8EI}\right)(L)\frac{L}{2} = \frac{wL^4}{24EI} = \nabla_b$$

$$\tan \theta_a = \theta_a = \frac{\nabla_b}{L} = \frac{wL^4/24EL}{L}$$

$$\theta_a = \frac{wL^3}{24EI}$$

(The sign of $\theta_a$ is not evident from the computations but is negative by inspection.)

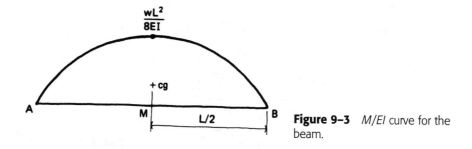

$\dfrac{wL^2}{8EI}$

+ cg

A  M  L/2  B

**Figure 9–3**  *M/EI* curve for the beam.

To find the midspan deflection, $\Delta_m$ (also negative by inspection), we will find the values of $\nabla_m$ and $t_{m/a}$ as shown in Figure 9–2. A subtraction will then give us $\Delta_m$. First find $\nabla_m$, the triangle side $MC$, by using $\theta_a$.

$$\theta_a = \tan \theta_a = \frac{\nabla_m}{L/2}$$

or

$$\nabla_m = \theta_a\left(\frac{L}{2}\right) = \frac{wL^3}{24EI}\left(\frac{L}{2}\right)$$

$$= \frac{wL^4}{48EI}$$

(Note that $\nabla_m$ could also have been found by using the similar triangles $AMC$ and $ABD$.)

The value $t_{m/a}$ can be found using the second moment-area theorem. From Figure 9–4,

$$t_{m/a} = \frac{2}{3}\left(\frac{wL^2}{8EI}\right)\frac{L}{2}\left(\frac{3}{8}\right)\frac{L}{2}$$

$$= \frac{wL^4}{128EI}$$

Then

$$\Delta_m = \nabla_m - t_{m/a}$$

$$= \frac{wL^4}{48EI} - \frac{wL^4}{128EI}$$

$$= \frac{5wL^4}{384EI}$$

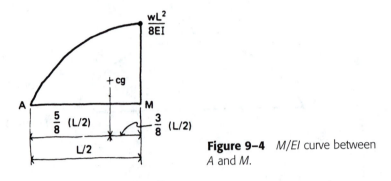

**Figure 9-4**  *M/EI* curve between *A* and *M*.

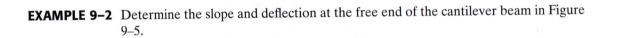

**EXAMPLE 9-2**  Determine the slope and deflection at the free end of the cantilever beam in Figure 9–5.

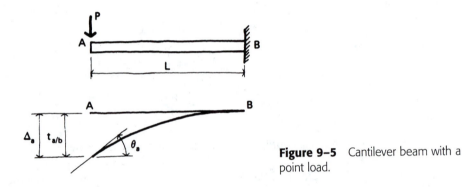

**Figure 9–5**  Cantilever beam with a point load.

### Solution

Notice that the slope of the elastic curve is zero at the fixed end. Knowing that $\theta_b$ is zero, $\theta_a$ can easily be found using the first moment-area theorem. The area under the *M/EI* curve between points *A* and *B* is equal to the change in $\theta$ between those same two points (Figure 9–6).

$$\Delta\theta_{ab} = \theta_a - \theta_b = \frac{1}{2}\left(\frac{PL}{EI}\right)L$$

Since $\theta_b = 0$,

$$\theta_a = \frac{PL^2}{2EI}$$

The sign is plus by inspection.

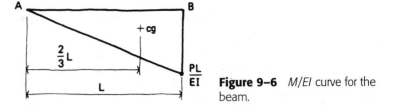

**Figure 9–6** *M/EI* curve for the beam.

To find $\Delta_a$, notice that a tangent line drawn from $B$ will be coincident with the initial position of the beam before loading. This means that deflections can be found directly because they are geometrically equal to tangential deviations.

$$\Delta_a = t_{a/b} = \frac{1}{2}\left(\frac{PL}{EI}\right)L\left(\frac{2}{3}L\right)$$

$$= \frac{PL^3}{3EI}$$

The selection of the tangent line location in this example gives us a hint as to how we could have set up the simple beam of Example 9–1 so as to reduce the numerical work involved.

In Figure 9–7, we see that the slope of the beam is zero at midspan due to symmetry. With $\theta_m$ equal to zero, $\theta_a$ can be found directly as the difference between $\theta_m$ and $\theta_a$. The midspan deflection $\Delta_m$ can also be found directly by noting that it is geometrically equal to $t_{a/m}$, which can be found by one application of the second moment-area theorem. Referring to Figure 9–4 and taking the statical moment about a vertical line through $A$, we get

$$\Delta_m = t_{a/m} = \frac{2}{3}\left(\frac{wL^2}{8EI}\right)\frac{L}{2}\left(\frac{5}{8}\right)\frac{L}{2}$$

$$= \frac{5wL^4}{384EI}$$

Moment-area computations can often be simplified through judicious selection of tangent line locations.

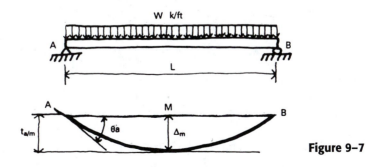

**Figure 9–7**

**EXAMPLE 9–3** Determine $\theta_b$ and $\Delta_m$ for the 4 × 16 timber beam in Figure 9–8. Let $E = 1.7(10)^6$ psi and $I = 1034$ in.$^4$.

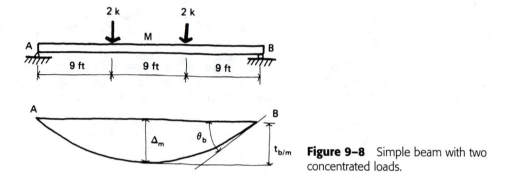

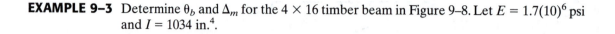

**Figure 9–8** Simple beam with two concentrated loads.

## Solution

(It is easier to work with the symbols $E$ and $I$ in the computations and replace them with numerical values as a final step; see Figure 9–9.)

$$\Delta\theta_{bm} = \theta_b - \theta_m$$

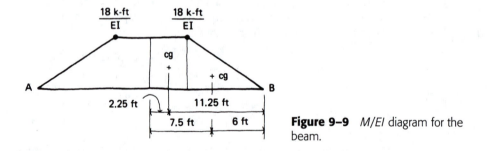

**Figure 9–9** $M/EI$ diagram for the beam.

The slope at midspan is zero; therefore,

$$\theta_b = \Delta\theta_{bm}$$

$$= \frac{18 \text{ kip-ft}}{EI}(4.5 \text{ ft}) + \frac{1}{2}\left(\frac{18 \text{ kip-ft}}{EI}\right)(9 \text{ ft})$$

$$= \frac{162 \text{ kip-ft}^2}{EI}$$

The deflection at midspan is equal to $t_{b/m}$.

$$t_{b/m} = \frac{18 \text{ kip-ft}}{EI}(4.5 \text{ ft})(11.25 \text{ ft}) + \frac{1}{2}\left(\frac{18 \text{ kip-ft}}{EI}\right)(9 \text{ ft})(6 \text{ ft})$$

$$\Delta_m = \frac{1400 \text{ kip-ft}^3}{EI}$$

The problem is completed by replacing $E$ and $I$ by their numerical values. Notice that the units of the numerator in each case must be converted from kips and feet to pounds and inches.

$$\theta_b = \frac{162 \text{ kip-ft}^2(1000 \text{ lb/kip})(12 \text{ in/ft})^2}{1.7(10)^6 \text{ psi}(1034 \text{ in}^4)}$$

$$= 0.0133$$

This is the slope of the beam at $B$ in radians. If degrees are desired, the value can be multiplied by $180/\pi$ to get

$$\theta_b = 0.76 \text{ degree}$$

To find the deflection,

$$\Delta_m = \frac{1400 \text{ kip-ft}^3(1000 \text{ lb/kip})(12 \text{ in/ft})^3}{1.7(10)^6 \text{ psi}(1034 \text{ in}^4)}$$

$$= 1.37 \text{ in}$$

---

**EXAMPLE 9–4M**

Determine $\theta_a$ and $\Delta_m$ for the W460 × 52 beam in Figure 9–10M. $E$ is 200 GPa, and from Appendix K, $I = 212(10)^6$ mm$^4$.

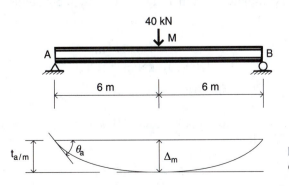

**Figure 9–10M** Steel beam with concentrated load.

### Solution
(It is easier to work with the symbols $E$ and $I$ in the computations and replace them with numerical values as a final step; see Figure 9–11M.)

$$\Delta\theta_{am} = \theta_a - \theta_m$$

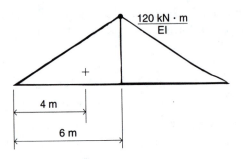

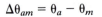

**Figure 9–11M**  *M/EI* diagram for the beam.

The slope at midspan is zero, therefore,

$$\theta_a = \Delta\theta_{am}$$

$$= \frac{1}{2}\frac{(120 \text{ kN} \cdot \text{m})}{EI}(6\text{m})$$

$$= \frac{360 \text{ kN} \cdot \text{m}^2}{EI}$$

The deflection at midspan is equal to $t_{a/m}$.

$$t_{a/m} = \frac{1}{2}\frac{(120 \text{ kN} \cdot \text{m})(6\text{m})(4\text{m})}{EI}$$

$$= \frac{1440 \text{ kN} \cdot \text{m}^3}{EI}$$

The problem is completed by replacing $E$ and $I$ by their numerical values. Notice that the $I$ value is in $\text{m}^4$.

$$\theta_a = \frac{360 \text{ kN} \cdot \text{m}^2}{[200(10)^6 \text{ kN/m}^2][212(10)^{-6} \text{ m}^4]}$$

$$= 0.0085$$

This is the slope of the beam at $A$ in radians. If degrees are desired, the value can be multiplied by $180/\pi$ to get

$$\theta_a = 0.49 \text{ degree}$$

To find the deflection,

$$\Delta_m = \frac{1440 \text{ kN} \cdot \text{m}^3}{[200(10)^6 \text{ kN/m}^2][212(10)^{-6} \text{ m}^4]}$$

$$= 0.034 \text{ m}$$

$$= 34 \text{ mm}$$

(*Note*: In the problems provided in this chapter, the self-weight of members has intentionally been ignored in the interest of simplicity.)

## PROBLEMS

**9–1.** Determine the free-end slope and deflection values in terms of $w, L, E,$ and $I$ for a cantilever beam with a uniform load.

**9–2.** Determine the maximum slope and deflection values in terms of $P, L, E,$ and $I$ for a simple beam with a midspan point load.

**9–3.** The beam in Figure 9–12 is a nominal 4 × 12 of Douglas fir No. 2. Will the deflection meet a code limitation of $L/240$?

**9–3M.** The beam in Figure 9–12M is an 89 × 285 of Douglas fir No. 2. Will the deflection meet a code limitation of $L/240$?

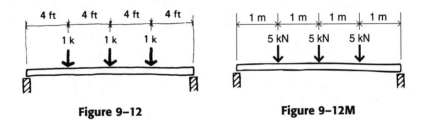

**Figure 9–12**                    **Figure 9–12M**

**9–4.** Determine the midspan deflection in terms of $EI$ for the beam of Figure 9–13.

**9–4M.** Determine the midspan deflection in terms of $EI$ for the beam of Figure 9–13M.

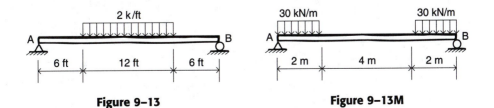

**Figure 9–13**                    **Figure 9–13M**

**9–5.** Assume that the beam in Figure 9–14 is a W16 × 57. Determine the deflection in inches at one of the free ends.

**9–5M.** Assume that the beam in Figure 9–14M is a W410 × 85. Determine the deflection in mm at one of the free ends.

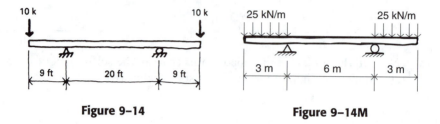

Figure 9–14                                                Figure 9–14M

# 9–3 PRINCIPLE OF SUPERPOSITION

Many structures, including simple beams, are acted upon by more than one load. It may be of advantage to treat the effects of such loads separately and add the results obtained to arrive at a final answer. This is referred to as using the *principle of superposition*. We must remember that building structures are (hopefully) never loaded such that any material reaches its yield limit or passes out of the region of elasticity. Therefore, the design loads could be placed on the structure one at a time or all at once and the resulting stresses and deflections would total the same. This idea can also be utilized from the opposite standpoint.

Suppose, for example, that we were asked to obtain the midspan deflection of the beam in Figure 9–15. Since there is no known point of zero slope, we could not obtain the answer with a single application of the second moment-area theorem. However, if we added a fictitious load of 2 kips at a point 9 ft from the right end, the loading would then be symmetrical and the problem is simply solved as in Example 9–3. The answer thus obtained would be exactly twice the true deflection that would result from the original 2-kip load.

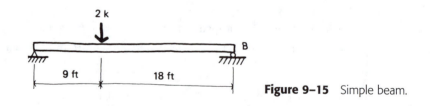

**Figure 9–15**   Simple beam.

# PROBLEMS

**9–6.** A simply supported W24 × 76 is 50 ft long and carries a uniform load of 2 kips/ft over the left half of its span. Determine the midspan deflection.

**9–6M.** A simply supported W610 × 113 is 15 m long and carries a uniform load of 30 kN/m over the right half of its span. Determine the midspan deflection.

**9–7.** The beam in Figure 9–16 is a nominal 4 × 16 with an $E$ value of $1.42(10)^6$ psi. Determine its midspan deflection.

**9–7M.** The beam in Figure 9–16M is an 89 × 385 with an $E$ value of 8900 MPa. Determine its midspan deflection.

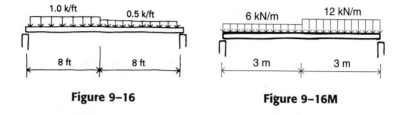

| **Figure 9–16** | **Figure 9–16M** |

**9–8.** Determine the deflection at a point 10 ft from the left support for the W16 × 57 shown in Figure 9–17. (*Hint:* It will be helpful to find the midspan deflection for the beam of Figure 9–14 first.)

**9–8M.** Determine the deflection at a point 3 m from the right support for the W410 × 85 shown in Figure 9–17M. (*Hint:* It will be helpful to find the midspan deflection for the beam of Figure 9–14M first.)

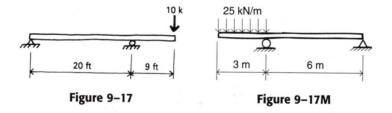

| **Figure 9–17** | **Figure 9–17M** |

**9–9.** Determine the free-end deflection of the beam in Figure 9–18. Let $E = 29(10)^6$ psi and $I = 300$ in.$^4$. (*Hint:* Use the principle of superposition to determine the deflection in parts. First find the upward deflection due to the load between the supports; then find the downward deflection due to the two overhanging loads. Algebraically add these values and then substitute for $E$ and $I$.)

**9–9M.** Determine the free-end deflection of the beam in Figure 9–18M. Let $E = 200$ GPa and $I = 100(10)^6$ mm$^4$. (*Hint:* Use the principle of superposition to determine the deflection in parts. First find the upward deflection due to the load between the supports; then find the downward deflection due to the two overhanging loads. Algebraically add these values and then substitute for $E$ and $I$.)

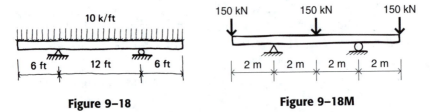

**Figure 9–18**                    **Figure 9–18M**

## 9–4   USE OF DEFLECTION FORMULAS

Many loading patterns and support conditions occur so frequently in construction that reference manuals and engineering handbooks tabulate the appropriate formulas for their deflections. A few such cases are given in Appendix K. More often than not, the required deflection values in a beam design situation can be obtained via these formulas, and one does not have to resort to deflection theory. Even when the actual loading situation does not match one of the tabulated cases, it is sufficiently accurate for most design situations to approximate the maximum deflection by using one or more of the formulas.

For such purposes it is helpful to know that in the case of beam simply supported at its ends, the point of greatest deflection will always be *very* close to midspan. This is true regardless of the pattern or placement of loads along the beam. The curvature of beams is very slight, and even when a point load is placed near one end, the maximum deflection is just a little bit greater than the deflection at midspan.

**EXAMPLE 9–5** Determine the approximate maximum deflection for the beam in Figure 9–19.

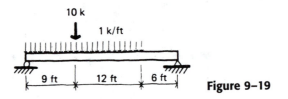

**Figure 9–19**

### Solution

Because of the relatively complex loading pattern, the determination of the real value of the maximum deflection could become somewhat involved. However, we know that the midspan deflection will be very close to the maximum. Furthermore, it would be conservative in this case to treat this beam as though it had a uniform load over the *full* span and had the concentrated load located at *midspan*. The midspan deflection for this fictitious situation can easily be computed by using Cases 3 and 4 in Appendix K.

$$\Delta_{max} = \frac{PL3}{48EI} + \frac{5wL^4}{384EI}$$

or

$$\Delta_{max} = \frac{10 \text{ kips}(27 \text{ ft})^3}{48EI} + \frac{5(1 \text{ kip/ft})(27 \text{ ft})^4}{384EI}$$

$$= \frac{4100 \text{ kip-ft}^3}{EI} + \frac{6900 \text{ kip-ft}^3}{EI}$$

$$= \frac{11\,000 \text{ kip-ft}^3}{EI}$$

The value thus obtained will be slightly larger than the actual maximum deflection. If this value falls within the code limitation or is even close to it, no further deflection investigation is warranted.

---

**EXAMPLE 9–6M**

Determine the approximate maximum deflection of the 10-m portion of the beam in Figure 9–20M.

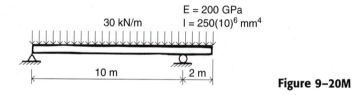

E = 200 GPa
I = 250(10)⁶ mm⁴
30 kN/m

10 m
2 m

**Figure 9–20M**

**Solution**
In this case, it would be "safe" to ignore the overhang and assume a 10-m uniformly loaded simple span.

$$\Delta_{max} = \frac{5wL^4}{384EI}$$

$$= \frac{5(30 \text{ kN/m})(10 \text{ m})^4}{[384(200)(10)^6 \text{ kN/m}^2][250(10)^{-6} \text{ m}^4]}$$

$$= 7.81(10)^{-2} \text{ m}$$

$$= 78 \text{ mm}$$

**EXAMPLE 9–7**  The W21 × 44 roof beam shown in Figure 9–21 is adequate in moment. Will it meet a deflection limitation of $L/180$?

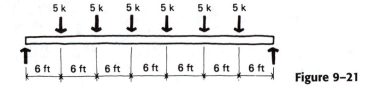

**Figure 9–21**

### Solution

The total load on the beam is 6 times 5 or 30 kips. If this load were uniformly spread out, the unit load would be $w = 0.71$ kip/ft. The deflection due to this fictitious load will be less than the actual deflection. If the loads were all gathered into one load of 30 kips and applied at midspan, the deflection thus generated would be considerably greater than the actual.

$$\Delta_{min} = \frac{5wL^4}{384EI}$$

$$= \frac{5(0.71 \text{ kip/ft})(42 \text{ ft})^4(1000 \text{ lb/kip})(12 \text{ in/ft})^3}{384[29(10)^6 \text{ psi}](843 \text{ in}^4)}$$

$$= 2.0 \text{ in}$$

$$\Delta_{max} = \frac{PL^3}{48EI}$$

$$= \frac{30 \text{ kips}(42 \text{ ft})^3(1000 \text{ lb/kip})(12 \text{ in/ft})^3}{48[29(10)^6 \text{ psi}](843 \text{ in}^4)}$$

$$= 3.3 \text{ in}$$

We now know that the actual beam deflection is between 2.0 and 3.3 in. and, in view of the loading pattern, is probably closer to the lesser value. The code limit is

$$\Delta_{code} = \frac{L}{180} = \frac{42 \text{ ft}(12 \text{ in/ft})}{180} = 2.8$$

Examining the upper and lower limits of the actual deflection versus the code allowable, this beam is probably adequate in deflection.

## PROBLEMS

**9–10.** A W16 × 31 serves as a simple beam 30 ft long. It supports a uniformly varying load that varies linearly from zero at one end to 1.5 kips/ft at the other. Determine the maximum deflection.

**9–10M.** A W410 × 46 serves as a simple beam 9 m long. It supports a uniformly varying load that varies linearly from zero at one end to 20 kN/m at the other. Determine the maximum deflection.

**9–11.** A Douglas fir No. 2 4 × 12 beam spans 20 ft and is loaded only over its central 10 ft by a uniform load of 200 plf. Determine the approximate maximum deflection.

**9–11M.** A doubled 38 × 235 of southern pine No. 2 is 6 m long and loaded at its quarter points by three 4-kN point loads. Determine the midspan deflection.

**9–12.** Assume that the beam of Problem 9–10 carries a point load of 10 kips at midspan in addition to the uniformly varying load. Determine the midspan deflection.

**9–12M.** Assume that the beam of Problem 9–10M carries a point load of 50 kN at midspan in addition to the uniformly varying load. Determine the midspan deflection.

**9–13.** For the cantilever beam of Figure 9–22, what fraction of the maximum deflection is caused by the left half of the load?

**9–13M.** For the cantilever beam of Figure 9–22M, what fraction of the maximum deflection is caused by the left half of the load?

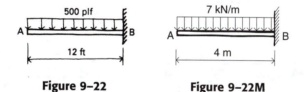

**Figure 9–22**          **Figure 9–22M**

## 9–5   SUPERPOSITION AND INDETERMINATE STRUCTURES

A proper investigation of statically indeterminate structures is beyond the scope of this book. However, certain structures can be readily approached using only the ideas of superposition and the equations of statics. For example, beams that are indeterminate to the first degree (those having only one redundant support component) are easily analyzed.

**EXAMPLE 9–8** Determine the vertical reactions for the indeterminate beam in Figure 9–23.

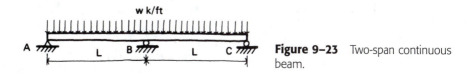

**Figure 9–23**   Two-span continuous beam.

### Solution

From statics we know that $A_y + B_y + C_y = 2wL$. Moment equations, however, cannot be used directly because any selected moment center will only eliminate one of the three forces and still leave two independent unknowns in each equation (Figure 9–24). It is noted that if any one of the three forces could be obtained by some other means, the remaining two can be easily evaluated. For example, if we denote reaction $B_y$ as the redundant force and remove it, the beam will deflect as shown in Figure 9–25.

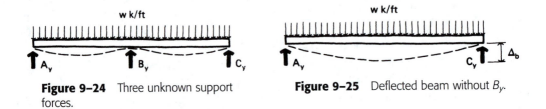

**Figure 9–24**  Three unknown support forces.

**Figure 9–25**  Deflected beam without $B_y$.

If we now apply a force $P$ vertically upward at $B$, the beam will be pushed back toward its original position. Indeed, if we apply just the right amount of $P$, say equal in magnitude to $B_y$, we will then have the deflected shape as shown in Figure 9–26, having reduced $\Delta_b$ to zero. In other words, the amount of $P$ necessary to remove the deflection $\Delta_b$ is called $B_y$.

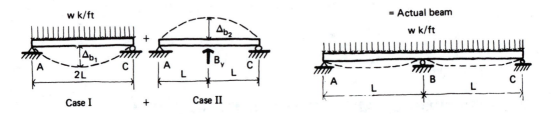

**Figure 9–26**  Loading cases for the beam.

The procedure is first to remove the redundant force and calculate the deflection at that point, $\Delta_{b_1}$. The deflection must be equal to $\Delta_{b_2}$, the upward deflection due to $B_y$ acting alone, if we are to get back to the real beam situation of zero deflection at $B$.

$$\Delta_{b_1} = \Delta_{b_2}$$

Using the appropriate deflection equations from Appendix K, we get

$$\frac{5w(2L)^4}{384EI} = \frac{B_y(2L)^3}{48EI}$$

$E, I,$ and $L^3$ will drop out, leaving us with

$$B_y = 1.25wL$$

Through symmetry and statics we can then find that

$$A_y = C_y = 0.375wL$$

It should be noted that any one of the three vertical support forces could have been declared as the redundant in this example. Owing to a lack of symmetry, the arithmetic would be a bit more involved if we had chosen $A_y$ or $C_y$.

---

The concept of equating deflections (really superposition of loads) is a very useful tool in structural analysis. The following example will illustrate its application to a very different kind of indeterminate structure.

---

**EXAMPLE
9–9M**

Figure 9–27M shows two beams crossed at midspan and having the same $EI$ value. Beam $A$, however, is twice as long as beam $B$. How much of the 90 kN is carried by each beam?

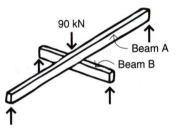

90 kN

Beam A

Beam B

**Figure 9–27M**   Crossed simple beams.

**Solution**
The key to the solution is to recognize that the midspan deflection of the two beams will be equal. From statics the amount of load carried by the two beams must sum to 90 kN. If the load carried by beam $B$ is called $P_b$, the *net* load carried by beam $A$ is $90 - P_b$, as shown in Figure 9–28M. Beam $A$ is simultaneously acted upon by 90 kN down and the contact force $P_b$ up, whereas beam $B$ is loaded only by $P_b$ downward.

$$\Delta_1 = \Delta_2$$

$$\frac{(90 - P_b)(2L)^3}{48EI} = \frac{P_b(L)^3}{48EI}$$

Solving for $P_b$ yields

$$P_b = 80 \text{ kN}$$

This indicates that the long beam is very lightly loaded, carrying only 10 kN. This is not surprising once we realize that the short beam is much stiffer; consequently it takes considerably more load to deflect than does the long beam. The support reactions will be 40 kN and 5 kN for the short and long beams, respectively.

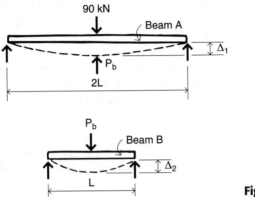

**Figure 9–28M**   Free-body diagrams.

Example 9–9M is often used to explain the behavior of a rectangular monolithic concrete slab supported on all four edges. The more rectangular or less square the slab, the greater is the fraction of the load taken by the long edge supports that make the short span. While the bending and torsional forces in the monolithic slab are more involved than this, the idea is essentially correct.

# PROBLEMS

**9–14.** Determine the reactions and the shear and moment diagrams for the beam in Figure 9–29. (*Hint: EI* need not be known.)

**9–14M.** Determine the reactions and the shear and moment diagrams for the beam in Figure 9–29M. (*Hint: EI* need not be known.)

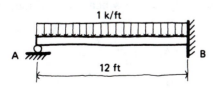 

**Figure 9–29**   Propped cantilever.        **Figure 9–29M**   Propped cantilever.

**9–15.** A beam of constant *EI* is continuous over three walls, making two equal spans. There is a concentrated load *P* applied at the center of each span. Determine the amount of reactive force provided by each wall in terms of *P*.

**9–16.** Determine the magnitude of the downward reaction at *B* for the overhanging beam in Figure 9–30.

**9–16M.** Determine the magnitude of the downward reaction at *B* for the overhanging beam in Figure 9–30M.

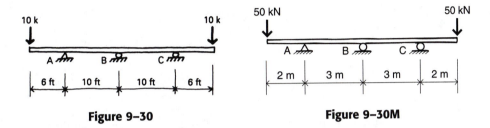

| **Figure 9–30** | **Figure 9–30M** |

**9–17.** In Figure 9–31, beams *A* and *B* are crossed at 90° in plan. The two beams are made of the same wood, but beam *A* has twice the *I* value of beam *B*. Determine how much of the load is taken by each beam.

**9–17M.** In Figure 9–31M, beams *A* and *B* are crossed at 90° in plan. The two beams are made of the same wood, but beam *A* has twice the *I* value of beam *B*. Determine how much of the load is taken by each beam.

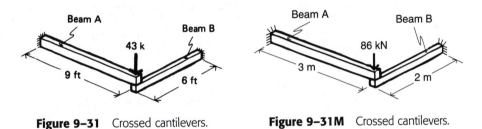

**Figure 9–31**  Crossed cantilevers.          **Figure 9–31M**  Crossed cantilevers.

**9–18.** If both beams in Figure 9–32 have the same *EI* value, determine the contact force at *A*. Construct the moment diagram for each beam.

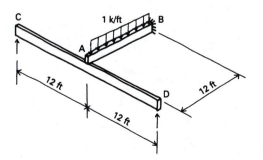

**Figure 9–32**

**9–18M.** If both beams in Figure 9–32M have the same $EI$ value, determine the contact force at $A$. Construct the moment diagram for each beam.

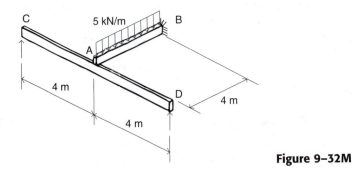

**Figure 9–32M**

**9–19.** Referring to Figure 9–32, assume that beam $AB$ has four times the $I$ value of beam $CD$, and determine the contact force at $A$. (Assume that the $E$ value remains the same for both beams.) Construct the moment diagram for each beam.

**9–19M.** Referring to Figure 9–32M, assume that beam $AB$ has four times the $I$ value of beam $CD$, and determine the contact force at $A$. (Assume that the $E$ value remains the same for both beams.) Construct the moment diagram for each beam.

**9–20.** (a) Compare the answers to Problems 9–14, 9–18, and 9–19 and then describe what will happen to the moment at the wall of the propped cantilever of Figure 9–29 if the support at $A$ encounters foundation problems and settles slightly downward.

(b) If $E = 29(10)^6$ and $I = 154$ in.$^4$, how much settlement at $A$ would result in a moment of 36 kip-ft at the wall?

**9–20M.** (a) Compare the answers to Problems 9–14M, 9–18M, and 9–19M and then describe what will happen to the moment at the wall of the propped cantilever of Figure 9–29M if the support at $A$ encounters foundation problems and settles slightly downward.

(b) If $E = 200$ GPa and $I = 38(10)^6$mm$^4$, how much settlement at $A$ would result in a moment of 20 kN $\cdot$ m at the wall?

**9–21.** With reference to Figure 9–23, let $W = 0.2$ k/ft and $L = 10$ ft. If $E = 1.5(10)^6$ psi and $I = 415$ in$^4$, how much would the supports at A and C have to settle downward before $A_y = C_y = 0$?

**9–22M.** Determine the reactions and construct the shear and moment diagrams for the hinged beam in Figure 9–33M. Assume that $EI$ is constant.

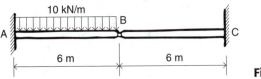

**Figure 9–33M**

## 9–6 THEOREM OF THREE MOMENTS

Example 9–8 illustrated how superposition could be used to find the reactions for a two-span continuous beam. This technique becomes cumbersome when more than two spans are present because the removal of one of the supports leaves a structure that is still indeterminate. The situation is further complicated if the various spans are not equal, because the deflection equations needed are difficult to derive or obtain.

Fortunately, continuous beams of many spans can be easily solved by simple computer programs using matrix algebra. A classical approach using only simple arithmetic is also available and is called the *theorem of three moments*. It is convenient for solving very simple continuous beams (such as those in the examples and the problems of this section) and for verifying the output of computer analyses.

In Figure 9–34, two spans of a uniformly loaded continuous beam have been sketched as free bodies to expose the internal moments at the supports $A$, $B$, and $C$. It is seen that for each span, these support moments are redundant in terms of equilibrium. If these moments were known, each beam segment would then become determinate, and the vertical support forces could be found using simple statics. The task of the theorem of three moments is to find these unknown support moments. One version of the theorem is:

$$M_a\frac{L_1}{I_1} + 2M_b\left(\frac{L_1}{I_1} + \frac{L_2}{I_2}\right) + M_c\frac{L_2}{I_2} = -\frac{w_1L_1^3}{4I_1} - \frac{w_2L_2^3}{4I_2}$$

$$-\frac{P_1a_1b_1}{I_1L_1}(L_1 + a_1) - \frac{P_2a_2b_2}{I_2L_2}(L_2 + b_2)$$

$$(9\text{–}1)$$

It can be derived rather easily using moment-area theory, and assumes that uniform loads, when present, act upon a given span for its full length. (This is the most common condition for uniformly loaded continuous members.) Figure 9–35 explains the notation of the theorem.

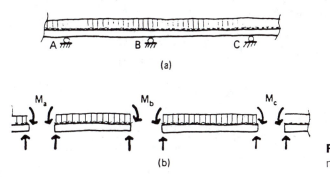

(a)

(b)

**Figure 9–34** Redundant support moments.

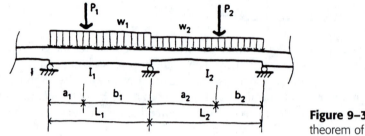

**Figure 9–35** Notation for the theorem of three moments.

If the beam is of constant cross section for all its spans, the theorem can be simplified by the deletion of the moment of inertia symbols, and we get Equation 9–2:

$$M_a L_1 + 2M_b(L_1 + L_2) + M_c L_2 = -\frac{w_1 L_1^3}{4} - \frac{w_2 L_2^3}{4}$$

$$-\frac{P_1 a_1 b_1}{L_1}(L_1 + a_1) - \frac{P_2 a_2 b_2}{L_2}(L_2 + b_2)$$

(9–2)

**EXAMPLE 9–10**

Determine the unknown support moment at $B$ and construct the shear and moment diagrams for the beam in Figure 9–36.

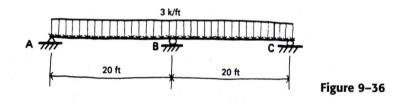

**Figure 9–36**

**Solution**

In this case, $M_a$ and $M_c$ are both zero and there are no concentrated loads, so the theorem of three moments reduces to

$$2M_b(L_1 + L_2) = -\frac{w_1 L_1^3}{4} - \frac{w_2 L_2^3}{4}$$

Substituting in the appropriate values, we get

$$2M_b(20 + 20) = -\frac{3(20)^3}{4} - \frac{3(20)^3}{4}$$

$$M_b = -150 \text{ kip-ft}$$

Figure 9–37 shows how the beam can now be treated as two determinate spans that share a common support at *B*. The 150-kip-ft support moment is applied to each span with negative sense (i.e., tension in the top fiber) as indicated by the sign of $M_b$. (*A support moment is a point of known moment when constructing moment diagrams.*)

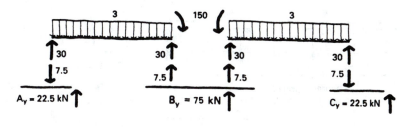

**Figure 9–37**

For convenience, the reactions have been determined in parts. The row of upward-acting 30-kip forces is due to the uniform load acting alone, and the couples formed by the 7.5-kip forces are in response to the 150-kip-ft support moment. These component forces can be algebraically summed to get the final reaction values at each support. Notice that the center support carries *more than half the total load*. Figure 9–38 illustrates the resulting increase in shear force present in the center support. The student should ascertain the full effects of continuity in this example by comparing the *V* and *M* diagrams to those that would result if the beam were constructed in two separate simple spans.

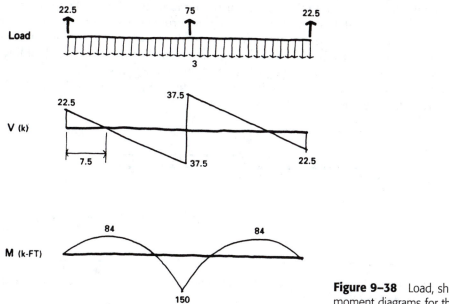

**Figure 9–38** Load, shear, and moment diagrams for the beam.

The theorem of three moments can be used to analyze continuous beams of any number of spans by applying it successively to each pair of adjacent spans. Making the appropriate extensions of the notation, as illustrated in Figure 9–39, we get a set of equations equal in number to the number of unknown support moments.

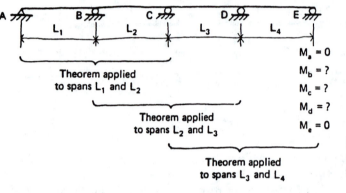

$M_a = 0$

$M_b = ?$

$M_c = ?$

$M_d = ?$

$M_e = 0$

**Figure 9–39**   Three unknown support moments and three equations.

The theorem can also be used for fixed-end beams of one or more spans. Here it is necessary to make use of the fact that the slope of the elastic curve is zero at such supports. This means that a fixed end can be simulated, in concept, by the central support of a set of continuous spans having mirror-image symmetry. The two-span beam could be fixed at support $B$ with no change in the elastic curve and thus no change in shear and moment. Conversely, the fixed-end beam in Figure 9–40 could be solved using the theorem of three moments by creating an imaginary $L_1$ span out to the left support $B$. Selecting the proper imaginary load and length for this span to achieve symmetry, the beam is then identical to the one in Example 9–10. Mathematically, the use of imaginary spans has the effect of doubling both sides of the three-moments equation, and thus the lengths and loads of such spans may be taken as zero when writing the equations. Examples 9–11M and 9–12 illustrate this procedure.

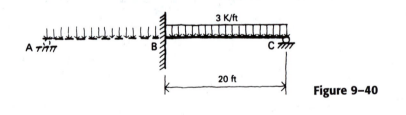

**Figure 9–40**

---

**EXAMPLE 9–11M**   Determine the unknown support moments and construct the shear and moment diagrams for the beam of Figure 9–41M.

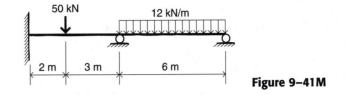

**Figure 9–41M**

### Solution

As indicated in Figure 9–42M, the theorem need not be applied to spans $AB$ and $BC$ because for those two spans all the load terms will be zero. Writing equations for the other two pairs, we get, for spans $BC$ and $CD$

$$2M_c(0 + 5) + (M_d)(5) = -\frac{50(2)(3)}{5}(5 + 3)$$

and for spans $CD$ and $DE$

$$M_c(5) + 2M_d(5 + 6) = -\frac{50(2)(3)}{5}(5 + 2) - \frac{12(6)^3}{4}$$

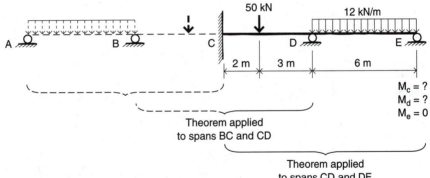

**Figure 9–42M**  Two unknown support moments and two equations.

Notice that the span $CD$ must be considered an $L_2$ (right-hand) span in the first equation and an $L_1$ (left-hand) span in the second equation. Solving these two simultaneously gives us

$$M_c = -27 \text{ kN} \cdot \text{m}$$

$$M_d = -42 \text{ kN} \cdot \text{m}$$

These values can then be used to determine the vertical reactions at $C$, $D$, and $E$, shown in Figure 9–43M.

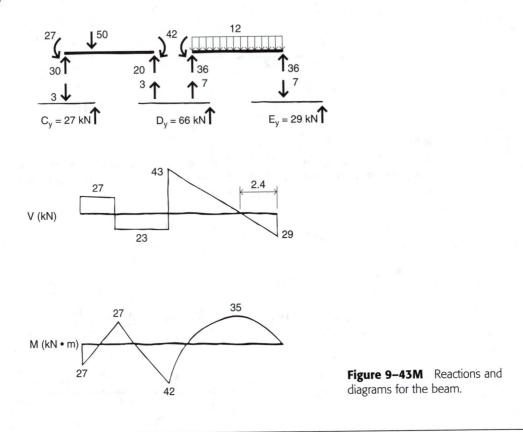

**Figure 9–43M** Reactions and diagrams for the beam.

---

**EXAMPLE 9–12**    Construct the moment diagram (in terms of $w$ and $L$) for the beam in Figure 9–44.

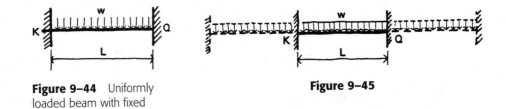

**Figure 9–44** Uniformly loaded beam with fixed ends.

**Figure 9–45**

**Solution**
Referring to Figure 9–45, for the left two spans

$$2M_k(O + L) + M_q(L) = -\frac{wL^3}{4}$$

and for the right two spans

$$M_k(L) + 2M_q(L + O) = -\frac{wL^3}{4}$$

Since by symmetry, $M_k = M_q$, either equation may be used to find the value.

$$M_k = M_q = -\frac{wL^2}{12}$$

Because the end moments are equal, the vertical reactions will each be $wL/2$, which means that the shear diagram will be the same as that of a simply supported beam. The moment diagram appears in Figure 9–46, and one should notice that it is quite different from that of a simple beam. End fixity not only reduces the positive moment by a factor of $3(wL^2/8$ to $wL^2/24)$ but also causes the maximum moment to be changed in location, sense, and magnitude. The absolute value of the maximum moment is, in fact, reduced by 50%.

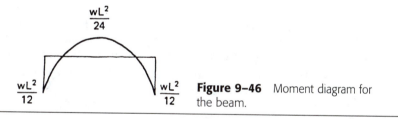

**Figure 9–46**  Moment diagram for the beam.

## PROBLEMS

**9–23.** Determine the support moment at $B$ and construct the $V$ and $M$ diagrams for the beam in Figure 9–47.

**9–23M.** Determine the support moment at $B$ and construct the $V$ and $M$ diagrams for the beam in Figure 9–47M.

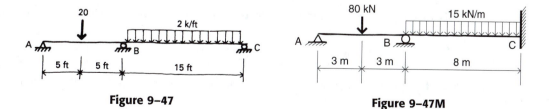

**Figure 9–47**                    **Figure 9–47M**

**9–24.** Determine the support moment at *P* and construct the *V* and *M* diagrams for the beam in Figure 9–48. (*Hint:* There will be two $P_1$ terms.)

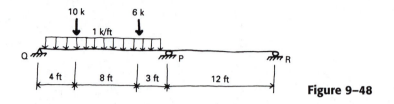

**Figure 9–48**

**9–24M.** Determine the support moment at *P* and construct the *V* and *M* diagrams for the beam in Figure 9–48M. (*Hint:* There will be two $P_2$ terms.)

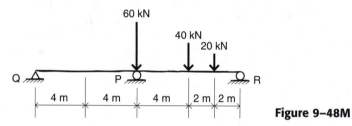

**Figure 9–48M**

**9–25.** Determine the moment at the wall and construct the *V* and *M* diagrams for the beam in Figure 9–49. (*Hint:* The moment at *B* is known by cutting a section located a differential distance to the right of that support.)

**9–25M.** Determine the moment at the wall and construct the *V* and *M* diagrams for the beam in Figure 9–49M. (*Hint:* The moment at *B* is known by cutting a section located a differential distance to the right of that support.)

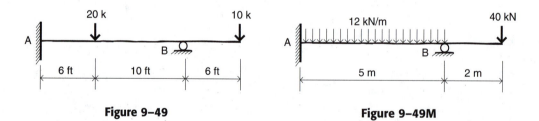

**Figure 9–49**                                    **Figure 9–49M**

**9–26.** Use the theorem of three moments to solve Problem 9–16. (*Hint:* There will be no load terms.)

**9–26M.** Use the theorem of three moments to solve Problem 9–16M. (*Hint:* There will be no load terms.)

**9–27.** Determine the unknown support moments for the beam of Figure 9–50 and construct the *V* and *M* diagrams.

**9–27M.** Determine the unknown support moments for the beam of Figure 9–50M and construct the *V* and *M* diagrams.

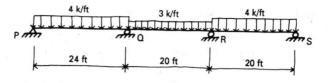

**Figure 9–50**

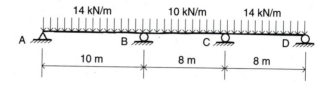

**Figure 9–50M**

**9–28.** Determine the support moments and construct the $V$ and $M$ diagrams for the fixed-end beam of Figure 9–51. (*Hint:* Make use of symmetry to reduce the number of equations.)

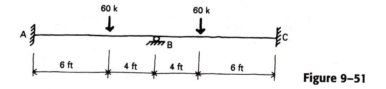

**Figure 9–51**

**9–29M.** Determine the unknown support moments for the beam of Figure 9–52M and construct the $V$ and $M$ diagrams.

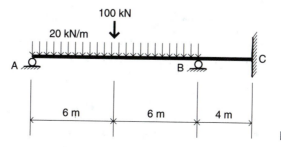

**Figure 9–52M**

## 9–7   LOADING PATTERNS

Continuous and fixed-end beams, like those on simple supports, are designed for the maximum stresses and deflections. Although deflection can be important, it is less likely to be the controlling design factor because of the inherently reduced angles of slope at the supports. Continuity or fixity serves to reduce the deflection between

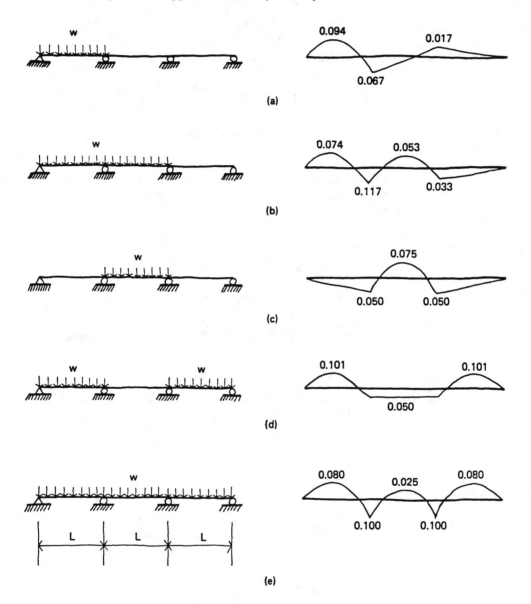

**Figure 9–53**   Moment coefficients for three-span continuous beams ($M$ = coefficient $\times wL^2$).

the supports. Bending and shearing stresses, however, may be increased or decreased by this continuity, and it is sometimes necessary to study more than one loading pattern to determine which combination will generate the maximum values. With continuous beams, the critical design loading might be one of less than full load. While the dead load is normally present on all spans, the analyst must consider the presence *and* absence of live loads. The addition or removal of a load will always change one or more moments elsewhere. As an illustration of this, look at the beam in Example 9–11M. Under the given loading conditions, the maximum moment is negative and occurs at support D, but if we remove the 50-kN concentrated load, this value will reduce and the largest moment will be positive and near the middle of the right-hand span.

It is difficult to make generalizations about continuous beams with moments of inertia that vary from span to span, but for those of constant $I$ (and $E$, of course), maximum negative moments will usually occur when adjacent spans are fully loaded. Maximum positive moments will be found when alternate spans are loaded. This is illustrated for uniform loads in Figure 9–53, where decimal coefficients of $wL^2$ are given for the moments in beams of three spans.

In a similar manner, it can be shown that the maximum shearing forces will generally result from loading adjacent spans. More important, these shearing forces, unlike bending moments, will generally be increased by the presence of continuity (see Example 9–10). This can be critical when designing with wood, where the low shear strength will sometimes favor noncontinuous or determinate spans.

# 10

# Beam Design and Framing

## 10-1 INTRODUCTION

Most conventional building structures involve a vertical support system and a horizontal spanning system. The configuration of such systems and the determination of the required sizes for elements in these systems is an integral part of building design. The horizontal framing or deck system is of particular importance because it almost always involves bending as the primary structural behavior. Its members can become quite large relative to those of the support system, which carries loads in an axial manner. Proper attention to the framing directions of joists, beams, and girders can save both depth and weight in a structural deck. Any excessive structural depth is to be avoided, particularly in multistoried structures where the overall building height and square area of exterior wall can be significantly increased by repetition of the floors. The location and shape of openings in a floor system can present unusual loading patterns and may require special attention structurally. The planning of such openings with the framing system in mind will usually avoid unnecessary conflicts. Similarly, the selection of structural materials is best made with due consideration for the proper span range for such materials and their relative capabilities to resist different kinds of stresses.

Novice designers all too frequently move too far along in the architectural planning process without attention to the structural planning process. Most attempts to "structure" building designs after the major spatial relationships and configurations have been determined result in contorted constructions and/or unsatisfactory compromises. The situation becomes even more disastrous when the preliminary design work was done by manipulating plans and elevations without sufficient attention to what was happening in building section. It has been the author's experience that many framing problems can be easily detected at an early stage by frequent sketches of building sections showing major structural elements.

283

## 10–2   SHAPE OF BEAM CROSS SECTIONS

Beams are designed for moment and shearing forces and deflections, all of which are brought about by the loads and spans of a given building structure. In designing bending members, it is helpful to remember that stresses and deflections are developed as functions of cross-sectional shape as well as in response to externally applied loads. The equations developed in Chapters 7, 8, and 9 illustrate the relationships between beam shape and the magnitude of stresses and deflections.

The general flexure formula, $f_b = Mc/I$, indicates that bending stress is directly proportional to the extreme fiber distance and inversely proportional to the moment of inertia. Since $I$ may be considered a product of cross-sectional area and distance to the neutral axis squared, the approximate net effect is that bending stress decreases linearly as either area or depth is increased. Bending stress is not an inverse function of depth squared, as it is often misrepresented. It *is* true, of course, that bending resistance is increased more by adding material that is remote to the neutral axis rather than near it.

The general shear formula, $f_v = VQ/Ib$, involves $Q$ (a function of area and depth) and $I$ (area and depth squared), which partially offset one another, and $b$, the section width. These factors tell us that shearing stress is inversely proportional to the area, as measured by $b$ and $d$. As either $b$ or $d$ gets smaller, the shearing stresses increase. An adequate depth is usually maintained to control flexural stresses, so with shear it becomes critical that an adequate width is maintained.

Deflection is inversely proportional to both $E$ and $I$. $E$ is independent of cross-sectional shape, but $I$ is, once again, a product of area and its location squared. Deflection will be greatly decreased by adding material to the section at points far from the neutral axis but will be decreased to an even greater degree by increasing the section depth.

In conclusion, it is important to notice that the moment of inertia occurs in the denominator in all three cases. It can be concluded that maintaining a large $I$ value per unit of cross-sectional area will almost always result in a structurally efficient beam design.

## 10–3   "IDEAL" BEAMS

Over the years, various structural analysts have attempted to determine the theoretically most efficient or "ideal" beam shapes to meet certain loading conditions. Although such exercises have little practical value, they are useful in understanding how forces in beams vary and how materials could be shaped to respond in purely structural terms.

The two beams in Figure 10–1(a) and (b) have been shaped so that the depth responds to moment diagrams for uniform and concentrated loads, respectively. As the depth reduces, the width must increase at the ends in order to maintain enough shear capacity. Actually, the uniformly loaded beam should increase its area in a linear manner from midspan to support in an ideal response to shear alone. A varia-

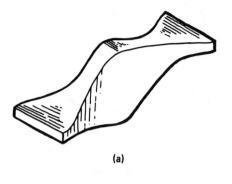

(a)

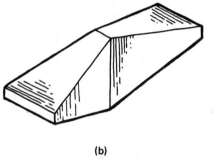

(b)

**Figure 10–1** "Ideal" simple beams.

tion on this same theme would have a wide-flange section using flanges that increase in thickness slightly as the overall beam depth decreases. In each case, deflection could be controlled by maintaining a high moment of inertia in regions of high bending moment.

The cantilever beams of Figure 10–2 are shaped so they have constant bending stress under concentrated free end loads. The beam of constant depth in Figure 10–2(a) must taper so the section modulus, $S = bd^2/6$, will change in direct proportion to the linear change in moment. To have the same uniform bending stress, the beam of constant width must also have a section modulus that varies with the moment diagram. Since the moment varies with the distance $x$, as in Figure 10–2(b), the depth of the cross section must vary with the square root of $x$. In other words, if $d$ is some constant times the square root of $x$, the section modulus will vary directly with $x$.

(a)

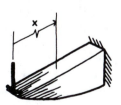

(b)

**Figure 10–2**

Any number of similar examples can be devised and manipulated to suit particular loading situations and shape constraints. As an exercise, the reader may wish to determine the proper shapes dictated by uniform rather than point loads acting on the cantilevers.

## 10–4   PROPERTIES OF MATERIALS

In designing framing systems or the individual elements of such systems, it is well to keep in mind the particular characteristics of the material being used. Such knowledge can help prevent irrational schemes and enable the designer to predict areas of probable high stress before any computations are done.

A brief review of the structural nature of each of the three major materials follows. There are always exceptions, of course, to these very general comments.

*Wood* is quite light and soft. It is generally weak perpendicular to the grain and strong parallel to it. It is very weak in shear, and this stress should always be checked. It is also subject to creep deformations.

Bending stress usually controls member size except for short, heavily loaded spans and overhangs, where shear can govern. The span range for *dimension lumber,* that which is nominally 2 to 4 in. thick (1½ to 3½ in. or 38 to 89 mm), extends to about 20 ft (6 m), and deflection will often control the design when the span is this great. Glued laminated sections, plywood box beams, and large timbers can span much farther, of course. Wood is a versatile framing material, and minor changes on the job are relatively easy to accomplish. Structure self-weight is seldom a design factor except for large members.

*Steel* is the strongest of all building materials and is both homogeneous and isotropic. Bending stress will usually control member size, but deflection should be checked. On longer simple spans and cantilevers, deflection limitations can readily govern. The increasing use of high-strength steels, requiring smaller $I$ values for flexure, also increases the proportion of beams designed for deflection. Shear almost never controls member size, but excess shear at selected points can require web stiffeners.

The normal span range for rolled steel sections is about 20 to 45 ft (6 to 14 m). Longer spans frequently employ trusses or plate girders. Steel building frames erect rapidly because of prior shop fabrication, but field modifications are not easily accommodated. Structure self-weight can be a significant part of the design load on longer spans.

*Concrete,* being an artificial stone, is quite heavy. Although not as dense as steel, much larger cross sections are required in reinforced concrete, and the self-weight of slabs and beams is always significant. Bending stress almost always controls member size. Excess shear is generally resisted by stirrups rather than increased member size. Concrete is subject to creep deformations, and members that are highly stressed on a constant basis may be governed by deflection limitations. The same is true of cantilevers, where deflection is naturally large.

Connections are inherently moment resistant, and advantage should be taken of the consequent indeterminacy of concrete structures. Continuous members require smaller sections than simple ones. The usual span range for reinforced concrete beams is about 15 to 35 ft (5 to 11 m). Longer spans become less efficient but can be accomplished using large member depths. Prestressing is a feasible alternative for longer spans. Reinforced concrete is formable, and cross-sectional shape is less restricted than in wood or steel. It is also naturally fire-resistant.

(The examples and problems of this chapter do not include the use of reinforced concrete because the behavior and analysis of this material is not addressed elsewhere in this book. Reinforced concrete is unlike wood or steel in that it is a composite of two materials. Framing patterns and load distribution are influenced by its continuous or monolithic nature, and the techniques for preliminary design can be quite different. Treatment of these topics logically occurs in a different textbook.)

## 10–5  TRIBUTARY AREA

Every structural element in a framing system has a *tributary area,* sometimes called a *contributing area.* In conventional structures, it is an area in horizontal plan from which a given member receives all its load. In previous examples and problems, this concept has been introduced by reference to parallel beams or joists spaced a certain distance apart.

Figure 10–3 illustrates how one beam in a parallel system must accept the loads on that portion of floor or roof deck that extends halfway to the neighboring beams. (The assumption is made here that the deck is constructed such that it acts structurally as shown by the arrows. For our purposes, these arrows may be assumed to represent plank or bar joists or the primary reinforcing of a one-way slab.) The larger, lightly shaded area constitutes the total load on the beam, W. The smaller and more heavily shaded area, called a *tributary strip,* would generate w, the load per running foot or meter of beam length. W, of course, equals wL.

In such a framing system, the girders are essentially point-loaded by the beam reactions. The only uniform load acting on the girder would be the girder self-weight. The small strip of dead and live load directly above the space occupied by the girder is usually included in the tributary areas of the beams. Exterior beams

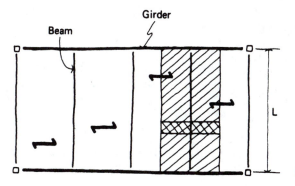

**Figure 10–3**  Tributary area.

have half the deck loading of interior beams but often end up being just as large, for reasons of repetition, economy, or exterior wall loads and torsional effects.

As shown in Figure 10–4, angled beams in plan will generate triangular loading patterns. In this case, both *AB* and *AC* have angled tributary areas.

The open well is framed in Figure 10–5(a) so that the girder *AB* will have the loading pattern shown in Figure 10–5(b). The concentrated load at midspan is the sum of two equal beam reactions from *CM* and *DM*. The student should investigate what happens to this loading pattern if the deck structure acts in the other direction (i.e., parallel to *AB*).

Any sort of opening in the horizontal system can usually be framed in more than one manner. The tributary areas and loading patterns on the framing elements will change, depending on the alternative selected. Figure 10–6 demonstrates two ways to frame a rectangular opening in a structural bay. The loading patterns that result are also shown. Such framing decisions should be made considering all of the influencing factors. If the structure is exposed, the beam directions can have a strong effect upon the character of the space. Mechanical duct runs may require beams to be shallow in one direction but not in the other. Structural efficiency, however, might best be served by a uniformity of depth.

Smaller beam sizes are usually possible if large concentrated loads can be avoided. When this is not practical, it is best to keep loads away from midspan regions, if possible.

The proper location of any type of opening can depend on the framing mate-

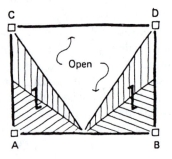

(a)

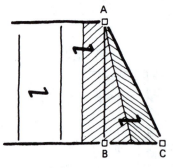

**Figure 10–4**  Theoretical end.

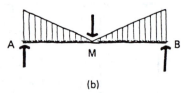

(b)

**Figure 10–5**  Tributary areas caused by a triangular opening.

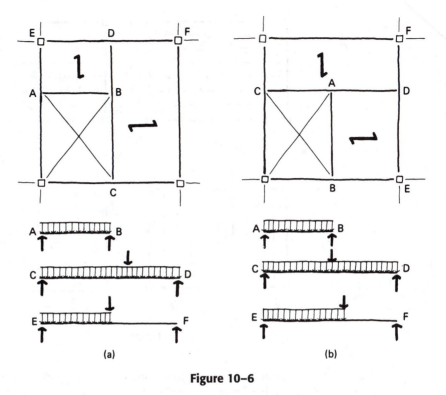

**Figure 10–6**

rial. For example, with reinforced concrete, it is very important not to interrupt the structure where continuity is required. In steel, it is important to avoid laterally unsupported spans which can require heavier sections.

Deck construction can also help to determine opening placement. In reinforced concrete slabs with beams framing between columns, an opening can be placed in the slab right next to a column. If there are no beams, as in a flat plate deck, all the material around a column is needed for shear and should not be interrupted.

## 10–6 FRAMING DIRECTION

One of the most frequently asked questions concerning framing refers to the optimum way to structure a rectangular bay. Figure 10–7 shows four possibilities without considering dimensions. In some cases, one or two possibilities may be ruled out by calling for unreasonably long or short spans for the material. For example, running the joists in the long direction as in Figure 10–7(a) would not be feasible in wood for spans more than about 20 ft (6 m). Common joist sizes are unavailable or expensive in longer lengths. Similarly, the long, heavily loaded beams of Figure 10–7(b) may not be feasible in reinforced concrete at distances greater than about 35 ft (11 m). Without prestressing, long concrete members become very deep.

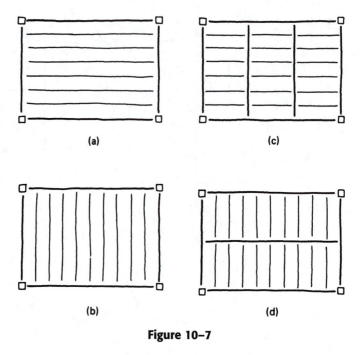

**Figure 10–7**

As mentioned in Section 10–5, one of the foremost determinants of framing direction should be the mechanical system. With a little advance planning, beam directions and large duct runs can be compatibly placed, reducing the overall finish ceiling to finish floor dimension. Smaller ducts and pipes can sometimes pass through holes in beams or girders at acceptable locations. Figure 10–8 shows the better locations for removal of material in a uniformly loaded simple beam, for example.

The writer's experience has been that many architects make the mechanical system "fit around" the structure by considering it too late in the design process. This is probably a mistake, as the structure should be made to compromise in the interest of total building design and economy. It should be recognized that there are alternative methods to structure a space, just as there are alternative ways to heat and cool it.

There is no universally correct way, from a structural standpoint, to frame such rectangular bays. It is easy to observe, however, that long, lightly loaded joists delivering their loads to shorter beams will lead to a uniformity of structural depth. Conversely, short joists creating heavy loads on the longer members will mean shallow joists and deep beams. It can also be noted that some of the load on certain

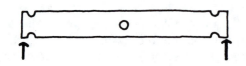

**Figure 10–8** Places of low stress in a uniformly loaded simple beam.

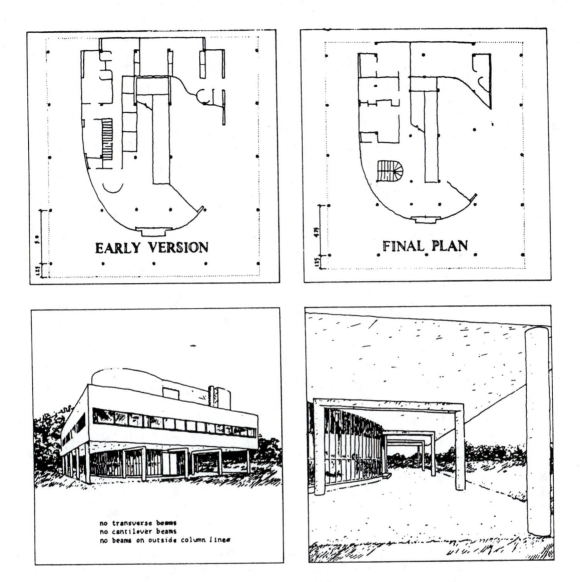

**EARLY VERSION**

**FINAL PLAN**

no transverse beams
no cantilever beams
no beams on outside column lines

Adaptability and cleverness in framing a design, whether in order to integrate structure properly with other building systems or to acknowledge, as a matter of principle, that a single perfect geometric system can seldom accommodate all the empirical demands of a given program, have been hallmarks of many good architects. It is instructive, for example, to look at the early and the as-built versions of Le Corbusier's Villa Savoye. Externally, the building presents a very simple structural system, a 4 × 4 m grid of equally spaced round columns, with cantilevers whose proportion to bay span is such that minimum moment is imparted to the end columns. In both the early and as-built versions, matters on the interior become less diagrammatic and more responsive. The early plan shows the central column line split to make room for a central ramp, and a column in the garage is simply dropped, but the house as a whole still presents an almost regular pattern. The Villa as built is more complicated, but the complexity is *systematic*. To cite a few items: A whole extra row of columns has been added inside, cantilever beams along the main axis of the house have been eliminated (presumably for the sake of a clearer volume above the piloti), and the external bay size has been lessened without a corresponding reduction of the cantilever. The final plan suggests that what appears to be a platonically ideal, two-way grid of columns and beams is actually a quite pragmatic one-way system, whose columns can be shifted and whose beams can be expressed or suppressed as circumstances, including aesthetic circumstances, demand.

beams can be reduced by one of the techniques of Figure 10–7(c) or (d), should this be beneficial. Such intermediate beams must be used, of course, whenever the deck or slab span becomes too great.

In "framing out" a building plan, it is helpful to remember the effects of span lengths and load types on the parameters that govern member size. Any increase in span almost always increases load by gathering more tributary area for the member, so the effects of load and span are difficult to differentiate. Assuming, for purposes of this discussion, however, that change in span does not necessarily mean change in load, we can state the following with respect to simple beams:

1. Increasing the span alone will cause a proportional change in moment, no change in shear, and will increase deflection as the cube of the relative span change.

2. Increasing a load while making no other changes will cause a proportional increase in moment, shear, and deflection.

3. Changing a uniform load to a concentrated one of the same magnitude will cause the moment to double, no change in shear, and a deflection increase of about 50%.

## 10–7   SELECTING WOOD BEAMS

Selecting member sizes of rectangular wood sections consists of:

1. Providing enough section modulus so that the allowable bending stress is not exceeded.

2. Providing enough cross-sectional area so that the allowable shearing stress is not exceeded.

3. Providing enough moment of inertia so that the permissible deflection is not exceeded.

The writer prefers to size the beam to meet the requirements for moment and then check to see that the area provided is sufficient for shear and that the moment of inertia provided is enough to control the deflection. (For a heavily loaded short span, where shear is likely to control, an alternative procedure of sizing for shear and then checking for moment and deflection might save time.)

---

**EXAMPLE 10–1**

Design the residential floor system of Figure 10–9. The live load is 40 psf and the total dead load, including an allowance for the self-weight of the members, is 10 psf. Use hem-fir No. 2 and assume that the deflection due to live load is limited to $L/360$.

(a) Select a typical joist.
(b) Design a lintel beam for the 10-ft opening using two elements nailed side by side of the same size as the joist.

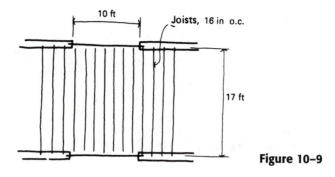

**Figure 10–9**

### Solution

(a) The total load = 40 + 10 = 50 psf. Since the joists are 16 in. on center, the load per linear foot on each joist will be

$$w = 50 \text{ psf} \left( \frac{16}{12} \text{ ft} \right) = 66.7 \text{ plf}$$

$$M = \frac{wL^2}{8} \qquad\qquad V = \frac{wL}{2}$$

$$= \frac{(66.7 \text{ lb/ft}) (17 \text{ ft})^2}{8} \qquad\qquad = \frac{(66.7 \text{ lb/ft}) (17 \text{ ft})}{2}$$

$$= 2410 \text{ lb-ft} \qquad\qquad = 567 \text{ lb}$$

As noted in Appendix H, joists usually qualify as repetitive members. Therefore, if we assume a trial section of 2 × 10, the tabled value of the allowable bending stress will be 935 psi. We will multiply it by 1.15 to provide the permitted repetitive member increase of 15%:

$$F_b = 935 (1.15) = 1095 \text{ psi}$$

Then

$$S_r = \frac{M}{F_b}$$

$$= \frac{(2410 \text{ lb-ft}) (12 \text{ in/ft})}{1075 \text{ psi}}$$

$$= 26.9 \text{ in}^3$$

Appendix I gives the $S_x$ value of a 2 × 10 as 21.4 in.$^3$, so we will have to try a 2 × 12, which has an $S_x$ value of 31.6 in.$^3$ From Appendix H, the new allowable stress is

$$F_b = 850 (1.15) = 977 \text{ psi}$$

Then

$$S_r = \frac{(2410 \text{ lb-ft}) (12 \text{ in/ft})}{977 \text{ psi}}$$

$$= 29.6 \text{ in}^3$$

Therefore, $2 \times 12$ will suffice for moment.

Checking shear, the required area will be

$$A_r = \frac{3V}{2F_v}$$

$$= \frac{3(567 \text{ lb})}{2(75 \text{ psi})}$$

$$= 11.3 \text{ in}^2$$

Since a $2 \times 12$ has an area of 16.9 in.$^2$, it will be OK in shear.

To check deflection we need the live load per foot.

$$w_{11} = 40 \text{ psf} \left( \frac{16}{12} \text{ ft} \right) = 53.3 \text{ plf}$$

From Appendix K, the maximum deflection will be

$$\Delta_{\max} = \frac{5wL^4}{384EI}$$

From Appendix H, $E = 1.3(10)^6$ psi, and from Appendix $I$, the moment of inertia is 178 in.$^4$.

$$\Delta_{11} = \frac{5(53.3 \text{ lb/ft}) (17 \text{ ft})^4 (12 \text{ in/ft})^3}{384[1.3(10)^6 \text{ psi}] (178 \text{ in}^4)}$$

$$= 0.43 \text{ in}$$

The permissible deflection due to live load is

$$\Delta_{\text{code}} = \frac{L}{360}$$

$$= \frac{17 \text{ ft} (12 \text{ in/ft})}{360}$$

$$= 0.57 \text{ in}$$

Therefore, the 2 × 12 will be OK in deflection.

(b) The discrete load points provided by the joists are sufficiently close to be considered as acting in a uniform manner upon each lintel. (This is a judgment call and is up to the designer. The writer prefers to treat point loads from repetitive members, such as joists, as being uniform when there are at least five such loads. When there are fewer, they should be treated as individual loads. At a larger scale, when long-span girders support a number of beams, it is best always to treat these as individual discrete loads.)

Since the tributary strip for either of the lintels is one-half the joist span or 8.5 ft, the uniform load per running foot on each lintel will be

$$w = 50 \text{ psf}(8.5 \text{ ft}) = 425 \text{ plf}$$

$$M = \frac{wL^2}{8} \qquad\qquad V = \frac{wL}{2}$$

$$= \frac{(425 \text{ lb/ft}) (10 \text{ ft})^2}{8} \qquad\qquad = \frac{(425 \text{ lb/ft}) (10 \text{ ft})}{2}$$

$$= 5310 \text{ lb-ft} \qquad\qquad = 2130 \text{ lb}$$

As per the problem statement, the lintels will be made up of 2 × 12 members nailed side by side. (In general, beams do not qualify as repetitive members.* Therefore, the allowable stress is *not* increased by 15%.) The allowable stress in bending will be the tabled value of 850 psi.

$$S_r = \frac{(5310 \text{ lb-ft}) (12 \text{ in/ft})}{850 \text{ psi}}$$

$$= 75.0 \text{ in}^3$$

Since each 2 × 12 provides a section modulus of 31.6 in.$^3$, the number needed to provide 75.0 in.$^3$ will be 75.0 ÷ 31.6 = 2.37. Therefore, three will be needed.

Checking shear, the area needed will be

$$A_r = \frac{3V}{2F_v}$$

---

* This is a preference of the writer. The National Design Specification permits the use of repetitive member bending stresses for built-up beams when there are *three or more* pieces and each piece runs the *full length* of the span.

$$= \frac{3(2130 \text{ lb})}{2(75 \text{ psi})}$$

$$= 42.6 \text{ in}^2$$

Using Appendix I again, each $2 \times 12$ provides an area of 16.9 in.$^2$, so *three* members will be needed for shear as well.

To check deflection, we will need the live load:

$$w_{11} = 40 \text{ psf}(8.5 \text{ ft}) = 340 \text{ plf}$$

The deflection will be

$$\Delta_{\text{max}} = \frac{5wL^4}{384EI}$$

The $I$ value will be $3 \times 178$ in.$^4$ = 534 in.$^4$.

$$\Delta_{11} = \frac{5(340 \text{ lb/ft}) \ (10 \text{ ft})^4 (12 \text{ in/ft})^3}{384[1.3(10)^6 \text{ psi}] \ (534 \text{ in}^4)}$$

$$= 0.11 \text{ in}$$

The code limit on live-load deflection is

$$\Delta_{\text{code}} = \frac{L}{360}$$

$$= \frac{10 \text{ ft}(12 \text{ in/ft})}{360}$$

$$= 0.33 \text{ in}$$

Therefore, the lintel will deflect only about one-third of the amount permitted.

---

**EXAMPLE 10–2M**

Design the garage roof system of Figure 10–10M. The live load is 1.5 kN/m$^2$ and the total dead load, including an allowance for the self-weight of the members, is 0.5 kN/m$^2$. Use hem-fir No. 2 and assume that the deflection due to live load is limited to $L/240$.

(a) Select a typical joist.
(b) Design a lintel beam for one of the 3 m openings using 38-mm-wide members nailed side by side.

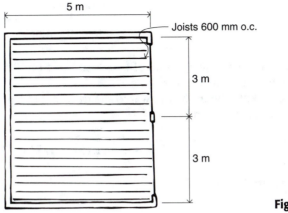

5 m

Joists 600 mm o.c.

3 m

3 m

**Figure 10–10M**

### Solution

(a) The total load $= 1.5 + 0.5 = 2.0$ kN/m². Since the joists are 0.6 m on center, the load per linear meter on each joist will be

$$w = 2.0 \text{ kN/m}^2(0.6 \text{ m}) = 1.2 \text{ kN/m}$$

$$M = \frac{wL^2}{8} \qquad\qquad V = \frac{wL}{2}$$

$$= \frac{(1.2 \text{ kN/m}) (5 \text{ m})^2}{8} \qquad\qquad = \frac{(1.2 \text{ kN/m}) (5 \text{ m})}{2}$$

$$= 3.75 \text{ kN} \cdot \text{m} \qquad\qquad = 3 \text{ kN}$$

As noted in Appendix H, joists usually qualify as repetitive members. Therefore, if we assume a trial section of 38 × 285, the tabled value of the allowable bending stress will be 5860 kPa. We will multiply it by 1.15 to provide the permitted repetitive member increase of 15%. We can also increase it by another 15% if we assume the live load on the roof to be of short duration. (See Appendix H again.)

$$F_b = 5860(1.15) \, (1.15) = 7750 \text{ kPa}$$

Then

$$S_r = \frac{M}{F_b}$$

$$= \frac{3.75 \text{ kN} \cdot \text{m}}{7750 \text{ kN/m}^2}$$

$$= 0.000\ 484\ \text{m}^3$$

$$= 484(10)^3\ \text{mm}^3$$

Appendix I gives the $S_x$ value of a 38 × 285 as $514(10)^3\ \text{mm}^3$, so the trial section is adequate in bending.

Checking shear, the allowable shearing stress can also be increased for snow. The $F_v$ value will then be

$$F_v = 500(1.15) = 575\ \text{kPa}$$

The required area will be

$$A_r = \frac{3V}{2F_v}$$

$$= \frac{3(3\ \text{kN})}{2\ (575\ \text{kN/m}^2)}$$

$$= 0.007\ 83\ \text{m}^2$$

$$= 7830\ \text{mm}^2$$

Since a 38 × 285 has an area of 10 800 $\text{mm}^2$, it will be more than adequate in shear.

To check deflection, we need the live load per meter:

$$w_{11} = (1.5\ \text{kN/m}^2)(0.6\ \text{m}) = 0.9\ \text{kN/m}$$

From Appendix K, the maximum deflection will be

$$\Delta_{\text{max}} = \frac{5wL^4}{384EI}$$

From Appendix H, $E = 9000$ MPa, and from Appendix I, the moment of inertia is $73.3(10)^6\ \text{mm}^4$.

$$\Delta_{11} = \frac{5(0.9\ \text{kN/m})\ (5\ \text{m})^4}{384[(9000)(10)^3\ \text{kN/m}^2](73.3)(10)^6\ \text{mm}^4}$$

$$= 0.011\ \text{m}$$

$$= 11\ \text{mm}$$

The permissible deflection due to live load is

$$\Delta_{code} = \frac{L}{240}$$

$$= \frac{5000 \text{ mm}}{240}$$

$$= 21 \text{ mm}$$

Therefore, the $38 \times 285$ will be OK in one deflection.

(b)  Refer to the note in part (b) of the solution to Example 10–1.

Since the tributary strip for either of the lintels is one-half the joist span or 2.5 m, the uniform load per running meter on each lintel will be

$$w = (2.0 \text{ kN/m}^2) \, (2.5 \text{ m}) = 5 \text{ kN/m}$$

$$M = \frac{wL^2}{8} \qquad\qquad V = \frac{wL}{2}$$

$$= \frac{5 \text{ kN/m}(3 \text{ m})^2}{8} \qquad\qquad = \frac{5 \text{ kN/m } (3 \text{ m})}{2}$$

$$= 5.63 \text{ kN} \cdot \text{m} \qquad\qquad = 7.5 \text{ kN}$$

In general, beams do not qualify as repetitive members. (See the footnote in Example 10–1.) The allowable stresses will be increased for duration of load, however. Select two $38 \times 285$s nailed together as a trial section. Using Appendix H, we get

$$S_r = \frac{M}{F_b}$$

$$= \frac{5.63 \text{ kN} \cdot \text{m}}{5860(1.15) \text{ kN/m}^2}$$

$$= 0.000 \, 835 \text{ m}^3$$

$$= 835(10)^3 \text{ mm}^3$$

From Appendix I, each $38 \times 285$ provides a section modulus of $514(10)^3 \text{ mm}^3$, so two will suffice for the lintel for bending. (The reader should verify that two $38 \times 235$s will almost suffice.)

Checking shear, the area needed will be

$$A_r = \frac{3V}{2F_v}$$

$$= \frac{3(7.5 \text{ kN})}{2(500 \text{ kN/m}^2)(1.15)}$$

$$= 0.0196 \text{ m}^2$$

$$= 19\,600 \text{ mm}^2$$

Since each $38 \times 285$ provides $10\,800 \text{ mm}^2$, two will be needed for shear. To check deflection, the live load will be

$$w_{11} = 1.5 \text{ kN/m}^2 \ (2.5 \text{ m}) = 3.75 \text{ kN/m}$$

The deflection will be

$$\Delta_{max} = \frac{5\,wL^4}{384EI}$$

The $I$ value will be $2 \times 73.3(10)^6 \text{ mm}^4 = 147(10)^6 \text{ mm}^4$.

$$\Delta_{11} = \frac{5(3.75 \text{ kN/m}) \ (3 \text{ m})^4}{384[(9000)(10)^3 \text{ kN/m}^3] \ 147(10)^{-6} \text{ m}^4}$$

$$= 0.003 \text{ m}$$

$$= 3 \text{ mm}$$

The code limit on line load deflection is

$$\Delta_{code} = \frac{L}{240}$$

$$= \frac{3000 \text{ mm}}{240}$$

$$= 12 \text{ mm}$$

Therefore, the lintel will have no deflection problems.

---

**EXAMPLE 10–3**

Design the roof members of Figure 10–11 using Douglas fir No. 2. The snow load is 25 psf, acting on the horizontal projection, and the dead load is 15 psf along the slope. Assume that an allowance has been included in the dead load for member self-weight. The arrows in the plan sketch indicate the direction of the deck spanning from beam to beam.

(a) Select a nominal $4 \times$ section for the sloping beams, assuming that deflection may be ignored.

(b) Select a nominal 6 × for the girder *AB*. Assume that the total deflection must not exceed *L*/180.

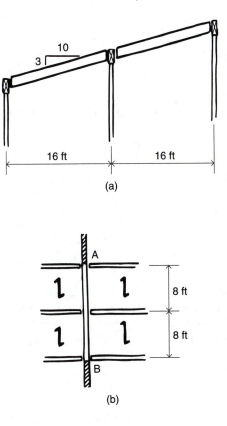

**Figure 10–11**  (a) Section; (b) plan.

### Solution

(a) Sloping roof beams are usually designed like horizontal members (i.e., taking the *span* as the *horizontal projection* rather than the sloping length).

Snow loads are always given as acting on the horizontal projection, but the dead load must be converted from the load along the sloped length to load acting on the horizontal projection, as explained in Section 2–11. Then the dead and live loads may be added to get the total load. (This procedure will result in accurate moment values and slightly conservative shearing values, and is much simpler than working with the sloped length, which would mean converting loads so they act perpendicularly and parallel to the member.)

The 15-psf dead load along the slope will be slightly larger when converted so it acts on the horizontal projection. This effect increases with the magnitude of the slope. In this case the hypotenuse of the 3-on-10 slope can be found, using the Pythagorean theorem, as 10.44 ft. The horizontal projection is related to the hypotenuse as the cosine of the roof slope, as shown in Figure 10–12. The projected load will be 15 psf ÷ cos θ or 15 psf ÷ (10.0/10.44) = 15.7 psf. The total load on the horizontal projection is then 25 psf live plus 15.7 psf dead for a total of 40.7 psf.

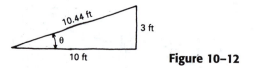

**Figure 10–12**

Since the tributary strip for a typical beam is 8 ft, the uniform load per foot will be

$$w = 40.7 \text{ psf}(8 \text{ ft}) = 326 \text{ plf}$$

$$M = \frac{wL^2}{8} \qquad\qquad V = \frac{wL}{2}$$

$$= \frac{(326 \text{ lb/ft}) (16 \text{ ft})^2}{8} \qquad\qquad = \frac{(326 \text{ lb/ft}) (16 \text{ ft})}{2}$$

$$= 10\ 400 \text{ lb-ft} \qquad\qquad = 2610 \text{ lb}$$

Let us select a trial section of 4 × 16. Appendix H gives us a base value of 875 psi. Since these beams are far apart, they do not qualify as repetitive members. However, Appendix H indicates that when snow is involved as the controlling design load, the tabled values can be increased 15% due to the reduced duration of the load. Therefore,

$$F_b = 875(1.15) = 1006 \text{ psi}$$

(As indicated by the factors, wood exhibits increased strength as the loading period becomes shorter. Had these members also been repetitive, two 15% increases could have been applied cumulatively!) Using the modified allowable bending stress,

$$S_r = \frac{M}{F_b}$$

$$= \frac{(10\ 400 \text{ lb-ft}) (12 \text{ in/ft})}{1006 \text{ psi}}$$

$$= 124 \text{ in}^3$$

From Appendix I, a 4 × 16 was a good trial choice because $S_x = 136$ in.$^3$.

Checking shear, the permissible stress will be 95 psi times 1.15 (15% increase for duration of load) = 109 psi. So the required area for shear is

$$A_r = \frac{3V}{2F_v}$$

$$= \frac{3(2610 \text{ lb})}{2(109 \text{ psi})}$$

$$= 35.9 \text{ in}^2$$

The area of a 4 × 16 is 53.4 in.$^2$, so shear will not be a problem. As specified in the problem statement, deflection need not be calculated. (This is often done on roofs with significant slopes since "ponding" will not be a problem.)

(b) As shown in Figure 10–13, girder $AB$ is loaded by a single concentrated load at midspan, which is equal in magnitude to two beam reactions. In such cases all the load on the girder is assumed to come into it from the beams, even that from the small strip of roof directly above it.

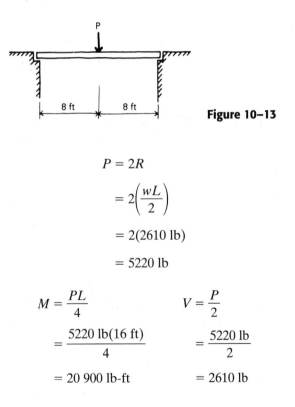

**Figure 10–13**

$$P = 2R$$

$$= 2\left(\frac{wL}{2}\right)$$

$$= 2(2610 \text{ lb})$$

$$= 5220 \text{ lb}$$

$$M = \frac{PL}{4} \qquad\qquad V = \frac{P}{2}$$

$$= \frac{5220 \text{ lb}(16 \text{ ft})}{4} \qquad\qquad = \frac{5220 \text{ lb}}{2}$$

$$= 20\,900 \text{ lb-ft} \qquad\qquad = 2610 \text{ lb}$$

A 6 × 16 will be tried for the girder. From Appendix H, the allowable stress is 850 psi. For the snow-load condition,

$$F_b = 850 \text{ psi}(1.15) = 978 \text{ psi}$$

Then

$$S_r = \frac{M}{F_b}$$

$$= \frac{(20\ 900\ \text{lb-ft})\ (12\ \text{in/ft})}{978\ \text{psi}}$$

$$= 256\ \text{in}^3$$

Since a $6 \times 16$ has a section modulus of only 220 in.$^3$, a $6 \times 18$ having $S_x = 281$ in.$^3$ will be tried. The new allowable stress will be

$$F_b = 839\ \text{psi}(1.15) = 965\ \text{psi}$$

Then

$$S_r = \frac{(20\ 900\ \text{lb-ft})\ (12\ \text{in/ft})}{965\ \text{psi}}$$

$$= 260\ \text{in}^3$$

So the $6 \times 18$ will be OK in moment.

Checking shear, the permissible stress will be 85 psi times 1.15 or 98 psi. Therefore,

$$A_r = \frac{3V}{2F_v}$$

$$= \frac{3(2610\ \text{lb})}{2(98\ \text{psi})}$$

$$= 39.9\ \text{in}^2$$

Since the $6 \times 18$ has an area of 96.3 in.$^2$, shear is easily handled.

The deflection can be checked by using Case 3 from Appendix K. We can get the $E$ value from Appendix H and the $I$ value from Appendix I.

$$\Delta_{\text{max}} = \frac{PL^3}{48EI}$$

$$= \frac{5220\ \text{lb}(16\ \text{ft})^3\ (12\ \text{in/ft})^3}{48[1.3(10)^6\ \text{psi}]\ (2456\ \text{in}^4)}$$

$$= 0.24\ \text{in}$$

As specified, the permissible deflection is

$$\Delta_{code} = \frac{L}{180}$$

$$= \frac{16 \text{ ft}(12 \text{ in/ft})}{180}$$

$$= 1.07 \text{ in}$$

So the deflection limit will not be close.

**EXAMPLE 10–4M**

A gabled roof system with collar beams is shown in Figure 10–14M. Design (a) the rafters and (b) the supporting lintels. Use 139 x members of southern pine No. 2. The snow load is 1.0 kN/m² and the dead load is 0.5 kN/m² along the slope. Assume that an allowance has been included for member self-weight. Assume that deflection will not control.

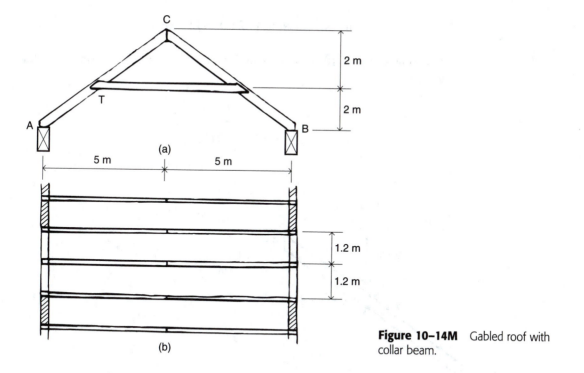

**Figure 10–14M** Gabled roof with collar beam.

### Solution

(a) As explained in Section 2–11, the dead load along the slope can be converted to load on the horizontal projection by dividing by the cosine of the angle of the slope. Using Figure 10–15M, we get

$$DL = \frac{0.5 \text{ kN/m}^2}{5/6.4}$$

$$= 0.64 \text{ kn/m}^2$$

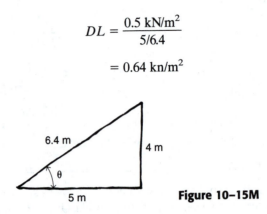

**Figure 10–15M**

Adding this to the snow load, we get 0.64 kN/m² plus 1.0 kN/m² for a total load on the horizontal projection of 1.64 kN/m². Since the rafters are placed 1.2 m on center, the load per meter will be

$$w = 1.64 \text{ kN/m}^2 \ (1.2 \text{ m}) = 1.97 \text{ kN/m}$$

Referring to Figure 10–16M, we can solve for the collar beam force $T$ by taking moments about $C$:

$$\Sigma M_c = 0$$

$$- 9.8(5) + 9.8(2.5) + T(2) = 0$$

$$T = 12.3 \text{ kN}$$

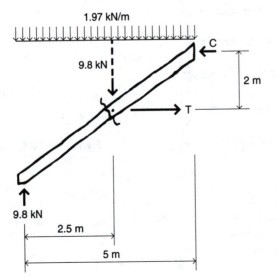

**Figure 10–16M**  Free-body diagram.

The maximum moment will be at midspan. From Figure 10–17M, we get

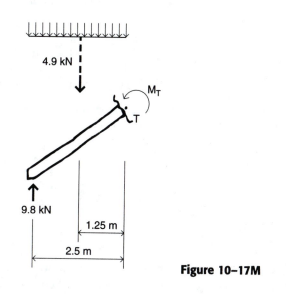

**Figure 10–17M**

$$\Sigma M_T = 0$$

$$-9.8(2.5) + 4.9(1.25) + M_T = 0$$

$$M_T = 18.4 \text{ kN} \cdot \text{m}$$

Try a 139 × 390 section. From Appendix H, the allowable stress will be 5700 kPa. (The rafters are too far apart to be considered as repetitive members, but $F_b$ will be modified for snow.)

$$F_b = 5700(1.15) \qquad = 6550 \text{ kPa}$$

$$S_r = \frac{18.4 \text{ kN} \cdot \text{m}}{6550 \text{ kN/m}^2}$$

$$= 0.002 \ 81 \text{ m}^3$$

$$= 2810(10)^3 \text{ mm}^3$$

From Appendix I, the 139 × 390 proved to be a good choice because $S_x = 3520(10)^3$ mm$^3$. (The student should verify that a 139 × 340 section will *not* be adequate.)

Checking shear, the permissible stress will be 689 kPa times 1.15 (15% increase for duration of load) = 792 kPa. The required area for shear is then

$$A_v = \frac{3V}{2F_v}$$

$$= \frac{3(9.8 \text{ kN} \cdot \text{m})}{2(792 \text{ kN/m}^2)}$$

$$= 0.0186 \text{ m}^2$$

$$= 18\ 600 \text{ mm}^2$$

Since the area of a 139 × 390 is 54 200 mm², shear will not be a problem.

(b) Each lintel beam must support a concentrated load of 9.8 kN (the rafter reaction) at its midspan.

$$M = \frac{PL}{4} \qquad\qquad V = \frac{P}{2}$$

$$= \frac{9.8 \text{ kN} (2.4 \text{ m})}{4} \qquad\qquad = \frac{9.8 \text{ kN}}{2}$$

$$= 5.88 \text{ kN} \cdot \text{m} \qquad\qquad = 4.9 \text{ kN}$$

Try a 139 × 240. From Appendix H, the allowable stress will be 5860 kPa. As we substitute in the equation, we will increase the value by 15% for snow, and then

$$S_r = \frac{M}{F_b}$$

$$= \frac{5.88 \text{ kN} \cdot \text{m}}{(5860 \text{ kN/m}^2)(1.15)}$$

$$= 0.000\ 873 \text{ m}^3$$

$$= 873(10)^3 \text{ mm}^3$$

This will be OK, because from Appendix I, the section modulus provided is 1330(10)³ mm³. (There is no reason to check out a 139 × 190 because the allowable stress will be the same. It has an $S_x$ value of only 836 (10)⁶ mm³.)

Checking shear, we find $F_v$ from Appendix H to be 689 kPa. The area required will be

$$A_r = \frac{3V}{2F_v}$$

$$= \frac{3(4.9 \text{ kN})}{2(689 \text{ kN/m}^2)(1.15)}$$

$$= 0.009\ 28 \text{ m}^2$$

$$= 9280 \text{ mm}^2$$

Shear is no problem, since the area provided by a 139 × 240 is 32 700 mm².

## PROBLEMS

**10–1.** Design the members for the balcony floor in Figure 10–18. The live load is 30 psf and the dead load is 10 psf. Use southern pine No. 2 members of nominal 2-in. thickness. Two or more should be used side by side to make the header and lintel. The maximum depth allowed is a nominal 10 in., and the live load deflection is limited to $L/300$ by local code.

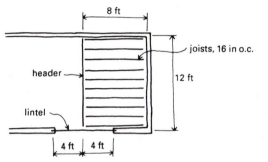

**Figure 10–18** Balcony framing.

**10–1M.** Design the members for the balcony floor in Figure 10–18M. The live load is 1.5 kN/m² and the dead load is 0.5 kN/m². Use hem-fir members of 38-mm stock. Two or more side by side may be used to make beams as needed. The maximum depth allowed is 235 mm, and the live load deflection is limited to $L/300$ by local code.

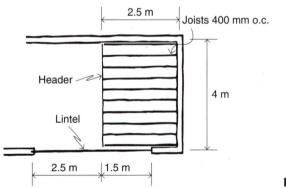

**Figure 10–18M** Balcony framing.

**10–2.** Beams $AB$ shown in Figure 10–19 are spaced 4 ft apart. The girder at $C$ is 12 ft long and supports two beams at its third points. Select a 4 × member suitable for the beams and a 6 × for the girder. The snow load is 20 psf on the horizontal projection and the dead load is 10 psf along the slope, which includes an allowance for the structural members. Assume that deflection will not be a concern. Use Douglas fir No. 2.

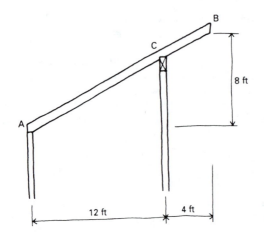

Figure 10-19    Timber shed roof.

**10-2M.** Beams *AB* shown in Figure 10–19M are spaced 1 m apart. The girder at *C* is 3 m long and supports two of the beams at its third points. Select an 89-mm member for the beams and a 139-mm member for the girder. The snow load is 1.2 kN/m$^2$ and the dead load is 0.8 kN/m$^2$, both taken on the horizontal projection. Assume that deflection is not a concern for the beam but that the total girder deflection should be limited to *L*/180. Use Douglas fir No. 2.

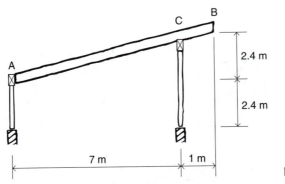

Figure 10-19M    Timber shed roof.

**10-3.** With reference to Figure 10–20, size the joists (for the 15-ft span) and determine how many 2 × 10s are needed for the built-up beam *AB*. The floor load is 40 psf and the dead load is 15 psf, including all self-weight. Use southern pine No. 2, and assume a midspan deflection limit of *L*/240 due to total loads. (*Hint*: Obtain the deflection for the beam by assuming three equal point loads. Case 3 and 15 of Appendix K may be combined. Use the *actual* point load values when considering shear and moment.)

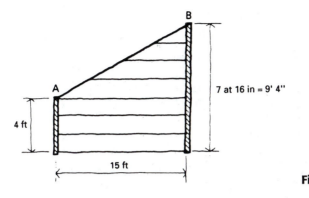

**Figure 10–20**   Floor framing plan.

**10–3M.** With reference to Figure 10–20M, size the joists (for the 4.5-m span) and determine how many $38 \times 285$ members are required for the beam $AB$. The snow load is 1.5 kN/m² and the dead load is 0.65 kN/m². Use southern pine No. 2 since "ponding" might be a problem, and, assume a total deflection limit of $L/480$. (*Hint*: Obtain the deflection for the beam by assuming three equal point loads. Cases 3 and 15 of Appendix K may be combined. Use the *actual* point load values when considering shear and moment.)

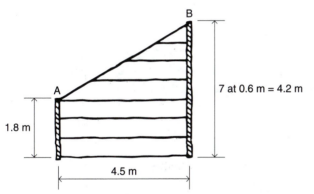

**Figure 10–20M**   Roof framing plan.

**10–4.** Figure 10–21 shows a schematic floor plan with a stairwell opening. Locate all joists and beams and size them using $2 \times$ pieces. No posts or walls may be added, but beams may be built up of parallel members, if needed. The live load is 30 psf and the total dead load is 10 psf. Joists are to be no farther apart than 16 in., and the live load deflection must be less than $L/360$. The dashed line represents the floor boundary. Use southern pine No. 2.

**10–4M.** Figure 10–21M shows a schematic floor plan with a stairwell opening. Locate all joists and beams and size them using 38-mm pieces. No posts or walls may be added, but beams may be built up of parallel members, if needed. The live load is 2.0 kN/m² and the total dead load is 0.5 kN/m². Joists are to be no farther apart than 400 mm, and live load deflection must be less than $L/360$. The dashed line represents the floor boundary. Use Douglas fir No. 2 throughout.

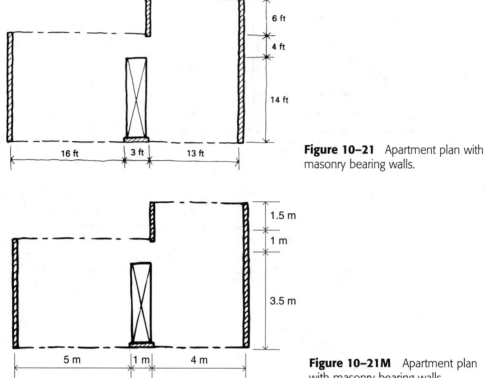

**Figure 10–21**  Apartment plan with masonry bearing walls.

**Figure 10–21M**  Apartment plan with masonry bearing walls.

## 10–8   SELECTING STEEL BEAMS

Selecting steel wide-flange beams is a simple matter of:

1. Providing enough section modulus so that the allowable bending stress is not exceeded.

2. Providing enough moment of inertia so that the permissible deflection is not exceeded.

Most designers select a trial size based on moment and then check deflection. Steel is so strong in shear that only rarely will this be a design concern. When shear is very high, for example when large concentrated loads exist near supports, web stiffeners are placed to keep the web from buckling (see Figure 8–5) or a beam is selected of the same overall dimensions but having a thicker web. Shear rarely causes us to use a larger beam as it sometimes does in wood.

As described in Section 7–5, narrow steel beams are prone to lateral buckling, and it is important to provide adequate lateral support for the compression flange. In the examples and problems of this section, adequate lateral support will be assumed to be provided by the floor deck. When this is the case the allowable bending stress can be taken as two-thirds of the yield stress.

There are fewer careless error possibilities when doing calculations in steel because all rolled steel has the same modulus of elasticity $[E = 29(10)^6 \text{ psi}]$ $(E = 200 \text{ GPa})$ regardless of its strength and because there are only two strengths of steel commonly used in building construction, $F_y = 36$ ksi (250 MPa) and $F_y = 50$ ksi (345 MPa). Some cross-sectional properties for only a few of the hundreds of different steel sections produced in the United States are tabled in Appendix J. As mentioned, the initial procedure involves selecting a member with enough section modulus. When selecting from a table such as that of Appendix J, one should notice that deeper sections are often lighter while providing more section modulus. For the beam shapes this has been emphasized by boldface type for the first member of each group. It is the deepest of its group and the strongest. It is also the lightest and is therefore the least expensive. (Steel is sold by the pound or kilogram.) These sections would be selected over the others except when factors such as mechanical, architectural, or construction detailing considerations would indicate otherwise. Appendix J serves as a list of available steel sections for use in the following examples and problems.

---

**EXAMPLE 10–5**

The floor system of Figure 10–22 consists of lightweight concrete on a steel deck spanning to beams spaced 10 ft o.c. It supports a live load of 60 psf and a dead load of 40 psf, which includes an allowance for the slab and the structural members. The deflection due to the live load is limited to $L/360$. Select the lightest A36 W shapes for typical members B–1 and G–1.

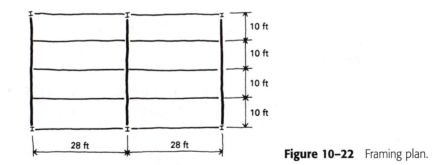

**Figure 10–22** Framing plan.

### Solution

Beam B–1 supports a tributary strip 10 ft long. The uniform load $w$ per linear foot can be obtained by multiplying the total load per square foot by 10 ft.

$$w = 100 \text{ psf}(10 \text{ ft}) = 1000 \text{ plf} = 1 \text{ klf}$$

The maximum moment is

$$M = \frac{wL^2}{8}$$

$$= \frac{(1 \text{ kip/ft}) (28 \text{ ft})^2}{8}$$

$$= 98 \text{ kip-ft}$$

As mentioned, if full lateral support is provided for the compression flange of the beam (in this case by the steel deck, which would normally be welded to the tops of the beams), the allowable bending stress is two-thirds of yield, or, for A36 steel, 24 ksi. Then the required section modulus can be found as

$$S_r = \frac{M}{F_b}$$

$$= \frac{98 \text{ kip-ft}(12 \text{ in/ft})}{24 \text{ ksi}}$$

$$= 49 \text{ in}^3$$

From Appendix J, a W18 × 35 is seen to be the lightest adequate one. Its $I$ value is 510 in.[4]. To check the live load deflection, we determine the live load as

$$w_{11} = 60 \text{ psf}(10 \text{ ft}) = 600 \text{ plf} = 0.6 \text{ klf}$$

From Appendix K, the maximum deflection will be

$$\Delta_{\text{max}} = \frac{5wL^4}{384EI}$$

The live-load deflection will be

$$\Delta_{11} = \frac{5(0.6 \text{ kip/ft}) (28 \text{ ft})^4 (1000 \text{ lb/kip}) (12 \text{ in/ft})^3}{384[29(10)^6 \text{ psi}] (510 \text{ in}^4)}$$

$$= 0.56 \text{ in}$$

The code limitation was given as

$$\Delta_{\text{code}} = \frac{L}{360}$$

$$= \frac{28 \text{ ft}(12 \text{ in/ft})}{360}$$

$$= 0.93 \text{ in}$$

The beam is well within the code limitations. Figure 10–23 shows the tributary strip and area for one of the point loads on the girder. In such cases it is customary to assume that there is no uniform load on the girder and that the load enters the girder from the beam reactions, even that from the floor directly above the girder and the allowance for its self-weight. The magnitude of each point load is equal to two beam reactions. Each reaction is, of course, equal to the uniform load per linear foot times one-half the beam span.

$$P = 2R$$

$$= 2\left(\frac{wL}{2}\right)$$

$$= 2\left[\frac{(1 \text{ kip/ft})(28 \text{ ft})}{2}\right]$$

$$= 28 \text{ kips}$$

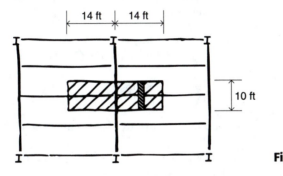

Figure 10–23 shows the tributary strip dimensions: 14 ft, 14 ft, and 10 ft.

**Figure 10–23**  Tributary area.

This value could also be obtained by multiplying the total load per square foot times the tributary area.

$$P = 0.1 \text{ ksf}(10 \text{ ft})(28 \text{ ft})$$

$$= 28 \text{ kips}$$

Figure 10–24 shows the loaded girder and its moment diagram. The maximum moment is 560 kip-ft.

The required section modulus is

$$S_r = \frac{M}{F_b}$$

$$= \frac{560 \text{ kip-ft}(12 \text{ in/ft})}{24 \text{ ksi}}$$

$$= 280 \text{ in}^3$$

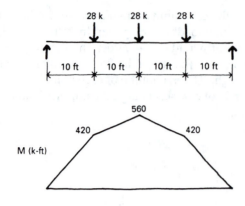

**Figure 10–24**   Moment diagram for the central girder.

From Appendix J, a W30 × 90 will be the lightest of those listed. Its $I$ value is 3620 in.[4]. To check the live load deflection, we find that since the live load given was 60 psf, the portion of each point load that is live is

$$P_{11} = 0.060 \text{ ksf}(10 \text{ ft})(28 \text{ ft})$$

$$= 16.8 \text{ kips}$$

(Since 60% of the total load was live, we would, of course, get the same value by taking 60% of 28 kips, the total load.)

Figure 10–25 illustrates how we can combine Cases 3 and 15 from Appendix K to get the deflection. The live load deflection is then

$$\Delta_{11} = \frac{PL^3}{48EI} + \frac{Pa}{24EI}(3L^2 - 4a^2)$$

$$= \left[ \frac{16.8(40)^3}{48} + \frac{16.8(10)[3(40)^2 - 4(10)^2]}{24} \right] \left[ \frac{(1000 \text{ lb/kip}) (12 \text{ in/ft})^3}{[29(10)^6 \text{ psi}] (3620 \text{ in}^4)} \right]$$

$$= 0.88 \text{ in}$$

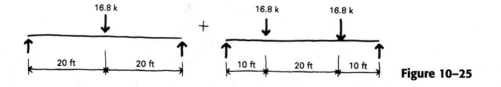

**Figure 10–25**

The code limitation is

$$\Delta_{code} = \frac{L}{360}$$

$$= \frac{40 \text{ ft}(12 \text{ in/ft})}{360}$$

$$= 1.33 \text{ in}$$

As with the beam, there will be no deflection problem.

---

**EXAMPLE 10–6M**

Select a steel W shape for the girder designated as *CD* in Figure 10–26M. The live load is 4 kN/m$^2$ and the dead load is 1 kN/m$^2$, which is sufficient to include the structural self-weight. The arrows indicate the direction of the bar joist floor system. Use $F_y$ = 250 MPa steel and a deflection limitation due to live load of $L/360$.

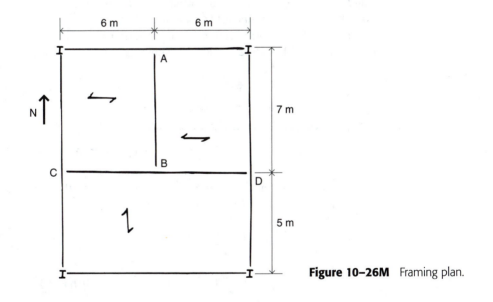

**Figure 10–26M**  Framing plan.

**Solution**

The total load is 5 kN/m$^2$. The allowable stress is $\frac{2}{3}F_y$ = 165 MPa, assuming that full lateral support is provided by the bar joists. The girder *CD* will carry a point load (the reaction at *B* from beam *AB*) at its midspan. It will also carry a uniform load from the floor area just south of it. These tributary areas are shown in Figure 10–27M.

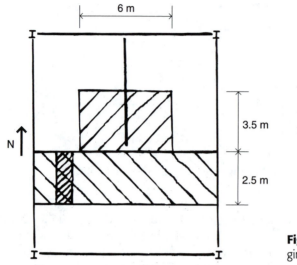

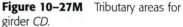

**Figure 10–27M** Tributary areas for girder *CD*.

The reaction at *B* for beam *AB* will result from the load acting upon the shaded area that is 3.5 m by 6 m in plan. The magnitude of this reaction will be the load per square meter times this area or 5 kN/m² times 21 m² = 105 kN. The tributary strip for the uniform load is 2.5 m long; therefore, the uniform load on girder *CD* will be $w = 5$ kN/m²(2.5 m) = 12.5 kN/m.

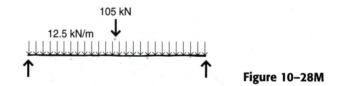

**Figure 10–28M**

The girder will then be loaded as shown in Figure 10–28M. The maximum moment from the combined loading is

$$M = \frac{PL}{4} + \frac{wL^2}{8}$$

$$= \frac{(105 \text{ kN})(12\text{m})}{4} + \frac{(12.5 \text{ kN/m})(12\text{m})^2}{8}$$

$$= 315 \text{ kN} \cdot \text{m} + 225 \text{ kN} \cdot \text{m}$$

$$= 540 \text{ kN} \cdot \text{m}$$

$$S_r = \frac{M}{F_b}$$

$$= \frac{540 \text{ kN} \cdot \text{m}}{165\,000 \text{ kN/m}^2}$$

$$= 0.003\,270 \text{ m}^3$$

$$S_r = 3270(10)^3 \text{ mm}^3$$

The lightest adequate section from Appendix J is a W690 × 125. To check this for deflection, the portion of the total load that is live can be obtained by ratio.

$$\frac{4}{5}(105 \text{ kN}) = 84 \text{ kN}$$

$$\frac{4}{5}(12.5 \text{ kN/m}) = 10 \text{ kN/m}$$

From Appendix K,

$$\Delta_{\text{max}} = \frac{PL^3}{48EI} + \frac{5\,wL^4}{384EI}$$

From Appendix J, the W690 × 125 has an $I$ value of $1190(10)^6 \text{ mm}^4$.

$$\Delta_{\text{max}} = \frac{(84 \text{ kN})\,(12 \text{ m})^3}{48[200(10)^6 \text{ kN/m}^2]\,[1190(10)^{-6} \text{ m}^4]}$$

$$+ \frac{5(10 \text{ kN/m})(12 \text{ m})^4}{384[200(10)^6 \text{ kN/m}^2]\,[1190(10)^{-6} \text{ m}^4]}$$

$$= 0.0127 \text{ m} + 0.0113 \text{ m}$$

$$= 0.024 \text{ m}$$

$$= 24 \text{ mm}$$

The code limit is

$$\Delta_{\text{code}} = \frac{L}{360}$$

$$= \frac{12\,000 \text{ mm}}{360}$$

$$= 33 \text{ mm}$$

Therefore, the girder meets the deflection criterion.

**EXAMPLE 10-7**

Select a W shape for the girder $AB$ in Figure 10–29. The live load on the floor is 100 psf and the dead load, including all structural elements, is 40 psf. There is a 10-ft-high masonry wall which bounds the opening from $A$ to $C$ to $D$. The wall weighs 80 lb per square foot of elevation. Assume that the total load deflection is limited to $L/240$ at any point and that the beam depth is limited to a nominal 18 in. Use high-strength steel having $F_y = 50$ ksi.

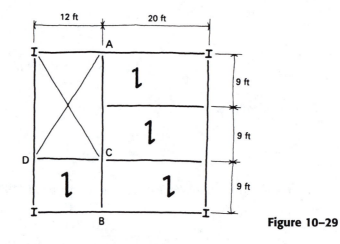

**Figure 10–29**

### Solution

The girder $AB$ is loaded by two unequal concentrated loads at the third points plus a uniform load from that portion of the wall from $A$ to $C$. The total floor load is 140 psf. The wall load per linear foot is 80 psf times the wall height.

$$w_{wall} = 80 \text{ psf}(10 \text{ ft}) = 800 \text{ plf}$$

The beam $DC$ carries the wall directly above it, plus a 4.5-ft tributary strip of floor. Its uniform load is

$$W_{DC} = 800 \text{ plf} + 140 \text{ psf}(4.5 \text{ ft}) = 1430 \text{ plf} = 1.43 \text{ klf}$$

As shown in Figure 10–30(a), one-half of the load on $DC$ is delivered as a point load on girder $AB$ at point $C$. This is actually the reaction at $C$ for beam $DC$ and is given as

$$(1.43 \text{ kips/ft})6 \text{ ft} = 8.6 \text{ kips}$$

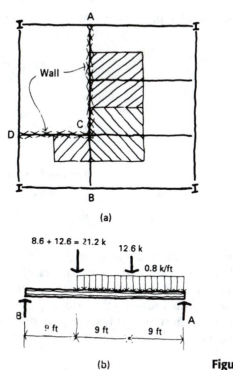

(a)

8.6 + 12.6 = 21.2 k    12.6 k

0.8 k/ft

B    A

9 ft    9 ft    9 ft

(b)    **Figure 10–30**

The two beams coming into $AB$ from the right each have a total tributary area of 9 ft times 20 ft, or 180 ft$^2$. One-half of each beam's load [represented by the shaded area in Figure 10–30(a)] will be delivered to $AB$, so each of these reactions will be

$$\tfrac{1}{2} (180 \text{ ft})^2 (140 \text{ psf}) = 12\,600 \text{ lb} = 12.6 \text{ kips}$$

The final loading pattern on the girder $AB$ is given to Figure 10–30(b). The only uniform load on the girder is from the wall. Figure 10–31 provides the $V$ and $M$ diagrams.

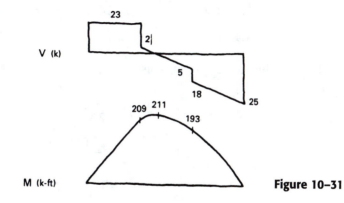

Figure 10–31

If full lateral support is provided by the floor deck, the allowable bending stress will be

$$F_b = \tfrac{2}{3} F_y = \tfrac{2}{3}(50 \text{ ksi}) = 33 \text{ ksi}$$

The required section modulus will be

$$S_r = \frac{M}{F_b}$$

$$= \frac{211 \text{ kip-ft}(12 \text{ in/ft})}{33 \text{ ksi}}$$

$$= 77 \text{ in}^3$$

From Appendix J, the lightest beam (if the depth must be less than or equal to a nominal 18 in.) is a W18 × 46. It has an $I$ value of 712 in.[4].

To make an approximate check on the maximum deflection (as described in Section 9–4), we shall simplify the loading pattern to that of Figure 10–32. This will result in a total deflection that is slightly greater than the actual one but will be very easy to calculate. From Cases 4 and 15 of Appendix K,

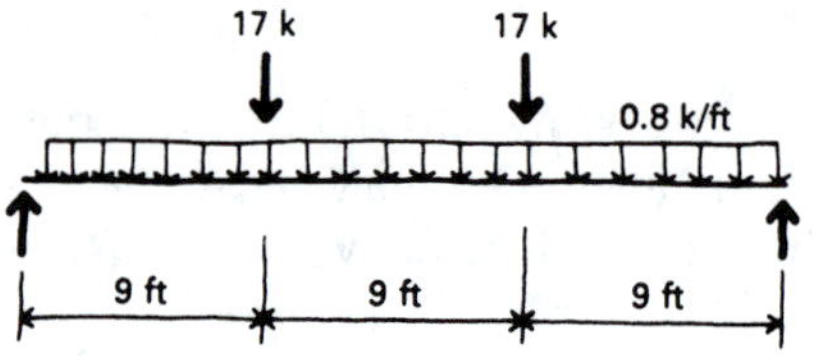

**Figure 10–32**

$$\Delta_{\max} = \frac{5wL^4}{384EI} + \frac{Pa}{24EI}(3L^2 - 4a^2)$$

$$= \left[ \frac{5(0.8)(27)^4}{384} + \frac{17(9)[3(27)^2 - 4(9)^2]}{24} \right] \left[ \frac{(1000 \text{ lb/kip})(12 \text{ in/ft})^3}{29(10)^6 \text{ psi}](712 \text{ in}^4)} \right]$$

$$= 1.46 \text{ in}$$

The code limitation was given as

$$\Delta_{\text{code}} = \frac{L}{240}$$

$$= \frac{27 \text{ ft}(12 \text{ in/ft})}{240}$$

$$= 1.35 \text{ in}$$

Our calculated approximate deflection exceeds the limit by 0.11 in., but sin
loads used were conservatively high, the actual deflection will almost surely be OK.

---

**EXAMPLE
10–8M**

Select a W shape for the beam $AB$ (Figure 10–33M). The live load on the floor is
5 kN/m$^2$ and the dead load, including all structural elements, is 1 kN/m$^2$. There is
also a 3-m-high masonry wall which bounds the opening from $A$ to $C$ to $D$. The wall
has a weight of 3.92 kN per square meter of elevation. The deflection due to live
load is limited to $L/360$ and that due to total load is limited to $L/240$. Use $F_y =$
345 MPa steel.

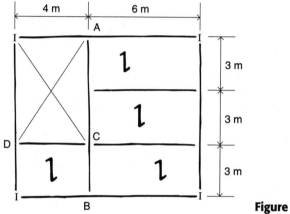

**Figure 10–33M** Framing plan.

### Solution

The beam $AB$ is loaded by two unequal concentrated loads at the third points plus
a uniform load from that portion of the wall from $A$ to $C$. The total floor load is
6 kN/m$^2$. Since the wall load is 3.92 kN/m$^2$ and the wall is 3 m tall, the wall load is

$$w = (3 \text{ m})(3.92 \text{ kN/m}^2)$$

$$= 11.8 \text{ kN/m}$$

The beam $DC$ carries the wall directly above it plus a 1.5-m tributary strip of
floor. Its uniform load is 11.8 kN/m plus 1.5 m(6 kN/m$^2$), for a total of 20.8 kN/m.
The reaction at $C$ is (20.8 kN/m)(2 m), or 41.6 kN.

The two beams coming into $AB$ from the right each have a tributary area of
18 m$^2$, and half of this gets to beam $AB$. The magnitude of each reaction upon $AB$
is, therefore, ½ (18 m$^2$)(6 kN/m$^2$), or 54 kN, so the loads end up as shown in Figure
10–34M(b). Figure 10–35M provides the $V$ and $M$ diagrams.

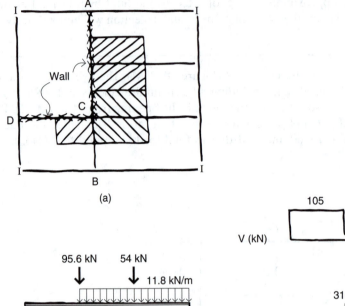

(a)

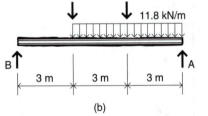

(b)

**Figure 10–34M** Tributary areas and final loading pattern.

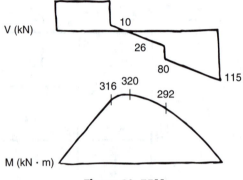

**Figure 10–35M**

With full lateral support, the allowable bending stress is $\frac{2}{3}F_y$, which is 230 MPa.

$$S_r = \frac{M}{F_b}$$

$$= \frac{320 \text{ kN} \cdot \text{m}}{230\,000 \text{ kPa}}$$

$$= 0.001\,39 \text{ m}^3$$

$$= 1390(10)^3 \text{ mm}^3$$

A W530 × 74 will provide a section modulus of $1550(10)^3 \text{ mm}^3$.

To make an approximate check on the total deflection, we shall simplify the loading pattern in the conservative direction as shown in Figure 10–36M.

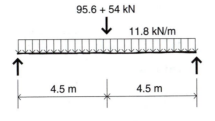

**Figure 10–36M**  Loading pattern for approximate deflection computation.

$$\Delta_{\substack{max \\ approx}} = \frac{5wL^4}{384EI} + \frac{PL^3}{48EI}$$

$$= \frac{5(11.8 \text{ kN/m})(9 \text{ m})^4}{384[200(10)^6 \text{ kN/m}^2][410(10)^{-6} \text{ m}^4]}$$

$$+ \frac{149.6 \text{ kN}(9 \text{ m})^3}{48[200(10)^6 \text{ kN/m}^2][410(10)^{-6} \text{ m}^4]}$$

$$= 0.0123 \text{ m} + 0.0277 \text{ m}$$

or

$$\Delta_{\substack{max \\ approx}} = 40 \text{ mm}$$

The permitted total deflection is

$$\Delta_{code} = \frac{L}{240}$$

$$= \frac{9000 \text{ mm}}{240}$$

$$= 38 \text{ mm}$$

Our approximate deflection exceeds the allowable value by 3 mm, but since the computation was in the conservative direction, the actual deflection will almost surely be OK.

We still need to check the deflection due to live load alone against $L/360$. The live loads are shown in Figure 10–37M(a) and can be combined into one load to over-estimate their effect, as we have done previously. Referring to Figure 10–37M(b), we get

$$\Delta_{\substack{max \\ approx}} = \frac{PL^3}{48EI}$$

$$= \frac{105 \text{ kN}(9 \text{ m})^3}{48[200(10)^6 \text{ kN/m}^2][410(10)^{-6} \text{ m}^4]}$$

$$= 0.0194 \text{ m}$$

or

$$\Delta_{\substack{\text{max} \\ \text{approx}}} = 19 \text{ mm}$$

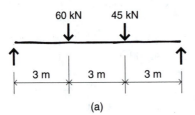

(a)

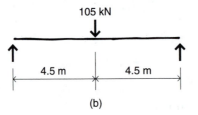

(b)

**Figure 10–37M**  Simplification of live load for approximate deflection computation.

This compares to a live load limitation of

$$\Delta_{\text{code}} = \frac{L}{360}$$

$$= \frac{9000 \text{ mm}}{360}$$

$$= 25 \text{ mm}$$

The W530 × 74 will be OK.

# PROBLEMS

**10–5.** Determine the lightest sections of A36 steel required for typical beams $AB$ and $AC$ in Figure 10–38. The dashed lines represent bar joists 24 in. on center. The spandrel beams $AC$ carry only a small strip of floor but must support a precast concrete curtain wall that is 13 ft high. The weight of the wall is 40 lb per square foot of elevation. The floor load is 70 psf live and 30 psf dead, including an allowance for all structural members. The live load deflection of beam $AB$ must be less than $L/360$, and the total load deflection of $AC$ must be less than $L/270$.

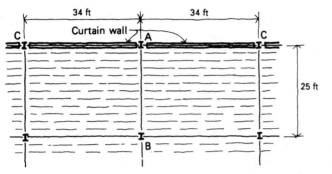

**Figure 10–38**   Portion of a bar joist floor system.

**10–5M.** Determine the lightest sections of $F_y = 250$ MPa steel required for the typical beams $AB$ and $AC$ in Figure 10–38M. The dashed lines represent bar joists 800 mm on center. The spandrel beams $AC$ carry only a small strip of floor but must support a precast concrete curtain wall that is 4 m high. The weight of the wall is 2 kN per square meter of elevation. The floor live load is 4 kN/m², and the dead load is 1 kN/m², including an allowance for all structural members. The live load deflection of beam $AB$ must be less than $L/360$, and the total load deflection of $AC$ must be less than $L/270$.

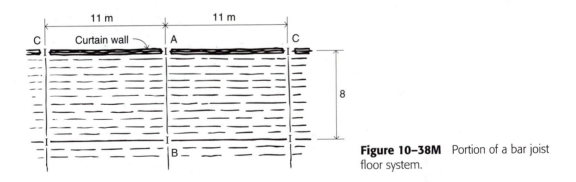

**Figure 10–38M**   Portion of a bar joist floor system.

**10–6.** Figure 10–39 shows a long-span, flat roof designed using bar joists (which span 20 ft), beams, and girders. The snow load is 20 psf and the dead load, including the weight of the joists (but *not* the beams and girders), is 20 psf. Since ponding could be a factor, the *total* load deflection is limited to $L/360$. Use $F_y = 36$ ksi steel for the beams and $F_y = 50$ ksi steel for the girders. (*Hint*: Be sure to account for member self-weight.)

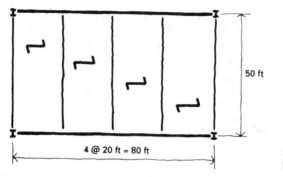

**Figure 10–39** Framing plan for a long-span steel frame.

**10–6M.** Figure 10–39M shows a long-span, flat roof structure designed using bar joists, beams, and girders. The snow load is 0.5 kN/m² and the dead load, including the weight of the joists (but not the beams and girders), is 0.5 kN/m². Since ponding could be a factor, the total load deflection of each member will be limited to $L/360$. Use $F_y = 250$ MPa steel for the beams and $F_y = 345$ MPa steel for the girders. (*Hint:* Use accurate deflection computations and include member self-weight.)

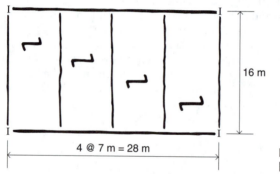

**Figure 10–39M** Framing plan for a long-span steel frame.

**10–7.** Beam B–1 in Figure 10–40 is limited in depth to a nominal 21 in. The live load is 60 psf and the dead load is 30 psf. The code limit on total deflection is $L/240$. Select a W shape for B–1 using $F_y = 50$ ksi steel.

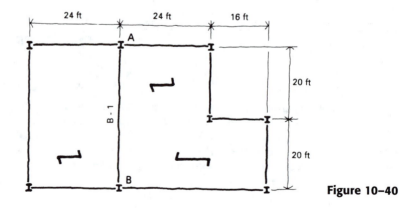

**Figure 10–40**

**10–7M.** Beam $AB$ in Figure 10–40M is limited in depth to a nominal 530 mm. The live load is 3 kN/m² and the dead load is 1.5 kN/m². The code limit on total deflection is $L/240$. Select a section for beam $AB$ using $F_y$ = 345 MPa steel.

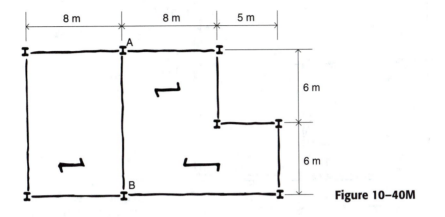

**Figure 10–40M**

**10–8.** Locate and size all the beams and girders for the office building floor plan in Figure 10–41. Note the absence of a column at the center of the east side. Closely spaced joists, which deliver uniform loads to beams, may be used for the floor deck, but these must span neither less than 12 ft nor more than 34 ft. The live load is 80 psf and the dead load is 35 psf, which includes an allowance for the self-weight of all structural elements. There is a masonry wall around each stairwell and along the periphery as shown. Its weight is 100 psf of elevation and it is 13 ft high. The live load deflection of the interior beams is limited to $L/360$, and the total load deflection of all spandrel beams must not exceed $L/240$. The dashed line represents the exterior face of the building. Assume that the floor deck provides full lateral support for all members, and use $F_y$ = 50 ksi steel throughout.

**10–8M.** Locate and size all the beams and girders for the office building floor plan in Figure 10–41M. Note the absence of a column at the center of the east side. Closely spaced joists, which deliver uniform loads to beams, may be used for the floor deck, but these must span neither less than 4 m nor more than 10 m. The live load is 4 kN/m² and the dead load is 1 kN/m², which includes an allowance for the self-weight of all structural elements. There is a masonry wall around each stairwell and along the periphery as shown. Its weight is 3 kN/m² of elevation and it is 4 m high. The live load deflection of the interior beams is limited to $L/360$, and the total load deflection of all spandrel beams must not exceed $L/270$. The dashed line represents the exterior face of the building. Assume that the floor deck provides full lateral support for all members, and use $F_y$ = 345 MPa steel throughout.

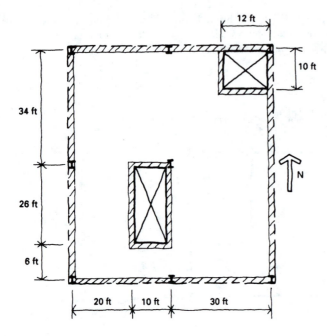

**Figure 10–41**  Office building schematic plan.

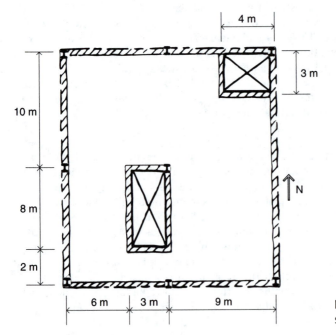

**Figure 10–41M**  Office building schematic plan.

## 10–9  DESIGN AIDS FOR WOOD AND STEEL

The process of beam sizing can be accelerated considerably by the proper use of design aids published by manufacturers' associations. Much of the data needed concerning member selection have been printed in the form of convenient graphs, charts, and tables. In many cases, beams can be selected with almost no computational work.

The American Institute of Steel Construction (AISC) publishes two valuable design manuals, one treating allowable stress design and the other, load and resistance factor design. As noted in Section 4–8, the two different approaches to design have primarily to do with how the factors of safety are applied. The use of one or the other of these manuals is an absolute necessity for any steel design work. Beam selection involves only a determination of the loads or the maximum moment. Even deflection magnitudes are provided by coefficient for common load situations.

For timber design, the *National Design Specification,* published by the American Forest and Paper Association, is absolutely essential. They also publish design handbooks, as do regional groups such as the Southern Forest Products Association and the Western Wood Products Association. Some have very complete tables, which enable the designer to select joists and beams often without a single written computation. For glued-laminated elements (beyond the scope of this book), the American Institute of Timber Construction publishes the very useful *Manual of Timber Construction.*

Open-web steel joists are so standardized that their design is almost always done by selection from allowable load tables. These tables are prepared by the Steel Joist Institute and include values of permissible uniform loads by section and span as limited by bending, end reaction, and deflection.

As with any design aid or shortcut, the designer must be aware of how the data were generated and when the actual situation departs from the one used to get the data. Many unfortunate errors have occurred through a misunderstanding or misapplication of tabled values. Similarly, the designer should always be wary of making a decision and leaving no record of how the choice was made or the options that were available. Good design is an iterative process, often accomplished over a substantial period of time, and unless good records are kept, much of the work ends up being done over and over again.

# 11

# Elastic Buckling of Columns

## 11–1  COLUMNS AS BUILDING STRUCTURAL ELEMENTS

Columns are probably the most important of the various structural elements in the conventional building frame. Stacked vertically, one on top of the other, they receive live and dead loads at each floor level and must transmit these loads to the foundation system below. The spacing of columns in plan usually determines what is referred to as the *structural bay*; for example, if columns are spaced 24 ft (7 m) on center in one direction and 32 ft (10 m) in the other, the structural bay size is said to be 24 × 32 ft (7 × 10 m). The size and shape of the structural bay has a great influence upon the type of framing system to be used in the floor structure.

Columns are usually designed with greater factors of safety than other structural elements, because any column failure would result in the catastrophic collapse of at least a major portion of the building frame. When a column fails, any beams or girders framing into it come down, as do all the other columns directly above, as shown in Figure 11–1.

Depending upon the skill of the designer in the initial stages of structural planning, columns can serve as valuable organizers or as ill-located hindrances to the architectural design process. Columns can be effective space dividers or modulators in large areas. Indeed, more columns than are needed stucturally are sometimes used to separate one space from another functionally while preserving visual continuity.

It is most important that the designer or structural planner understand the structural behavior of columns under various kinds of loading. How long and slender can a given column be and still have a useful load capacity? When is a rectangular cross section more appropriate than a square one? How can the column work together with the horizontal spanning members that frame into it? What happens when the center of a column is removed so that services can be run vertically through it? How do the top and bottom connections affect column capacity? Can column capacity always be increased by using a stronger material? The structural planner should not only be able to answer these questions and others but should also have a real understanding of the principles that generate those answers.

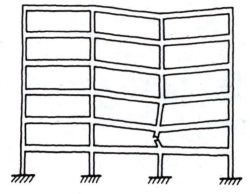

**Figure 11–1** Second-story column failure.

Many codes include a provision for reducing the total design live load on certain structural members that have large tributary areas. This reduction is based on the relatively low probability that the entire area will ever be loaded to the full design value. Although this can be applicable to lower-story columns, which often support many square feet of floor area, such load reductions have been purposely ignored in the examples and problems of this chapter. It is felt that such a provision could easily lead to large errors in the nonconservative direction during the preliminary design stages.

## 11–2 COLUMN FAILURE MODES

Columns are essentially compression elements and, when overloaded sufficiently, will fail by crushing or buckling or a combination of these two effects (Figure 11–2). Very short stout columns will fail by crushing, and long slender columns will fail by buckling. Actually, most columns in buildings are proportioned such that both effects would be involved.

Pure crushing is a relatively easy concept to understand, and a very simple design formula is available to prevent its occurrence. The designer merely provides

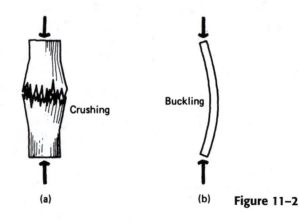

(a)  Crushing          Buckling  (b)    **Figure 11–2**

enough cross-sectional areas in the column so that the allowable compressive stress for the material is not exceeded.

$$A_r = \frac{P}{F_c}$$ (11–1)

where $A_r$ = area required (in.$^2$) (m$^2$)
$P$ = total load on the column (kips or lb) (kN)
$F_c$ = allowable compressive stress (ksi or psi) (MPa or kPa)

The allowable compressive stress is obtained by reducing the value of the actual crushing strength for a material by an appropriate factor of safety. (It should be noted that in the case of most steel column shapes, such crushing does not occur because a similar type of failure called *local crippling* or *buckling* occurs under a lesser load. An example of such a failure occurs when we "crush" a tin can vertically without actually crushing the material. Local buckling is different from elastic buckling, which is the primary subject of this chapter.)

## 11–3   THE EULER THEORY

Pure buckling or elastic buckling of long, slender columns is not so easy to understand. Here, column capacity is dependent upon the dimensions and shape of the column and upon the stiffness of the material. Surprisingly enough, pure buckling is totally independent of the strength of the material!

The basic theory of elastic buckling was successfully formulated over 200 years ago by Leonhard Euler (1707–1783), a Swiss mathematician. Essentially, such buckling occurs because there exists more than one position of equilibrium for a long, straight compression member. The slightly deflected column shown in Figure 11–3 could carry a load and be in equilibrium just like the straight one. Conversely, this could never happen in a tension member.

As you gradually increase the axial load on a long column that is initially straight, it will suddenly deflect laterally. If the load is removed, the column will return to its initial straight shape. This behavior is called *elastic buckling*. The particu-

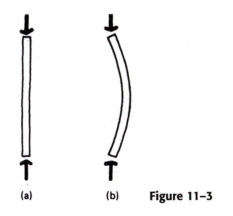

(a)              (b)        **Figure 11–3**

lar value of axial load (called the critical load) that causes buckling is given by the *Euler equation*:

$$P_{cr} = \frac{\pi^2 EI}{L^2} \tag{11-2}$$

where $P_{cr}$ = axial load necessary to cause buckling (lb) (kN)
$\quad\quad\quad E$ = modulus of elasticity of the column material (psi) (kPa)
$\quad\quad\quad I$ = moment of inertia of the column cross section (in.$^4$) (m$^4$)
$\quad\quad\quad L$ = length of the column (in.) (m)

(See Appendix C for the derivation of Equation 11–2.)

(Experimentally, most long columns buckle under loads that are somewhat less than the value given by this equation. This is due to a number of factors, such as slight irregularities in the straightness of the member, eccentricity of the load, and the material being nonhomogeneous. Load eccentricity is further addressed in Section 11–7.)

This equation is more useful if we make a minor modification using the radius-of-gyration concept from Section 3–5:

$$r = \sqrt{\frac{I}{A}}$$

where $r$ = radius of gyration of the column cross section (in.) (mm)
$\quad\quad\quad A$ = area of the column cross section (in.$^2$) (mm$^2$)

If we solve this expression for $I$ and substitute into Equation 11–2,

$$\left(\frac{P}{A}\right)_{cr} = \frac{\pi^2 E}{(L/r)^2} \tag{11-3}$$

where $(P/A)_{cr}$ is the stress caused by the critical load.

The parameter $L/r$ is called the *slenderness ratio*. The critical stress (and load) is inversely proportional to the square of this ratio.

Equations 11–2 and 11–3 are equally valid and can be used interchangeably. As stated previously, notice that neither one includes a term representing the strength of the material. Also notice that the equations give the failure loads and stresses. No factor of safety has been included.

Euler's theory assumes that the column has pinned ends that allow the column ends to rotate freely but not to translate. The following examples and problems assume pinned ends. When other end conditions occur, the column length must be adjusted by a special end conditions factor, and this is discussed in Section 11–4.

**EXAMPLE 11-1**

Determine the critical buckling stress and load for a wood 4 × 4 post that is 10 ft. long. Assume that $E = 1.2(10)^6$ psi.

**Solution**

Properties of timber sections may be found in Appendix I.

$$\left(\frac{P}{A}\right)_{cr} = \frac{\pi^2 E}{(L/r)^2}$$

$$A = 12.3 \text{ in}^2$$
$$I = 12.5 \text{ in}^4$$
$$r = 1.01 \text{ in (computed as } r = \sqrt{I/A})$$
$$\frac{L}{r} = \frac{10 \text{ ft}(12 \text{ in/ft})}{1.01 \text{ in}} = 119$$

$$\left(\frac{P}{A}\right)_{cr} = \frac{\pi^2(1.2)(10)^6 \text{ psi}}{(119)^2}$$

$$= 836 \text{ psi}$$

$$P_{cr} = \left(\frac{P}{A}\right)_{cr}(A)$$

$$= 836 \text{ psi}(12.3 \text{ in}^2)$$
$$= 10\ 300 \text{ lb}$$

**EXAMPLE 11-2M**

Determine the critical buckling stress and load for a wood column 89 × 89 mm in cross section and 3 m long. Assume that $E = 10\ 500$ MPa.

**Solution**

Properties of timber sections may be found in Appendix I.

$$\left(\frac{P}{A}\right)_{cr} = \frac{\pi^2 E}{(L/r)^2}$$

$$A = 7920 \text{ mm}^2$$
$$I = 5.23(10)^6 \text{ mm}^4$$
$$r = 25.7 \text{ mm (computed as } r = \sqrt{I/A})$$
$$\frac{L}{r} = \frac{3000}{25.7} = 117$$

$$\left(\frac{P}{A}\right)_{cr} = \frac{\pi^2(10\ 500)(10)^3\ \text{kPa}}{(117)^2}$$

$$= 7570\ \text{kPa}$$

$$P_{cr} = \left(\frac{P}{A}\right)_{cr}(A)$$

$$= (7570\ \text{kN/m}^2)(0.007\ 92\ \text{m}^2)$$

$$= 60.0\ \text{kN}$$

**EXAMPLE
11–3(a)**

Determine the critical buckling stress for a W8 × 35 steel column that is 30 ft. long.

**Solution**

Properties of selected steel shapes may be found in Appendix J. $E$ is taken as $29(10)^6$ psi for all rolled steel shapes. The radius of gyration values may be computed from $I$ and $A$.

$$\left(\frac{P}{A}\right)_{cr} = \frac{\pi^2 E}{(L/r)^2} \quad r_x = 3.51\ \text{in}$$

$$r_y = 2.03\ \text{in}$$

Compute the $L/r$ value for each of the two axes. Substitute the larger of the two values into the Euler equation because it will yield the smaller critical stress value.

$$\frac{L}{r_x} = \frac{30\ \text{ft}(12\ \text{in/ft})}{3.51\ \text{in}} = 103$$

$$\frac{L}{r_y} = \frac{30\ \text{ft}(12\ \text{in/ft})}{2.03\ \text{in}} = 177$$

$$\left(\frac{P}{A}\right)_{cr} = \frac{\pi^2[29(10)^6\ \text{psi}]}{(177)^2}$$

$$= 9140\ \text{psi}$$

The use of $L/r_x$ would clearly yield a much larger critical stress value. This indicates that the column would buckle about the y-axis (in the x direction) under a much smaller load than would be required to make it buckle the other way. In practical terms this means that, in case of overload, the column would not be able to reach the critical load necessary to make it buckle about its strong axis; it would have failed at a lower load value by buckling about its weak axis. Therefore, in computing critical load and stress values, always use the greater $L/r$ value.

**EXAMPLE**          Determine the critical buckling load for the column of Example 11–3(a).
**11–3(b)**

**Solution**

$$P_{cr} = \frac{\pi^2 EI}{L^2}$$

$$A = 10.3 \text{ in}^2$$

$$I_y = 42.6 \text{ in}^4$$

Since $L$ is the same for both axes, we need only $I_y$ for use in the equation (i.e., $L/r_y$ will be greater than $L/r_x$).

$$P_{cr} = \frac{\pi^2 [29(10)^6 \text{ psi}](42.6 \text{ in}^4)}{[(30 \text{ ft})(12 \text{ in/ft})]^2}$$

$$= 94\,100 \text{ lb}$$

The same answer could have been obtained using the critical stress value from Example 11–3(a).

$$P_{cr} = \left(\frac{P}{A}\right)_{cr} (A)$$

$$= 9140 \text{ psi}(10.3 \text{ in}^2)$$

$$= 94\,100 \text{ lb}$$

The writer prefers this latter approach to getting the critical load as opposed to the use of Equation 11–2. Computing the critical stress enables the analyst to watch its magnitude, which is necessary because the Euler equation is only valid for long slender columns having *low* critical stresses (see Section 11–6). Practical design methods developed by the wood and steel industries often involve calculating the stress and then multiplying by the area of the cross section to get the permissible load.

---

**EXAMPLE**          Determine the critical buckling stress for a W200 × 52 steel column that is 9 m long.
**11–4M(a)**

**Solution**
Properties of the steel shapes used in the examples and problems of this chapter may be found in Table 2 of Appendix J. Assume $E$ to be 200 000 MPa for all rolled steel shapes. The radius of gyration values may be computed from $I$ and $A$.

$$\left(\frac{P}{A}\right)_{cr} = \frac{\pi^2 E}{(L/r)^2}$$

$$r_x = 89.1 \text{ mm}$$

$$r_y = 51.6 \text{ mm}$$

Compute the $L/r$ value for each of the two axes. Substitute the larger of the two values into the Euler equation because it will yield the smaller critical stress value.

$$\frac{L}{r_x} = \frac{9000}{89.1} = 101$$

$$\frac{L}{r_y} = \frac{9000}{51.6} = 174$$

$$\left(\frac{P}{A}\right)_{cr} = \frac{\pi^2 (200)(10)^3 \text{ MPa}}{(174)^2}$$

$$= 65.2 \text{ MPa}$$

See the note at the end of Example 11–3(a).

---

**EXAMPLE
11–4M(b)**

Determine the critical buckling load for the column of Example 11–4M(a).

**Solution**

$$A = 6650 \text{ mm}^2$$

$$I_y = 17.7(10)^6 \text{ mm}^4$$

$$P_{cr} = \frac{\pi^2 EI}{L^2}$$

Since $L$ is the same for both axes, we need only $I_y$ for use in the equation.

$$P_{cr} = \frac{\pi^2 [(200)(10)^6 \text{ kN/m}^2](17.7)(10)^{-6} \text{ m}^4}{(9 \text{ m})^2}$$

$$= 431 \text{ kN}$$

The same answer could have been obtained using the critical stress value from Example 11–4M(a).

$$P_{cr} = \left(\frac{P}{A}\right)_{cr} (A)$$

$$= (65\,200 \text{ kN/m}^2)(0.006\,65 \text{ m}^2)$$

$$= 433 \text{ kN}$$

(The discrepancy between the two answers is due to rounding error and is less than one-half of 1%.)

See the note at the end of Example 11–3(b).

## PROBLEMS

**11–1.** Determine the critical buckling stress and load for a nominal $6 \times 6$ post that is 17 ft long. Assume that $E = 1.3(10)^6$ psi.

**11–1M.** Determine the critical buckling load for a timber column $139 \times 139$ mm and 5.5 m long. Assume that $E = 8000$ MPa.

**11–2.** Determine the critical buckling stress and load for a southern pine No. 2 nominal $4 \times 4$ that is 14 ft long. (See Appendix H for the modulus of elasticity.)

**11–2M.** Determine the critical buckling stress and load for a southern pine No. 2 $89 \times 89$ mm section that is 4.3 m long. (See Appendix H for the modulus of elasticity.)

**11–3.** Work Problem 11–2 if the column is a $4 \times 6$.

**11–3M.** Work Problem 11–2M if the column is an $89 \times 139$ section.

**11–4.** Work Problem 11–2 if the column is a $6 \times 6$. (*Hint:* Be sure to use the correct $E$ value.)

**11–4M.** Work Problem 11–2M if the column is a $139 \times 139$ section. (*Hint:* Be sure to use the correct $E$ value.)

**11–5.** Compare the results of Problems 11–3 and 11–4 and discuss why the $6 \times 6$ is able to carry so much more load even though it has a lesser $E$ value and only 50% more cross section than the $4 \times 6$.

**11–5M.** Compare the results of Problems 11–3M and 11–4M and discuss why the $139 \times 139$ section is able to carry so much more load even though it has a lesser $E$ value and only 50% more cross section than the $89 \times 139$.

**11–6.** A W6 $\times$ 25 is used as a column 20 ft long. Compute its critical buckling stress and load.

**11–6M.** A W150 $\times$ 37 is used as a column 6 m long. Compute its critical buckling stress and load.

**11–7.** Determine the critical buckling stress for a steel pipe column that is 16 ft long. The outside diameter is $4\frac{1}{2}$ in. and the wall thickness is $\frac{1}{4}$ in.

**11–7M.** Determine the critical buckling stress for a steel pipe column that is 5 m long. The outside diameter is 114 mm and the wall thickness is 6 mm.

**11–8.** Determine the critical buckling load for a single wood $2 \times 4$ stud that is 8 ft long. Assume that $E = 1.3(10)^6$ psi.

**11–8M.** Determine the critical buckling load for a single wood $38 \times 89$ mm stud 2.4 m long. Assume that $E = 9500$ MPa.

## 11–4    INFLUENCE OF DIFFERENT END CONDITIONS

How the column ends are connected to the rest of the structure has a large influence on the critical buckling load. If the column ends are restrained from rotation in some manner, the effective buckling length can be very different from the true length, as shown in Figure 11–4. True length will be called $L$ and the effective length $KL$, where $K$ is a theoretical modifier that accounts for the effect of different end conditions.

The effective length for a column with both ends fixed is just one-half that of a column with both ends pinned. The "flagpole" type of column has a $K$ value of 2.0, which is rationalized by noting the mirror image below the fixed end needed to obtain a full buckling curve. Fortunately, this case seldom appears in building columns.

The problems presented so far have all assumed pinned ends with $K = 1.0$. Many building columns have $K$ values that are in between the four cases shown, and judgment must be used in estimating a proper $K$ value.

The equations presented previously should be modified to include this end condition factor:

$$P_{cr} = \frac{\pi^2 EI}{(KL)^2} \tag{11–2a}$$

$$\left(\frac{P}{A}\right)_{cr} = \frac{\pi^2 E}{(KL/r)^2} \tag{11–3a}$$

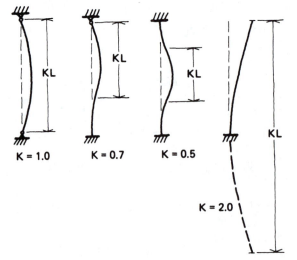

**Figure 11–4**

Relative $\frac{1}{r}$ increased during the evolution of the classical orders.

Euler's formula, which correlates the factors principally responsible for a column's resistance to buckling, is another abstraction seemingly distant from "design," whose manipulation in fact has immediate visual consequences. The type of *end connections—K* in *KL/r*—and *column slenderness*—the *L/r* in the formula—can be seen as partial rationales underlying what seem to be merely arbitrary formal gestures or superficial modifications due only to changes in taste.

For example, a compression strut in Frei Otto's pavilion for the Museum of Modern Art garden displays a quite conscious gradation in cross section, center to ends, reflecting the fact that with pinned ends the middle of a column must resist the tendency to buckle. Alvar Aalto may have had a similar structural logic less directly in mind when he designed the columns in his 1937 Finnish Pavilion in Paris. Each has six ribs, tapering center to ends, added to a cylindrical section; yet the end connections are certainly not pins, and the ribs may in fact add only marginally to the columns' buckling resistance. But Aalto was using our intuitive visual knowledge of behavior under load in order to involve us with the building; the ribs are a kind of plausible structural fairy tale, an invitation to empathy.

The evolution of the classical orders can be interpreted as another instance where patterns of structure coincide with patterns of visual sophistication. Despite numerous individual exceptions, the clear pattern is one of regular increase in the slenderness ratio, from the earlier Tuscan and Doric to the later Composite. To put it another way, the increase in *L/r* reflects both a greater technical confidence, the result of accumulated experience with columns and loads, and a greater affinity for visual lightness produced by changes in both proportion and ornamentation.

By noting that $K$ is squared along with $L$, we can see the big difference these end conditions will make. For example, if we hold all other parameters constant and vary only $K$, then

1. A fixed-end column will support four times the load of one with pinned ends.

2. A pinned-end column will support four times the load of one with one fixed and one free end (flagpole).

Some of the examples that follow indicate fixed ends for timber columns. In actual construction detailing, this is quite difficult to achieve. However, such examples are included here to illustrate the use of the $K$ factor.

**EXAMPLE 11–5**

The timber $6 \times 6$ column of Figure 11–5 is 24 ft long and can be considered pinned at the lower end and effectively fixed by deep trusses framing into it at the top. Determine the critical buckling stress and load. Assume that $E = 1.5(10)^6$ psi.

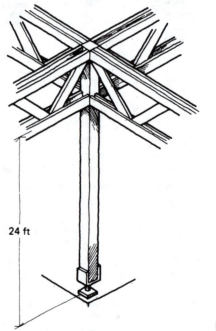

**24 ft**

**Figure 11–5**

*Solution*

$$\left(\frac{P}{A}\right)_{cr} = \frac{\pi^2 E}{(KL/r)^2}$$

$$A = 30.3 \text{ in}^2$$

$$r = 1.6 \text{ in (computed as } r = \sqrt{I/A})$$

$$K = 0.7$$

$$\frac{KL}{r} = \frac{0.7(24 \text{ ft})(12 \text{ in/ft})}{1.6 \text{ in}} = 126$$

$$= \frac{\pi^2[(1.5)(10)^6 \text{ psi}]}{(126)^2}$$

$$= 933 \text{ psi}$$

$$P_{cr} = \left(\frac{P}{A}\right)_{cr}(A)$$

$$= (933 \text{ psi})(30.3 \text{ in}^2)$$

$$= 28\,300 \text{ lb}$$

---

**EXAMPLE 11–6M**  The timber $139 \times 139$ column of Figure 11–6M is 7 m long and can be considered pinned at the top and effectively fixed by masonry walls at the bottom. Determine the critical buckling stress and load. Assume that $E = 10\,000$ MPa.

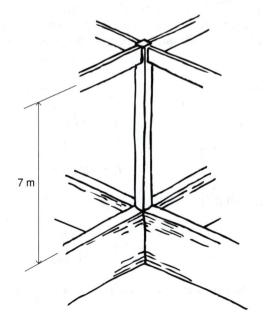

7 m

**Figure 11–6M**

### Solution

$$\left(\frac{P}{A}\right)_{cr} = \frac{\pi^2 E}{(KL/r)^2}$$

$$A = 19\,300 \text{ mm}^2$$

$$r = 40 \text{ mm (computed as } r = \sqrt{I/A})$$

$$K = 0.7$$

$$\frac{KL}{r} = \frac{0.7(7000 \text{ mm})}{40 \text{ mm}} = 123$$

$$= \frac{\pi^2(10\,000)(10)^3 \text{ kPa}}{(123)^2}$$

$$= 6520 \text{ kPa}$$

$$P_{cr} = \left(\frac{P}{A}\right)_{cr}(A)$$

$$= (6520 \text{ kN/m}^2)\,(0.0193 \text{ m}^2)$$

$$= 126 \text{ kN}$$

---

**EXAMPLE 11–7**

A W8 × 67 section is used for the column in Figure 11–7. The bottom clip angle connection is a pin. Deep plate girders frame into the web, which serve to fix the weak axis at the top. Small bracing beams are clipped to the flanges and provide a pinned condition. Determine the critical buckling stress and load.

### Solution
From Appendix J,

$$A = 19.7 \text{ in}^2$$

$$I_x = 272 \text{ in}^4$$

$$I_y = 88.6 \text{ in}^4$$

We can then find that

$$r_x = 3.72 \text{ in}$$

$$r_y = 2.12 \text{ in}$$

Next determine which axis is critical (i.e., which has the greater $KL/r$).

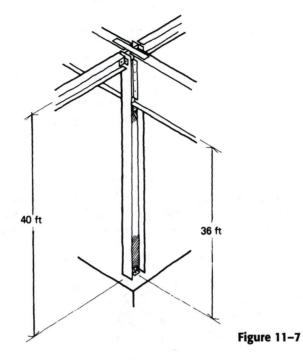

**Figure 11–7**

$$\left(\frac{KL}{r}\right)_x = \frac{1.0(40 \text{ ft})(12 \text{ in/ft})}{3.72 \text{ in}} = 129$$

$$\left(\frac{KL}{r}\right)_y = \frac{0.7(36 \text{ ft})(12 \text{ in/ft})}{2.12 \text{ in}} = 143$$

The weak axis is critical for this column. We can now determine the critical buckling stress and multiply by the area to get the critical load. We can also find the load directly by substituting the critical axis properties into

$$\left(\frac{P}{A}\right)_{cr} = \frac{\pi^2 E}{(KL/r)^2}$$

$$= \frac{\pi^2[29(10)^6 \text{ psi}]}{(143)^2}$$

$$= 14\,000 \text{ psi}$$

$$P_{cr} = \left(\frac{P}{A}\right)_{cr} (A)$$

$$= 14\ 000\ \text{psi}(19.7\ \text{in}^2)$$

$$= 276\ 000\ \text{lb}$$

$$= 276\ \text{kips}$$

---

**EXAMPLE 11–8M**

A W200 = 100 section is used for the column in Figure 11–8M. The bottom clip angle connection is a pin. Deep plate girders frame into the web, which serve to fix the weak axis at the top. Small bracing beams are clipped to the flanges and provide a pinned condition. Determine the critical buckling load.

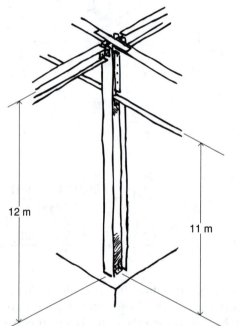

12 m

11 m

**Figure 11–8M**

**Solution**

From Appendix J,

$$A = 12\ 700\ \text{mm}^2$$

$$I_x = 113(10)^6\ \text{mm}^4$$

$$I_y = 36.9(10)^6\ \text{mm}^4$$

We can then find that

$$r_x = 94.2 \text{ mm}$$

$$r_y = 53.9 \text{ mm}$$

and

$$L_x = 12 \text{ m}$$

$$L_y = 11 \text{ m}$$

Next determine which axis is critical (i.e., which one has the greater $KL/r$).

$$\left(\frac{KL}{r}\right)_x = \frac{1.0(12\,000)}{94.2} = 127$$

$$\left(\frac{KL}{r}\right)_y = \frac{0.7(11\,000)}{53.9} = 143$$

$$\left(\frac{P}{A}\right)_{cr} = \frac{\pi^2 E}{(KL/r)^2}$$

$$= \frac{\pi^2[200(10)^6 \text{ kPa}]}{(143)^2}$$

$$= 96\,500 \text{ kPa}$$

$$P_{cr} = \left(\frac{P}{A}\right)_{cr}(A)$$

$$= (96\,500 \text{ kN/m}^2)\,(0.0127 \text{ m}^2)$$

$$= 1230 \text{ kN}$$

---

**EXAMPLE 11–9**    A steel pipe column has one end fixed and one end free. It has an outside diameter of 2.4 in. and an inside diameter of 2.0 in. It supports an axial load of 10 kips. Determine the actual length $L$ that this column can reach without buckling.

**Solution**

$$I = \frac{\pi}{4}(R_o^4 - R_i^4)$$

$$= 0.843 \text{ in}^4$$

$$K = 2.0$$

$L$ is defined as the length at which $P$ becomes critical.

$$P_{cr} = \frac{\pi^2 EI}{(KL)^2}$$

or

$$L = \frac{\sqrt{\pi^2 EI/P_{cr}}}{K}$$

$$= \frac{\sqrt{\dfrac{\pi^2[29(10)^3 \text{ ksi}](0.843 \text{ in}^4)}{10 \text{ kips}}}}{2.0}$$

$$= 78 \text{ in or } 6.5 \text{ ft}$$

---

**EXAMPLE 11-10M**

A steel pipe column has one end fixed and one end free. It has an outside diameter of 60.3 mm and an inside diameter of 52.5 mm. It supports an axial load of 45 kN. Determine the actual length $L$ that this column can reach without buckling.

**Solution**

$$I = \frac{\pi}{4}(R_o^4 - R_i^4)$$

$$= 0.277(10)^6 \text{ mm}^4$$

$$K = 2.0$$

$L$ is defined as the length at which $P$ becomes critical.

$$P_{cr} = \frac{\pi^2 EI}{(KL)^2}$$

or

$$L = \frac{\sqrt{\pi^2 EI/P_{cr}}}{K}$$

$$= \frac{\sqrt{\dfrac{\pi^2[200(10)^6 \text{ kN/m}^2](0.277)(10)^{-6} \text{ m}^4}{45 \text{ kN}}}}{2.0}$$

$$= 1.7 \text{ m}$$

**EXAMPLE 11–11M**

A timber section 89 × 139 mm is to be used as a column 4.5 m long. The $K$ value for the strong axis, $K_x$, is 1.0. At both ends, the weak axis is partially restrained, so $K_y$ is estimated to be 0.8. Determine the critical buckling stress. Assume that $E = 8000$ MPa.

**Solution**

$$r_x = 40.1 \text{ mm}$$

$$r_y = 25.7 \text{ mm}$$

First determine the larger $KL/r$.

$$\left(\frac{KL}{r}\right)_x = \frac{1.0(4500)}{40.1} = 112$$

$$\left(\frac{KL}{r}\right)_y = \frac{0.8/(4500)}{25.7} = 140$$

The weak axis is critical for this column.

$$\left(\frac{P}{A}\right)_{cr} = \frac{\pi^2 E}{(KL/r)^2}$$

$$= \frac{\pi^2(8000)(10)^3 \text{ kPa}}{(140)^2}$$

$$= 4030 \text{ kPa}$$

## PROBLEMS

**11–9.** A W12 × 50 is used as a column 50 ft long. If both ends are fixed, determine the critical buckling stress and load.

**11–9M.** A steel W310 × 74 is used as a column 17 m long. Both ends are fixed. Determine the critical buckling stress and load.

**11–10.** Work Problem 11–6 if the lower end is fixed and the upper end remains pinned.

**11–10M.** Work Problem 11–6M if the lower end is fixed and the upper end remains pinned.

**11–11.** Work Problem 11–6 if the lower end is fixed and the upper end becomes free (i.e., a flagpole condition).

**11–11M.** Work Problem 11–6M if the lower end is fixed and the upper end becomes free (i.e., a flagpole condition).

**11–12.** Work Problem 11–7 if both ends are partially restrained instead of pinned. Let $K = 0.85$.

**11–12M.** Work Problem 11–7M if both ends are partially restrained instead of pinned. Let $K = 0.85$.

**11–13.** A 6-in.-diameter (actual dimension) wood post is fixed into a large foundation pier at grade and is completely free at its upper end. How long can it be and still just support a load of 2 kips without failing? Assume that $E = 1.3(10)^6$ psi.

**11–13M.** A 150-mm-diameter wood post is fixed into a large foundation pier at grade and is completely free at its upper end. How long can it be and still just support a load of 9 kN without failing? Assume that $E = 10\,000$ MPa.

**11–14.** (a) Work Example 11–7 if the deep girders frame into the flanges of the W8 × 67 and the small bracing beams are clipped to its web.
(b) Read Section 11–7 and explain why a designer might prefer this framing scheme over the original one.

**11–14M.** (a) Work Example 11–8M if the deep girders frame into the flanges of the W200 × 100 and the small bracing beams are clipped to its web.
(b) Read Section 11–7 and explain why a designer might prefer this framing scheme over the original one.

**11–15.** Figure 11–9 shows a 4 × 2 in. structural tube with a 1/4-in. wall thickness serving as a column 16 ft long. Its upper end has pinned connections. The lower end is braced by a masonry wall so that its weak axis is fixed and the strong axis pinned. Determine the critical buckling load.

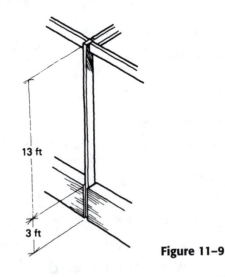

**13 ft**

**3 ft**

**Figure 11–9**

**11–15M.** An 89 × 139 mm wood post frames into stiff box beams, as shown in Figure 11–9M. Its strong axis may be considered pinned at the top by a floor deck. The lower end is a simple pin for both axes. Determine the critical buckling stress and load. Assume that $E = 9000$ MPa.

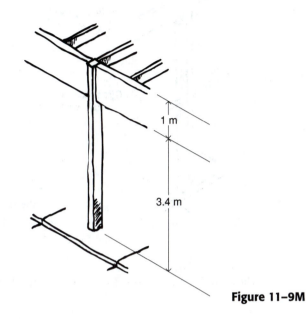

1 m

3.4 m

**Figure 11–9M**

# 11–5   INTERMEDIATE LATERAL BRACING

We now know that if the end conditions are the same for both axes, a column will always buckle about its weak axis. A rectangular timber post will buckle in a direction parallel to its least dimension. A steel wide-flange shape will buckle in a direction parallel to its flanges. With any asymmetrical shape, we have a situation in which the full capacity of the strong axis is not normally utilized. However, there are many situations in structural frames where we can increase the capacity of such asymmetrical shapes by decreasing the effective weak-axis length. In Example 11–7 and Example 11–8M, this occurred to a certain degree by virtue of the different end connections. We will have a structurally more efficient column if $(KL/r)_x$ and $(KL/r)_y$ have values that are similar in magnitude. Intermediate lateral bracing members are a most effective way of doing this. Often such elements occur rather naturally for other construction reasons.

In Figure 11–10, the column is braced against weak-axis buckling by a secondary wall element. Bracing can be provided by load-carrying beams and girders as well. It is important to realize that such members do not provide any bracing for the other axis of the column. In terms of the support provided for the

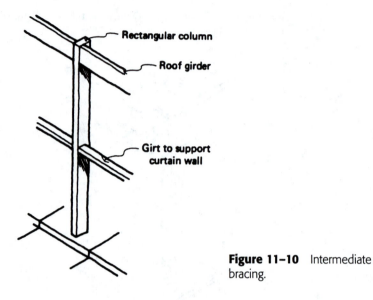

**Figure 11-10** Intermediate bracing.

column, they can be considered as two-force members (i.e., unable to resist non-axial forces).

---

**EXAMPLE 11-12**

A southern pine No. 2 nominal 4 × 8 section is used as a column 20 ft long. It has pinned ends and is braced against weak-axis buckling at midheight (Figure 11–11). Determine the critical buckling stress.

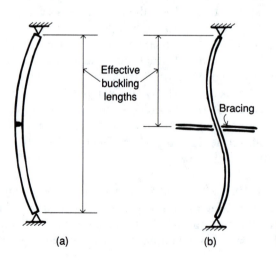

Effective buckling lengths

Bracing

(a)                              (b)

**Figure 11-11** (a) Strong-axis buckling; (b) weak-axis buckling.

### Solution

From Appendix H, $E = 1.6(10)^6$ psi and from Appendix I, the area and $I$ values can be found to determine the $r$ values.

$$r_x = 2.1 \text{ in}$$

$$r_y = 1.01 \text{ in}$$

$$\left(\frac{KL}{r}\right)_x = \frac{1.0(20 \text{ ft})(12 \text{ in/ft})}{2.1 \text{ in}} = 114$$

$$\left(\frac{KL}{r}\right)_y = \frac{1.0(10 \text{ ft})(12 \text{ in/ft})}{1.01 \text{ in}} = 119$$

The weak axis is critical, but not by much.

$$\left(\frac{P}{A}\right)_{cr} = \frac{\pi^2 E}{(KL/r)^2}$$

$$= \frac{\pi^2[(1.6)(10)^6 \text{ psi}]}{(119)^2}$$

$$= 1120 \text{ psi}$$

---

**EXAMPLE 11–13M**

A wood $89 \times 185$ mm section is used as a column 4.3 m long. It has pinned ends and is braced against weak-axis buckling at midheight (Figure 11–11). Determine the critical buckling stress and load. Assume that $E = 8000$ MPa.

### Solution

From Appendix I, the area and $I$ values can be found to determine the $r$ values.

$$r_x = 53.4 \text{ mm}$$

$$r_y = 25.7 \text{ mm}$$

$$A = 16\,500 \text{ mm}^2$$

$$\left(\frac{KL}{r}\right)_x = \frac{1.0(4300)}{53.4} = 80.5$$

$$\left(\frac{KL}{r}\right)_y = \frac{1.0(2150)}{25.7} = 83.7$$

The weak axis is critical, but not by much.

$$\left(\frac{P}{A}\right)_{cr} = \frac{\pi^2 E}{(KL/r)^2}$$

$$= \frac{\pi^2(8000)(10)^3 \text{ kPa}}{(83.7)^2} = 11\ 300 \text{ kPa}$$

$$P_{cr} = \left(\frac{P}{A}\right)_{cr} (A)$$

$$= (11\ 300 \text{ kN/m}^2)(0.0165 \text{ m}^2)$$

$$= 186 \text{ kN}$$

**EXAMPLE 11–14**

A W10 × 49 is used as a 40-ft-long column. It has pinned ends and its weak axis is braced at a point 22 ft up from the lower end (Figure 11–12). Determine the critical buckling stress and load.

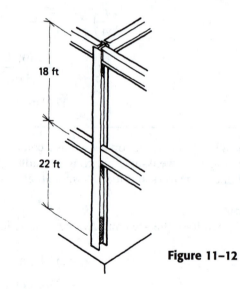

18 ft

22 ft

**Figure 11–12**

**Solution**

Values from Appendix J are

$$I_x = 272 \text{ in}^4$$
$$I_y = 93.4 \text{ in}^4$$
$$A = 14.4 \text{ in}^2$$

We can then find that

$$r_x = 4.35 \text{ in}$$
$$r_y = 2.54 \text{ in}$$

Next determine the larger $KL/r$.

$$\left(\frac{KL}{r}\right)_x = \frac{1.0(40 \text{ ft})(12 \text{ in/ft})}{4.35 \text{ in}} = 110$$

$$\left(\frac{KL}{r}\right)_y = \frac{1.0(22 \text{ ft})(12 \text{ in/ft})}{2.54 \text{ in}} = 104$$

The strong axis is critical.

$$\left(\frac{P}{A}\right)_{cr} = \frac{\pi^2 E}{(KL/r)^2}$$

$$= \frac{\pi^2[29(10)^3 \text{ ksi}]}{(110)^2}$$

$$= 23.6 \text{ ksi}$$

$$P_{cr} = \left(\frac{P}{A}\right)_{cr}(A)$$

$$= 23.6 \text{ ksi}(14.4 \text{ in}^2)$$

$$= 340 \text{ kips}$$

---

**EXAMPLE 11–15M**

A steel W250 × 73 is used as a 13.4-m-long column. It has pinned ends and its weak axis is braced at a point 7.3 m up from the lower end (see Figure 11–13M). Determine the critical buckling stress and load.

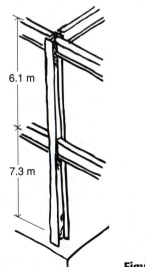

6.1 m

7.3 m

**Figure 11–13M**

### Solution

$$I_x = 113 \times 10^6 \text{ mm}^4$$

$$I_y = 38.9 \times 10^6 \text{ mm}^4$$

$$A = 9290 \text{ mm}^2$$

$$r_x = 110 \text{ mm}$$

$$r_y = 64.5 \text{ mm}$$

First determine the larger $KL/r$.

$$\left(\frac{KL}{r}\right)_x = \frac{1.0(13\ 400)}{110} = 121$$

$$\left(\frac{KL}{r}\right)_y = \frac{1.0(7300)}{64.5} = 113$$

The strong axis is critical.

$$\left(\frac{P}{A}\right)_{\text{cr}} = \frac{\pi^2 E}{(KL/r)^2}$$

$$= \frac{\pi^2 (200)(10)^3 \text{ MPa}}{(121)^2}$$

$$= 135 \text{ MPa}$$

$$P_{\text{cr}} = (135 \text{ MN/m}^2)(0.009\ 29 \text{ m}^2)$$

$$= 1.25 \text{ MN}$$

or

$$P_{\text{cr}} = 1250 \text{ kN}$$

# PROBLEMS

**11–16M.** A steel rectangular structural tube, TS152 × 76, with a wall thickness of 12.7 mm, is used as a 6-m-long column. It has pinned ends, and its weak axis is fully braced by a masonry curtain wall, as shown in Figure 11–14M. Determine the critical buckling stress. Let $r_x = 48.3$ mm and $r_y = 27.7$ mm.

**11–17.** A W8 × 35 is used as a column 36 ft long with pinned ends. If its weak axis is braced at midheight, compute its critical buckling stress and load.

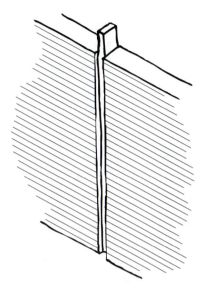

**Figure 11–14M**

**11–17M.** A W200 × 52 is used as a column 11 m long with pinned ends. If its weak axis is braced at midheight, compute its critical buckling stress and load.

**11–18.** In typical residential construction, the 2 × 4 studs are sheathed by drywall on both sides. This has the effect of fully bracing the weak axis but leaving the strong axis unbraced.
   (a) Determine the critical buckling load for a single 2 × 4 stud in such a wall if it is 8 ft tall. Assume that $E = 1.3(10)^6$ psi. Compare your answer to that of Problem 11–8.
   (b) Determine the critical load for the wall (*w* in lb per linear foot of wall) if the studs are spaced at 16 in. o.c.

**11–18M.** In typical residential construction, the 38 × 89 mm studs are sheathed by drywall on both sides. This has the effect of fully bracing the weak axis but leaving the strong axis unbraced.
   (a) Determine the critical buckling load for a single 38 × 89 stud in such a wall if it is 2.4 m tall. Assume that $E = 9500$ MPa. Compare your answer to that of Problem 11–8M.
   (b) Determine the critical load for the wall (*w* in kN per linear meter of wall) if the studs are spaced 400 mm o.c.

**11–19.** Work Problem 11–9 if the weak axis of the wall W12 × 50 is braced 30 ft up from its lower end.

**11–19M.** Work Problem 11–9M if the weak axis of the wall W310 × 74 is braced 10 m up from its lower end.

**11–20.** A W8 × 10 section is used as a 30-ft-long column. The upper end is pinned, the lower end fixed, and the weak axis braced at midheight by two angles as shown in Figure 11–15. Determine the critical buckling stress. (*Hint:* Watch the K values.)

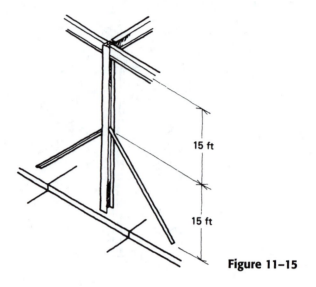

**Figure 11–15**

**11–20M.** A W200 × 15 section is used as a 10-m-long column. The upper end is pinned, the lower end fixed, and the weak axis braced at midheight by two angles, as shown in Figure 11–15M. Determine the critical buckling stress.

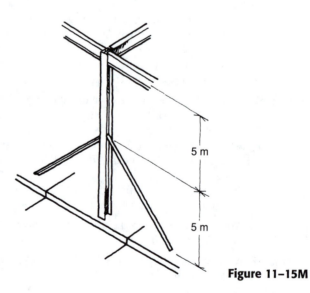

**Figure 11–15M**

**11–21.** Determine the critical buckling stress and load for a wood 2 × 6 if it is 20 ft long, has pinned ends, and has its weak axis braced at 5-ft intervals. Assume that $E = 1.2(10)^6$ psi.

**11–21M.** Determine the critical buckling stress and load for a wood 38 × 139 mm if it is 6 m long, has pinned ends, and has its weak axis braced at 1.5-m intervals. Assume that $E = 9500$ MPa.

**11–22.** Figure 11–16 shows a C10 × 30 channel used as a long pinned-end compression member of length $L$. Determine the optimum spacing $XL$ of intermediate bracing elements such that the critical buckling load will be the same for both axes. $X$ will be a fraction of $L$.

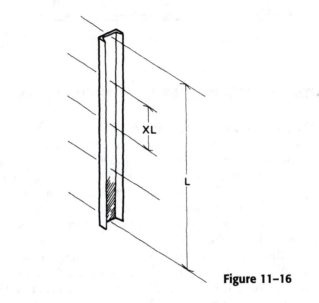

**Figure 11–16**

**11–22M.** Figure 11–16 shows a C250 × 45 channel used as a long pinned-end compression member of length $L$. Determine the optimum spacing $XL$ of intermediate bracing elements such that the critical buckling load will be the same for both strong and weak axes. $X$ will be a fraction of $L$.

**11–23.** The Douglas fir No. 2 stud bearing wall in Figure 11–17 is 8 ft tall and must support a load of 2 kips per running foot. The wall is not finished by sheathing or drywall on

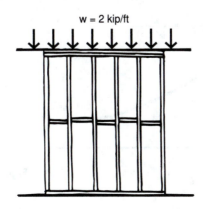

w = 2 kip/ft

**Figure 11–17** Elevation of a stud wall.

either side but it does have blocking at midheight. Will 2 × 4 studs 16 in. o.c. be sufficient if a factor of safety of 2.0 is required?

**11–24M.** The hem-fir No. 2 stud bearing wall of Figure 11–17M has 38 × 139 members spaced 600 mm o.c. Using a factor of safety of 2.5, how tall can the wall safely be and still support 22 kN per running meter?

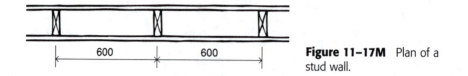

600                           600                           **Figure 11–17M** Plan of a
                                                            stud wall.

# 11–6   LIMITS TO THE APPLICABILITY OF THE EULER EQUATION

The plot of the Euler equation in Figure 11–18 shows that it is asymptotic to both axes. We can see that for very low values of $KL/r$ (i.e., for short stout columns), the critical stress becomes very high. Indeed, as $KL/r$ approaches zero, $(P/A)_{cr}$ goes to infinity. Obviously, this cannot be valid because the stresses in this region of the graph would be above the yield stress or crushing stress of the material. It is clear that the Euler equation cannot be valid if it predicts a buckling stress above the yield stress. A short column will fail by crushing under a load that is less than that predicted by the Euler equation.

The value $(P/A)_{cr}$ represents an average unit stress over the entire cross section. When a column is buckled into an arc, the stresses are not uniform over the cross section. The maximum fiber stress will be compressive, occurring on the inside of the arc. For some columns of relatively low slenderness, $(P/A)_{cr}$ may be large enough to cause a yield or crushing of these fibers. This implies that the Euler equation for elastic buckling is not completely valid except for long, thin columns. A col-

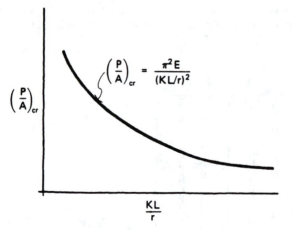

**Figure 11–18**   Euler equation.

umn with a moderately (often called "intermediate") low slenderness ratio will still basically buckle rather than crush, but as this occurs some of the fibers will reach their yield stress. This type of buckling is called *inelastic buckling* and is really a combination of buckling and crushing.

In terms of developing design equations for safe loads on columns, both the timber and steel industries have come up with appropriate lower limits on $KL/r$. Below these values, the Euler equation cannot be used, and empirically developed equations to handle these shorter columns are employed. Since this chapter is intended to treat only elastic buckling, most of the examples and problems provided involve relatively long and slender columns. Some of the problems even involve slenderness ratios that are above code limits.

## 11-7 ECCENTRIC LOADING AND BEAM-COLUMNS

In the previous sections of this chapter we have assumed that the loads acting on columns are axial. This represents an ideal condition seldom realized outside an experimental testing laboratory. It is true that loads on a column from columns stacked directly on top of it do produce axial loads. However, when beams and girders frame into the sides of a column (standard practice for construction stability and ease), as shown in several figures in the early part of this chapter, the loads that are delivered to the column by those members are not axial. They are eccentric to the column's axis by a significant distance. This results in an applied moment to the top of the column as illustrated in Figure 2–51.

Even if the framing system were fully symmetrical, as in Figure 11–5, an effective axial load would occur only if each of the four trusses carried exactly the same load. Since floor and roof members carry live load as well as dead, this is highly unlikely. Under such eccentric loads, a column has an increased tendency to buckle due to bending caused by the applied moments. These columns are more properly called *beam-columns*. In the examples and problems of this chapter, the bending due to such inevitable eccentricities has been ignored, with the result that the critical stresses and loads computed are optimistically high, often considerably so.

Bending also occurs more obviously in some columns. For example, almost every column located in an exterior wall must resist lateral loads due to wind pressures and suctions delivered to it from the wall. These members are also beam-columns. Special formulas beyond the scope of this book are used to handle both eccentric and lateral loads on columns.

# 12

---

# Trusses

---

## 12–1    INTRODUCTION

A truss is a lightweight frame generally used for relatively long spans in buildings and bridges. They are usually placed parallel to one another to make a one-way system for a floor or roof deck. Their lightness means they are deeper than beams would be if used on a similar span, and for this reason trusses are more frequently used in roof structures.

In the United States, trusses are almost always constructed of wood or steel, but in other countries they have also been precast in reinforced concrete. The light triangular wood truss, made up of nominal 2 × 4 and 2 × 6 (38 × 89 and 38 × 139) elements and placed 2 ft (600 mm) on center, is used almost exclusively to make residential gable roofs in many parts of the United States. It erects rapidly and enables the floor below to be free of interior bearing partitions. Steel trusses, both flat and curved, are used to span the large majority of long-span buildings such as field houses and gymnasiums. In such structures, self-weight can easily become a controlling design factor, and the small span/depth ratio of a truss (with its increased building envelope) becomes a welcome trade-off to minimize this dead load.

Trusses can be fabricated in almost any shape. In technical terms, a truss is a triangulated planar framework made up of linear elements that connect at pin joints. When actually constructed, these joints are seldom truly pinned, but the initial structural analysis makes this assumption anyway. (For many trusses, the members are thin and have relatively little bending resistance, so the pinned-joint assumption causes no great error.) A few of the more commonly used truss shapes are illustrated in Figure 12–1. Some have been named for the engineer or designer who popularized that particular type.

Loaded properly, each member of a truss is (in ideal terms) a two-force member. It is either in tension or compression, and if in compression, it behaves as a slender column and must be designed with buckling in mind. When trusses are used on simple spans, all the top chord members form a continuous line of compression and the entire top of the truss is subject to the lateral buckling phenomenon discussed in Section 7–5. Usually, the roof "skin" provides the required lateral bracing unless the trusses are exposed. Overhanging trusses will have compression in some bottom chord elements, and these are subject to the same buckling effects.

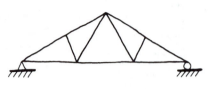

Fink

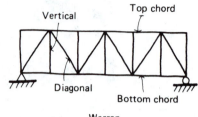

Warren

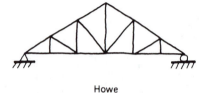

Howe

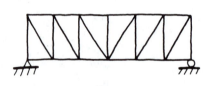

Howe

Pratt

Pratt

Bowstring

Trussed arch

**Figure 12–1**

For the purposes of preliminary design, it is assumed that trusses are loaded by concentrated loads that act only at the joints. In actuality, most floor and roof loads are uniform, and when the deck surface is attached directly to the top chord elements, these members are subjected to the combined action of axial and bending forces. The examples and problems of this chapter are concerned only with the analysis of trusses loaded through the joints, or "panel points," and all the members will be considered to carry only axial loads.

Almost all trusses are statically determinate with respect to the external reaction components. Depending on the manner of triangulation, trusses can be determinate or indeterminate with respect to the internal forces in the members. Trusses

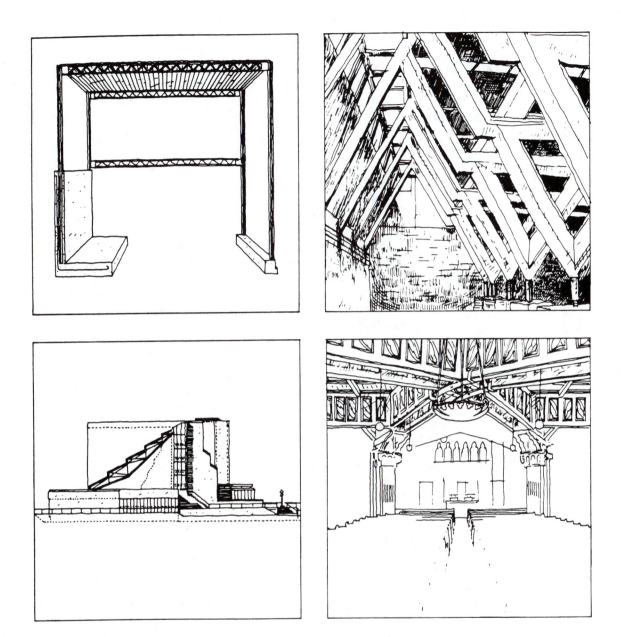

Be they straightforward or complex, trusses are relatively long-span building elements that offer considerable possibility for architectural effect along with the efficient performance of structural jobs. The variety of truss shapes, types, and adaptations is large, but even the simplest can have both visual and intellectual fascination: the attraction of a well-made puzzle, the paradox of a large strong thing made from many small weak things. Further, the usefulness of trusses in roof structures is an eternal invitation to experiments with the interacting effects of structure and light on interior space. From Charles Eames' unassumingly elegant use of stock steel bar joists in his own California house, to Frank Lloyd Wright's more spectacular (but really no more complicated) wood trusses over the drafting room at Taliesin East, the range of expression available with simple trusses is enormous. But complexity, too, has its places and desirable effects; the technically complex trusses and the soaring space shadowed by them in James Stirling's Cambridge History Faculty and the decoratively complex trusses and brooding interior of Bernard Maybeck's First Church of Christ Scientist are two diverse instances.

with redundant members are internally statically indeterminate, and the member forces cannot be resolved using statics alone. Whether or not a truss is internally determinate can be ascertained by Equation 12–1. Trusses without enough members to make triangles using every joint will be unstable, and those with excess members are indeterminate.

$$m + 3 = 2j \qquad (12\text{–}1)$$

where $m$ = number of members, assuming no member runs through a joint
$j$ = number of joints

In Equation 12–1, the constant 3 represents the usual three external reaction components. The concept here is that the number of unknowns equals the unknown member forces ($m$) plus the reaction components. At a planar joint, only two force equations of equilibrium, $\Sigma F_x = 0$ and $\Sigma F_y = 0$, can be written, and this means that the total number of available equations is twice the number of joints ($j$).

The trusses in Figure 12–1 are determinate, as are all of the trusses in the examples and problems. Figure 12–2 shows two indeterminate trusses. The one in Figure 12–2(b) has two diagonals, which cross without a joint. This type of truss becomes determinate if we assume that those two members are so slender and flexible as to be worthless in compression, in which case only one of them will be functional, depending upon the loading pattern. The diagonals are then called *counters*.

The geometry imposed by triangulation means that, under certain loading conditions, some of the members of a truss may have no internal force. In such cases, the member acts as a bracing element and is usually needed for stability. These *zero members* could also carry force under a different loading pattern.

## 12–2   ANALYSIS BY JOINT EQUILIBRIUM

If we assume that all joints are pinned and that loads and reactions act only at the joints, each joint becomes a small concurrent force system. It must be held in equilibrium by the known forces acting on it from the loads (including reactions) and by the unknown forces from the two-force members. Each joint can then be analyzed like the simple structures of Section 2–4. As pointed out previously, there are only

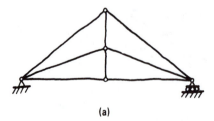

(a)

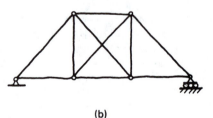

(b)

**Figure 12–2**  Intermediate trusses.

two equations available for each joint, so we must move from joint to joint over the truss in such a manner as to be always working with only two unknowns. In many cases, this means starting at one of the joints at the ends of the truss and progressing toward the center.

The external reaction components should be determined before isolating the joints, and this has been done in the examples that follow. After solving for the reactions of a given truss, the reader should attempt to guess which members are in tension and compression before continuing with the solution. The answers obtained from any numerical analysis can then be rationalized with the visual analysis, and arithmetical errors can often be caught before they accumulate.

---

**EXAMPLE 12–1**

Determine the forces in each of the members of the truss in Figure 12–3. A free-body diagram of the truss is shown in Figure 12–4.

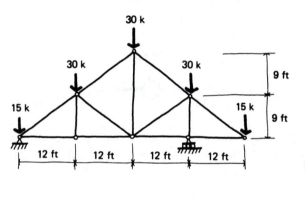

**Figure 12–3**

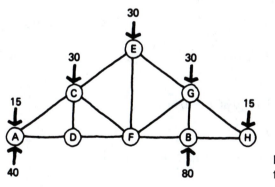

**Figure 12–4** Free-body diagram of the truss.

### Solution

Make a free-body diagram of each joint in turn, showing compressive arrows acting toward the joint and tensile arrows pointing away from the joint. As usual, incorrect sense assumptions will result in negative answers.

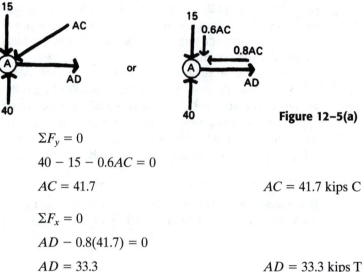

**Figure 12–5(a)**

$$\Sigma F_y = 0$$

$$40 - 15 - 0.6AC = 0$$

$$AC = 41.7 \qquad\qquad AC = 41.7 \text{ kips C}$$

$$\Sigma F_x = 0$$

$$AD - 0.8(41.7) = 0$$

$$AD = 33.3 \qquad\qquad AD = 33.3 \text{ kips T}$$

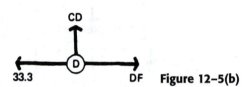 **Figure 12–5(b)**

(The presence of members having zero force is not unusual; such members will usually go into tension or compression under other loading patterns.)

$$\Sigma F_x = 0$$

$$DF - 33.3 = 0$$

$$DF = 33.3 \qquad\qquad DF = 33.3 \text{ kips T}$$

$$\Sigma F_y = 0 \qquad\qquad CD = 0$$

$$\Sigma F_x = 0$$

$$33.3 - 0.8CE - 0.8CF = 0$$

$$\Sigma F_y = 0$$

$$-30 + 25 - 0.6CE + 0.6CF = 0$$

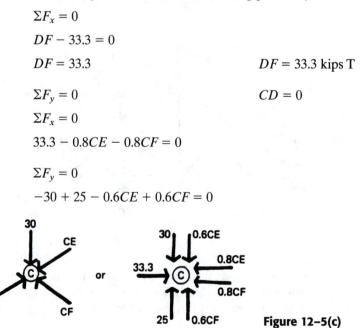

**Figure 12–5(c)**

Solving simultaneously gives us

$$CF = 25 \qquad\qquad CF = 25 \text{ kips C}$$
$$CE = 16.7 \qquad\qquad CE = 16.7 \text{ kips C}$$

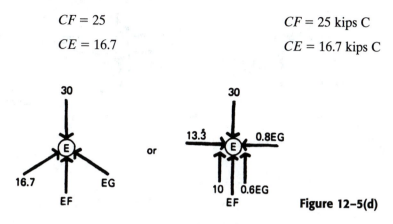

**Figure 12–5(d)**

$$\Sigma F_x = 0$$
$$13.3 - 0.8EG = 0$$
$$EG = 16.7 \qquad\qquad EG = 16.7 \text{ kips C}$$

$$\Sigma F_y = 0$$
$$-30 + 10 + EF + 0.6(16.7) = 0$$
$$EF = 10 \qquad\qquad EF = 10 \text{ kips C}$$

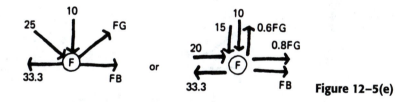

**Figure 12–5(e)**

$$\Sigma F_y = 0$$
$$-15 - 10 + 0.6FG = 0$$
$$FG = 41.7 \qquad\qquad FG = 41.7 \text{ kips T}$$

$$\Sigma F_x = 0$$
$$-33.3 + 20 + 0.8(41.7) + FB = 0$$
$$FB = -20 \qquad\qquad FB = 20 \text{ kips C}$$

Sense of *FB* was assumed incorrectly.

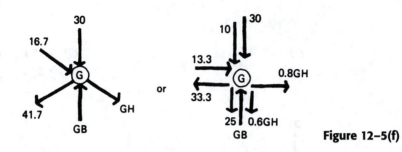

**Figure 12–5(f)**

$$\Sigma F_x = 0$$

$$-33.3 + 13.3 + 0.8GH = 0$$

$$GH = 25 \qquad\qquad\qquad\qquad GH = 25 \text{ kips T}$$

$$\Sigma F_y = 0$$

$$-10 - 30 - 25 - 0.6(25) + GB = 0$$

$$GB = 80 \qquad\qquad\qquad\qquad GB = 80 \text{ kips C}$$

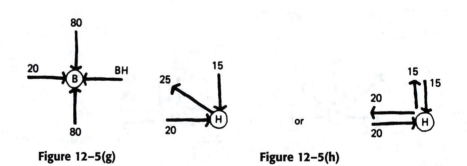

**Figure 12–5(g)**                              **Figure 12–5(h)**

$$\Sigma F_x = 0$$

$$-BH + 20 = 0$$

$$BH = 20 \qquad\qquad\qquad\qquad BH = 20 \text{ kips C}$$

Joint $H$ is isolated as a check in Figure 12–5(h).
    The member forces are shown in Figure 12–6.

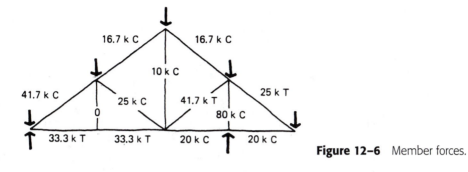

**Figure 12–6** Member forces.

**EXAMPLE
12–2M**

Determine the forces in each member of the wind bent shown in Figure 12–7M. A free-body diagram of the wind bent is shown in Figure 12–8M.

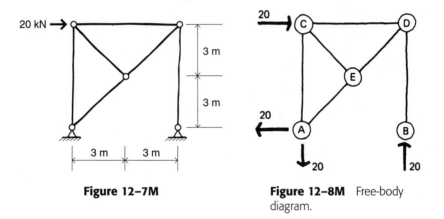

**Figure 12–7M**

**Figure 12–8M** Free-body diagram.

### Solution

Make a free-body diagram of each joint in turn, showing compressive arrows acting toward the joint and tensile arrows pointing away from the joint. As usual, incorrect sense assumptions will result in negative answers.

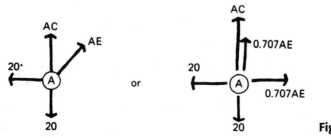

**Figure 12–9M(a)**

$$\Sigma F_x = 0$$

$$-20 + 0.707AE = 0$$

$$AE = 28.3 \hspace{4cm} AE = 28.3$$

kN T

$$\Sigma F_y = 0$$

$$AC + 0.707(28.3) - 20 = 0$$

$$AC = 0 \hspace{4cm} AC = 0$$

(The presence of members having zero force is not unusual; such members will usu-ally go into tension or compression under other loading patterns.)

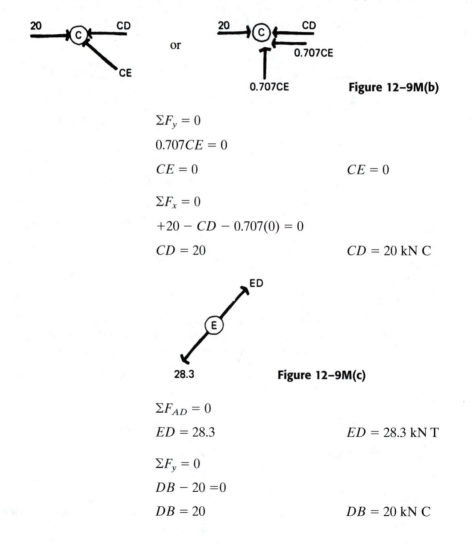

or

Figure 12–9M(b)

$$\Sigma F_y = 0$$

$$0.707CE = 0$$

$$CE = 0 \hspace{4cm} CE = 0$$

$$\Sigma F_x = 0$$

$$+20 - CD - 0.707(0) = 0$$

$$CD = 20 \hspace{4cm} CD = 20 \text{ kN C}$$

Figure 12–9M(c)

$$\Sigma F_{AD} = 0$$

$$ED = 28.3 \hspace{4cm} ED = 28.3 \text{ kN T}$$

$$\Sigma F_y = 0$$

$$DB - 20 = 0$$

$$DB = 20 \hspace{4cm} DB = 20 \text{ kN C}$$

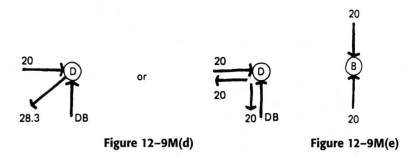

**Figure 12–9M(d)**      **Figure 12–9M(e)**

Joint *B* is isolated as a check in Figure 12–9M(e).
The member forces are shown in Figure 12–10M.

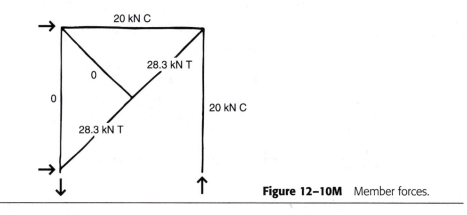

**Figure 12–10M**    Member forces.

## PROBLEMS

**12–1.** Determine the forces in the members of the truss in Figure 12–11.

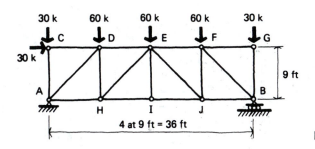

**Figure 12–11**    Flat Howe truss.

**12–1M.** Determine the forces in the members of the truss in Figure 12–11M.

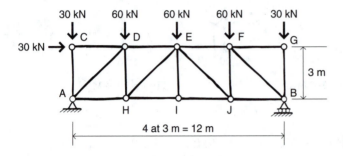

**Figure 12–11M**   Flat Howe truss.

**12–2.** Determine the forces in the members of the truss in Figure 12–12.

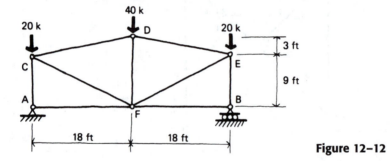

**Figure 12–12**

**12–2M.** Determine the forces in the members of the truss in Figure 12–12M.

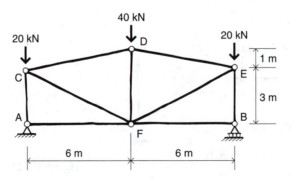

**Figure 12–12M**

**12–3.** Determine the forces in the members of the Fink truss in Figure 12–13.

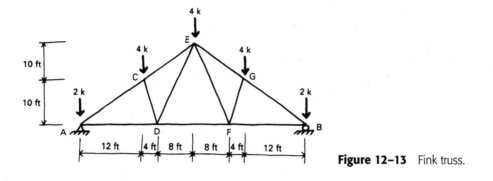

**Figure 12–13** Fink truss.

**12–3M.** Determine the forces in the members of the shed roof truss in Figure 12–13M.

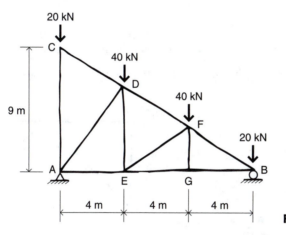

**Figure 12–13M** Shed roof truss.

**12–4.** Determine the forces in the members of the cantilevered roof truss in Figure 12–14.

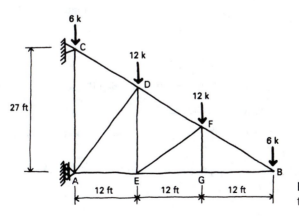

**Figure 12–14** Cantilevered roof truss.

**12–4M.** Determine the forces in the overhanging truss in Figure 12–14M.

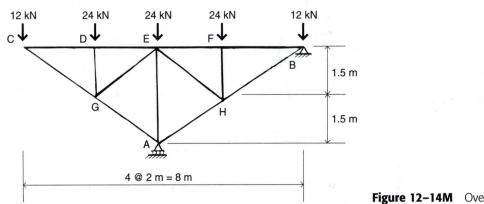

**Figure 12–14M**  Overhanging truss.

**12–5.** Determine the forces in the members of the truss in Figure 12–15.

**12–5M.** Determine the forces in the members of the truss in Figure 12–15M.

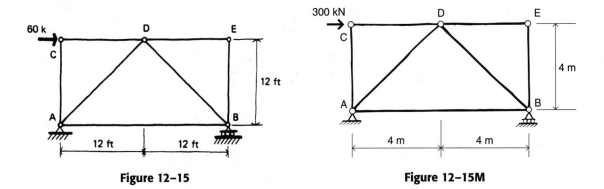

**Figure 12–15**

**Figure 12–15M**

**12–6.** Determine the forces in the members of the hillside truss in Figure 12–16.

**12–6M.** Determine the forces in the members of the hillside truss in Figure 12–16M.

**12–7.** Work Problem 12–6 after adding a 4-kip wind load acting to the right at joint *C* in Figure 12–16.

**12–7M.** Work Problem 12–6M after adding a 20-kN wind load acting to the right at joint *C* in Figure 12–16M.

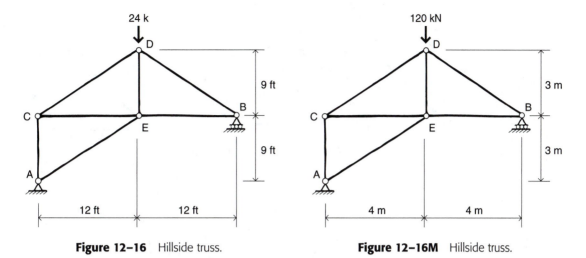

**Figure 12–16**   Hillside truss.          **Figure 12–16M**   Hillside truss.

## 12–3   METHOD OF SECTIONS

A second approach to truss analysis, called the *method of sections*, consists of cutting sections through the truss so that a free-body diagram of a portion of the truss will involve the desired unknown member forces. In general, it is faster than the method of joints because it makes use of all three equations of planar static equilibrium. The method of joints deals with concurrent forces and therefore no moment arms are available.

Realizing that the entire truss is in equilibrium and that each joint is in equilibrium, it follows that larger portions of the truss will also be in equilibrium. If a truss is cut through by an imaginary cutting plane and a portion to one side of that plane is isolated, it will be held in a state of balance by the external forces acting on the truss and the unknown forces in the cut members. Since the free-body diagram makes these internal forces external, the equations of statics can be used to find them. With three equations available, three unknown member forces can be determined with each free-body cut.

Successive cutting planes may be used to isolate increasingly larger portions of the truss, as shown in Figure 12–17. However, one of the advantages to this method is that all the member forces need not be found if we are interested only in those in one area of the truss. (Whenever a section cuts through two concurrent unknowns, as in Figure 12–17(b), this method reduces to a joint equilibrium problem. The two procedures, of course, can be used to work different parts of the same truss.)

The senses of the forces in the unknown members are assumed so that an arrow acting against the cutting plane is compressive and one pulling away from it is tensile. As usual, a negative sign in the answer will indicate an incorrect assumption. The isolated portion of the truss is treated as a rigid free body, and moment centers may be located on or off the body. The forces in cut sloping members are usually broken down into their rectangular components so that moment arms can be ob-

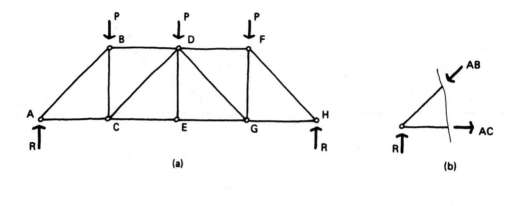

(a)                                                                                          (b)

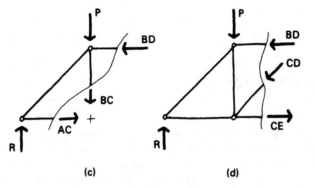

(c)                                          (d)

**Figure 12–17**  Cutting planes.

tained from known dimensions. It is helpful to realize that a resultant force can be translated forward or backward along its line of action (before being replaced by its components), and this can sometimes result in convenient moment-arm distances. This translation does not in any way affect the state of equilibrium.

**EXAMPLE 12–3**    Determine the forces in members *CE, DE,* and *DF* of the truss shown in Figure 12–18.

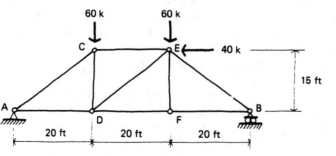

**Figure 12–18**

### Solution

First find the external reactions and then cut a section through the three members. Applying the equations of statics to the body in Figure 12–19(b), we get

$$\Sigma F_y = 0$$

$$-60 + 50 + 0.6DE = 0$$

$$DE = 16.7 \qquad\qquad\qquad DE = 16.7 \text{ kips C}$$

$$\Sigma M_E = 0$$

$$50(20) - DF(15) = 0$$

$$DF = 66.7 \qquad\qquad\qquad DF = 66.7 \text{ kips T}$$

$$\Sigma M_D = 0$$

$$-CE(15) - 60(20) + 40(15) + 50(40) = 0$$

$$CE = 93.3 \qquad\qquad\qquad CE = 93.3 \text{ kips C}$$

$$\Sigma F_x \overset{?}{=} 0 \text{ (check)}$$

$$CE + 0.8DE - DF - 40 = 0$$

$$93.3 + 0.8(16.7) - 66.7 - 40 \overset{\checkmark}{\approx} 0$$

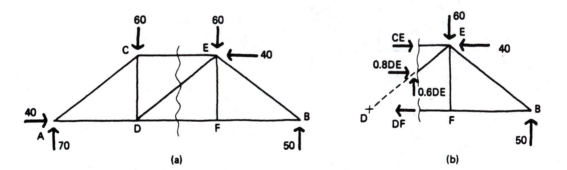

**Figure 12–19**  Section line and free-body diagram.

**EXAMPLE
12–4M**

Determine the forces in members *EG*, *FG*, and *FB* of the overhanging truss in Figure 12–20M.

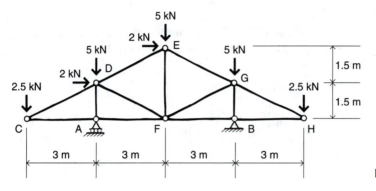

**Figure 12–20M**

### Solution

Using the isolated portion in Figure 12–21M(b), we get

$$\Sigma M_G = 0$$

$$-4(1.5) - 2.5(3) + FB(1.5) = 0$$

$$FB = 9 \qquad\qquad\qquad FB = 9 \text{ kN C}$$

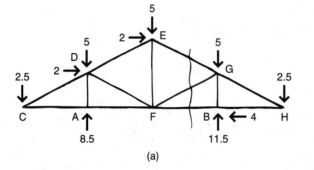

(a)

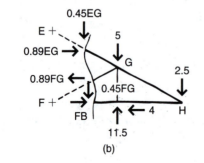

(b)

**Figure 12–21M**  Section line and free-body diagram.

$$\Sigma M_F = 0$$

(Let the force $EG$ be translated back to point $E$, where its vertical component will have no moment arm with respect to moment center $F$.)

$$-5(3) + 11.5(3) - 2.5(6) - 0.89EG(3) = 0$$

$$EG = 1.7 \qquad\qquad\qquad EG = 1.7 \text{ kN C}$$

$$\Sigma M_H = 0$$

(Let the force $FG$ be translated forward to point $F$, where its horizontal component will have no moment arm with respect to the moment center $H$.)

$$5(3) - 11.5(3) + 0.45FG(6) = 0$$

$$FG = 7.2 \qquad\qquad\qquad FG = 7.2 \text{ kN T}$$

Since neither force equation was used, either will be valid for a check.

$$\overset{?}{\Sigma F_x = 0} \text{ (check)}$$

$$0.89EG - 0.89FG - 4 + 9 = 0$$

$$0.89(1.7) - 0.89(7.2) + 5 \overset{\checkmark}{\approx} 0$$

## PROBLEMS

**12–8.** Use the method of sections combined with joint equilibrium (as necessary) to determine the forces in the members of the Warren truss in Figure 12–22.

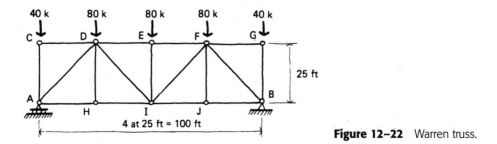

**Figure 12–22** Warren truss.

**12–8M.** Use the method of sections combined with joint equilibrium to determine the forces in the members of the Warren truss in Figure 12–22M.

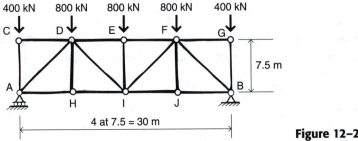

**Figure 12–22M**  Warren truss.

**12–9.**  Use a cut section to determine the force in member *DB* of the truss in Figure 12–15.

**12–9M.**  Use a cut section to determine the force in member *DB* of the truss in Figure 12–15M.

**12–10.**  Use a cut section to find the force in member *DE* of the truss in Figure 12–16.

**12–10M.**  Use a cut section to find the force in member *DE* of the truss in Figure 12–16M.

**12–11.**  Use the method of sections to determine the forces in members *DF, EF,* and *EG* of the cantilevered roof truss in Figure 12–14.

**12–11M.**  Use the method of sections to determine the forces in members *DF, EF,* and *EG* of the shed roof truss in Figure 12–13M.

**12–12.**  Use the method of sections combined with the method of joints (as necessary) to determine the forces in the members of the cantilevered truss in Figure 12–23.

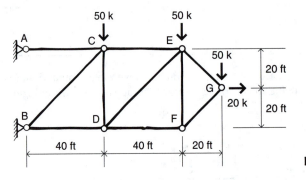

**Figure 12–23**  Cantilevered truss.

**12–12M.**  Use the method of sections combined with the method of joints (as necessary) to determine the forces in the members of the overhanging truss in Figure 12–14M.

**12–13.**  Determine the force in each member of the three-story wind bent in Figure 12–24.

**12–13M.**  Determine the force in each member of the three-story wind bent in Figure 12–24M.

**12–14.**  Determine the force in the tensile tie *AB* of the truss in Figure 12–25.

**12–14M.**  Determine the force in member *HI* of the arched truss in Figure 12–25M.

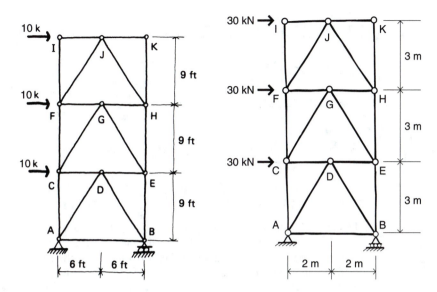

Figure 12–24   Wind bent.

Figure 12–24M   Wind bent.

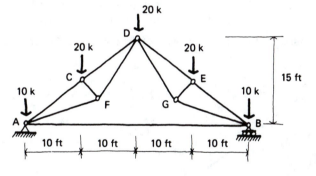

Figure 12–25

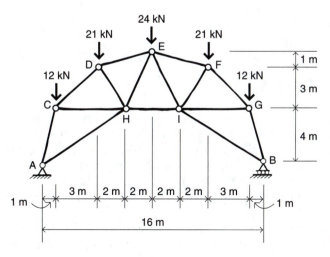

Figure 12–25M   Arched truss.

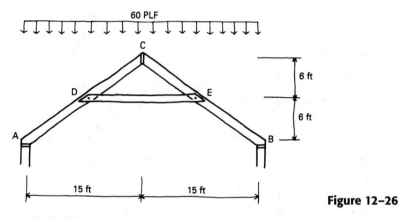

**Figure 12–26**

**12–15.** Figure 12–26 shows a schematic section through a residential roof system using a collar beam connecting two rafters, $AC$ and $BC$. The only load on the roof is the snow load, shown acting along the horizontal projection with a magnitude of 60 lb per foot. (Assume the weight of the members to be negligible, as in the other problems.)

(a) Assuming that the walls at $A$ and $B$ offer negligible resistance to lateral forces and that joint $C$ acts as a pin, determine the force in the collar beam $DE$.

(b) If the collar beam were relocated so that it connected joints $A$ and $B$ (like an attic floor joist), what force would it then have to carry?

**12–16M.** Determine the force in member $EF$ of the fan truss in Figure 12–27M.

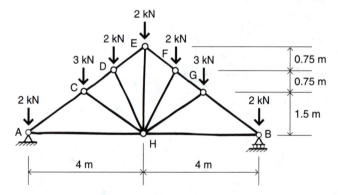

**Figure 12–27M** Fan truss.

## 12–4 SPECIAL TYPES OF TRUSSES

There are two categories of trusses that are used so frequently in construction today that they warrant special attention. One is the prefabricated light timber truss mentioned in the first section of this chapter. They are usually made up of pieces small in cross section and fastened at the joints so that all members can lie in the same plane. These joints are nailed together using special toothed light-gage-steel plates as seen in Figure 12–28. Whenever possible, chord members continue through joints for ease of fabrication.

When fully braced by a roof plane, these trusses are very stiff and much stronger than they appear. Specifications have been set in a standard manner so

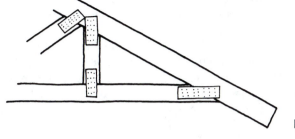

**Figure 12–28**  Nailing plates.

that, if no special loads or unusual support conditions are present, these trusses can be ordered almost as stock items, with obvious economy.

The second type of frequently used truss is the open-web steel bar joist, often referred to simply as a bar joist. They are usually fabricated from angles and round bar stock, with the larger ones using only angles. Shop-welded together, using continuous chord members, they are manufactured to meet certain load capacities as specified by the Steel Joist Institute. Without special detailing, they are not suitable for concentrated loads but can handle the uniform loads of most floors and flat (or nearly flat) roofs over a very wide range of spans. They are seldom designed in the usual sense. Like rolled steel beams, they are selected from load tables that have been developed with due consideration for the moments, shears, and deflections involved in simple spans.

Unlike wood joists and precast concrete joists usually used on shorter spans, open web joists easily permit the through passage of wires and pipes and even small ducts. The minimum on-site labor required for their erection means that steel joists can often provide a very economical deck system. They are particularly well suited for one-story structures with high ceilings, such as factories and gymnasiums, where fireproofing and acoustics needs are minimal.

A third type of special truss is not really a truss at all. As shown in Figure 12–29, the Vierendeel "truss" is really more like a rigid frame or a beam with large holes. The absence of triangulation and the presence of fully moment-resistant joints mean that this structure is grossly misnamed when called a truss. The Vierendeel frame takes its name from its designer, Arthur Vierendeel (1825–c.1930), a Belgian engineer and builder. It is usually made of reinforced concrete, which inherently provides the required joint fixity, but can also be fabricated from steel. It carries its load through the development of bending stresses in all or most of the segments. As

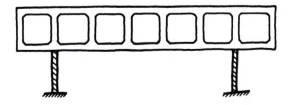
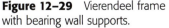

**Figure 12–29**  Vierendeel frame with bearing wall supports.

the frame bends, the members assume "S" patterns for their deflected shapes because the joints apply moments to the ends of the members as they rotate. This joint continuity makes the structure highly indeterminate.

While the Vierendeel frame is quite inefficient compared to a truss, it can be very useful in specific structural situations. The lack of diagonal members means that there are large clear openings in the frame that can be used functionally. Their best application occurs when the span and loads are such that a frame equal in depth to one story height is required. If the floor plan can accommodate the web verticals, the frame will become an integrated part of the architectural section. The Vierendeel frame of Figure 12–29 is used to provide a large column-free area beneath a heavily loaded second floor.

# Appendix A

## Derivation of Basic Flexural Stress Equation

The beam in Figure A–1 has a rectangular cross section, which has been specified here for the sake of convenience and clarity. The beam cross section can be any shape and the following derivation will still be valid. It is necessary, however, to make a number of other assumptions concerning the material and the geometry, and these will be listed at the end of this appendix. The most important of these is that "planes before bending remain plane after bending," which is stated graphically in Figure A–1(b). To understand this, visualize a straight unloaded beam and make a series of imaginary vertical slices through it quite close together. The planes made by the imaginary slicer should be parallel to each other and normal to the beam axis. When the beam is bent under load, as in Figure A–1(a), the planes will not warp or twist out of shape but will merely tilt toward one another. They will remain flat, getting closer together where the fibers are in compression and farther apart where tension occurs. This assumption is quite valid and can be proven visually using a material of low stiffness.

There will be a horizontal plane, designated ab in Figure A–1(b) and called the neutral axis, which will neither lengthen nor shorten. The fact that "planes remain plane" ensures that the unit strain, $\epsilon$, will be proportional to its distance from the neutral axis. The fiber located $y$ distance above the neutral axis in Figure A–1(b) has an original unit length of $dx$ and has a total strain of $\epsilon_y$. Similarly, the top fiber of the beam, $c$ distance from the neutral axis, has the same original unit length and a total strain of $\epsilon_c$. By similar triangles,

$$\frac{\epsilon_y}{\epsilon_c} = \frac{y}{c} \tag{A–1}$$

Since stress is proportional to strain by Hooke's law (if we keep the stresses in the elastic region for the material), then

$$\frac{f_y}{f_c} = \frac{y}{c} \tag{A–2}$$

as illustrated in Figure A–2(c) by the triangular, straight-line stress distribution.

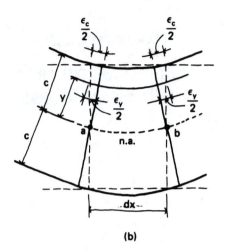

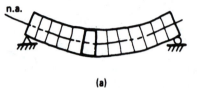

(a)                                                                (b)

**Figure A–1**  Flexural strain.

The value of $f_y$ varies from zero at the neutral axis to a maximum value of $f_c$ at the extreme fibers of the section. Each stress $f_y$ acts on a small area $dA$ of beam cross section and causes a small moment about the neutral axis.

$$dM = [f_y(dA)]y$$

The summation of these small moments must equal the couple $M$, which is caused by the external loads. (The value of $M$, of course, varies along the length of a beam as represented by a moment diagram.)

$$M = \int_0^A f_y y \, dA$$

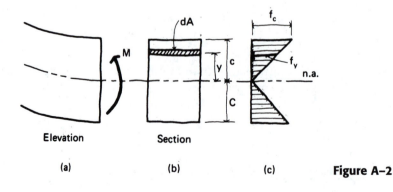

Elevation            Section

(a)                    (b)                    (c)                **Figure A–2**

The variable $f_y$ can be eliminated by using Equation A–2:

$$f_y = f_c\left(\frac{y}{c}\right)$$

giving us

$$M = \int_0^A \frac{f_c}{c} y^2 \, dA$$

Removing the constants, we get

$$M = \frac{f_c}{c} \int_0^A y^2 \, dA$$

in which the quantity $\int_0^A y^2 \, dA$ is the moment of inertia of the cross section taken with respect to the neutral axis. Making the change, we get

$$M = \frac{f_c}{c} I_{\text{n.a.}}$$

Solving for bending stress, we get

$$f_c = \frac{Mc}{I_{\text{n.a.}}} \tag{A–3}$$

where $f_c$ is the extreme fiber bending stress. (*Note:* $f_c$ is often written as $f_b$ with $b$ understood to mean "extreme fiber bending.")

It will be proven that the neutral axis is coincident with the centroidal axis, as implied by the dimensions labeled $c$ in Figure A–1(b). Therefore, $I_{\text{n.a.}} = I_{\text{c.g.}}$ and the subscripts are usually deleted.

$$f = \frac{Mc}{I} \tag{7–1a}$$

The stress at some fiber other than at the top or bottom locations is found as

$$f_y = \frac{My}{I} \tag{7–1}$$

To show that the neutral axis is a centroidal one, examine Figure A–3. The forces, $f_y dA$, must algebraically add to zero to ensure horizontal equilibrium ($\Sigma F_x = 0$) for the beam section on which they act.

$$\int_0^A f_y dA = 0$$

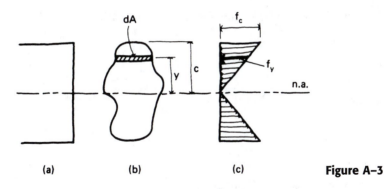

**(a)**              **(b)**              **(c)**              **Figure A–3**

However, $f_y = f_c(y/c)$ as before, so

$$\frac{f_c}{c}\int_0^A y\, dA = 0$$

The quantity $f_c/c$ is clearly not equal to zero; therefore,

$$\int_0^A y\, dA = 0$$

From Chapter 3 the reader should recall that this integral is the expression for the statical moment of the area in Figure A–3(b), using the neutral axis as the reference axis. The only way the statical moment of an area can be zero is if the reference axis is a centroidal axis.

The general formula for flexural stress developed here is subject to the following idealistic restrictions:

1. Transverse sections remain plane.
2. The beam is straight, of constant cross section, and does not twist under load.
3. The material is homogeneous and isotropic in the direction of stress.
4. The proportional limit is not exceeded.
5. The deformations are small.

# Appendix B

## Derivation of Basic Horizontal Shearing Stress Equation

As with the derivation of the flexural stress formula, a rectangular cross section will be used here. This is done only for simplicity, and the formula so developed is not restricted as to cross-sectional shape.

The beam shown with its moment diagram in Figure B–1(a) is typical in that its moment varies from one transverse section to another (e.g., from section 1 to section 2). In this case, $M_2$ is slightly greater than $M_1$, by the amount $dM$. Figure B–2 shows this difference in moment between sections 1 and 2 by showing the normal stresses that act on those sections. There is an increase in these stresses acting on the transverse sections containing the element $dA$ as we move from plane 1 to plane 2. This change in stress will result in an unbalanced force on the block that has dimensions $b$, $c - y'$, and $dx$. Therefore, a stress $f_v$ is developed to put this block back in equilibrium. The magnitude of this horizontal shearing stress at level $y$ can be determined from the basic equilibrium equation,

$$\Sigma F_x = 0$$

$$f_v b \, dx = C_2 - C_1$$

where $C_2$ and $C_1$ of Figure B–3 are resultants or summations of the normal stresses, shown in Figure B–2(b) acting on the elemental areas $dA$.

$$C_2 = \int_{y'}^{c} f_{y_2} dA$$

$$C_1 = \int_{y'}^{c} f_{y_1} dA$$

In these equalities, $f_y$ can be replaced by the appropriate values of $My/I$, where $M$ is a constant for a given section and $I$ is presumed constant for the entire beam.

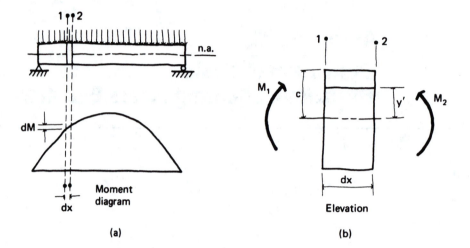

(a)                    (b)

**Figure B–1**

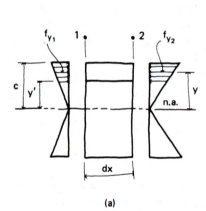

(a)

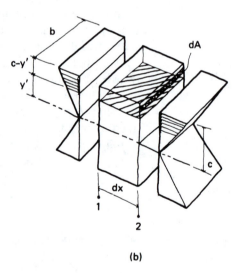

(b)

**Figure B–2**   Change in bending stress.

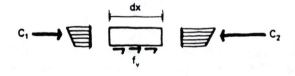

**Figure B–3**   Development of horizontal shearing stress.

$$C_2 = \frac{M_2}{I} \int_{y'}^{c} y \, dA$$

$$C_1 = \frac{M_1}{I} \int_{y'}^{c} y \, dA$$

Substituting into Equation B–1, we get

$$f_v b \, dx = \frac{M_2 - M_1}{I} \int_{y'}^{c} y \, dA$$

where $M_2 - M_1 = dM$, as indicated previously. Solving for the shearing stress at level $y$ yields

$$f_v = \frac{dM}{dx \, Ib} \int_{y'}^{c} y \, dA$$

where $dM/dx = V$, the vertical shearing force:

$$f_v = \frac{V}{Ib} \int_{y'}^{c} y \, dA$$

The expression $\int_{y}^{c} y \, dA$ has been given the symbol $Q$. It is the statical moment of that portion of the cross section which lies between level $y'$ (where we are finding the stress) and level $c$ (the edge of the section), taken with respect to the neutral axis. Using the symbol $Q$ for this statical moment, the complete formula is

$$f_v = \frac{VQ}{Ib} \tag{8-1}$$

Because the flexure formula, $f_y = My/I$, was used in this derivation, the shearing stress formula is subject to the same assumptions or restrictions listed at the end of Appendix A.

# Appendix C
## Derivation of Euler Column Buckling Equation

Under the application of a certain critical load, a column can be in equilibrium in the curved (buckled) position. In this buckled position, an increase in $P$ will cause an increase in the lateral deflection (leading to failure), and a decrease in $P$ will cause the column to return to its initially straight position. It is this critical value of $P$ that is quantified by the Euler equation.

Just as in a beam, the rate of change of slope of the column is directly proportional to the bending moment and inversely proportional to the stiffness. With respect to the free-body diagram in Figure C–1(b), this relationship is given by

$$-\frac{d^2y}{dx^2} = \frac{Py}{EI}$$

(The negative sign is present because of the selection of the origin for the coordinate axes.) As $x$ increases, the slope decreases; thus, the rate of change of slope is negative. The equation can be rewritten as

$$\frac{d^2y}{dx^2} + \frac{P}{EI}y = 0 \tag{C–1}$$

If we let $m = \sqrt{P/EI}$ such that

$$\frac{d^2y}{dx^2} + m^2y = 0 \tag{C–2}$$

the solution of the differential equation is of the form

$$y = A \cos mx + B \sin mx \tag{C–3}$$

which involves two arbitrary constants, $A$ and $B$. That this is a valid solution for $y$ may be verified by taking the second derivative of Equation C–3 and substituting it into Equation C–2.

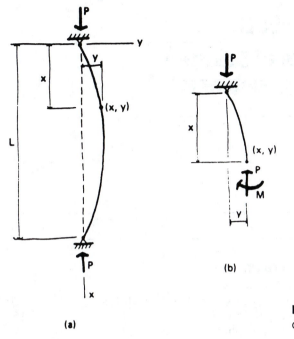

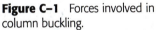

(b)

(a)

**Figure C–1** Forces involved in column buckling.

To evaluate the constants $A$ and $B$, we can use the boundary condition of $y = 0$ when $x = 0$, from which we get $A = 0$, and therefore $y = B \sin mx$. Likewise, $y = 0$ when $x = L$, or $B \sin mL = 0$. If $B$ is to have a value, then $\sin mL$ must be zero. This is only true if $mL = 0, \pi, 2\pi, 3\pi$, etc. The coefficient of $\pi$ represents the buckling mode. For the single-wave mode of our column, $mL = \pi$. Replacing $m$ by $\sqrt{P/EI}$ and solving for $P$, we get

$$P = \frac{\pi^2 EI}{L^2} \tag{11–2}$$

where $P$ is the critical buckling load.

# Appendix D

## Weights of Selected Building Materials*

|  | $lb\backslash ft^3$ | $kN/m^3$ |
|---|---|---|
| Aluminum | 160 | 25 |
| Brick masonry construction | 120 | 19 |
| Cement plaster | 120 | 19 |
| Concrete masonry construction, hollow blocks | 80 | 13 |
| Concrete, stone, reinforced | 150 | 24 |
| Concrete, structural lightweight, reinforced | 110 | 17 |
| Earth, sand, loose | 100 | 16 |
| Earth, topsoil, packed | 90 | 15 |
| Glass | 180 | 28 |
| Gypsum board | 50 | 8 |
| Insulation, rigid | 20 | 3 |
| Plywood | 40 | 6 |
| Steel | 490 | 77 |
| Stone | 160 | 25 |
| Wood, Douglas fir | 30 | 5 |
| Wood, oak | 45 | 7 |
| Wood, redwood | 25 | 4 |
| Water, fresh | 62 | 9.8 |

* Values in this appendix are intended to be representative rather than precise. Most material densities vary considerably, depending upon type and/or ambient conditions.

# Appendix E
## Properties of Selected Materials*

* Values given in the tables of this appendix are intended to be representative rather than precise. Many properties vary, depending on type, manufacturing process, and conditions of use.

### Table E–1    Values in Customary Units

| Material | Strength (psi) (Yield values except where noted) | | | Modulus of Elasticity (E) (ksi) | Coefficient of Thermal Expansion $(°F^{-1})(10^{-6})$ |
| --- | --- | --- | --- | --- | --- |
| | Tension | Compression | Shear | | |
| Wood (dry)[a] | | | | | |
|   Douglas fir | 6 000 | 3 500 | 500 | 1 500 | 2 |
|   Redwood | 6 500 | 4 500 | 450 | 1 300 | 2 |
|   Southern pine | 8 500 | 5 000 | 600 | 1 500 | 3 |
| Steel | | | | | |
|   Mild, low-carbon | 50 000 | 50 000 | 30 000 | 29 000 | 6.5 |
|   Cable, high-strength | 275 000[b] | — | — | 25 000 | 6.5 |
| Concrete | | | | | |
|   Stone | 200[b] | 3 500[b] | 180[b] | 3 500 | 5.5 |
|   Structural, lightweight | 150[b] | 3 500[b] | 130[b] | 2 100 | 5.5 |
| Brick masonry | 300[b] | 4 500[b] | 300[b] | 4 500 | 3.4 |
| Aluminum, structural | 30 000 | 30 000 | 18 000 | 10 000 | 12.8 |
| Iron, cast | 20 000[b] | 85 000[b] | 25 000[b] | 25 000 | 6 |
| Glass, plate | 10 000[b] | 36 000[b] | — | 10 000 | 4.5 |
| Polyester, glass-reinforced | 10 000[b] | 25 000[b] | 25 000[b] | 1 000 | 35 |

[a] Values given are for the parallel-to-grain direction.
[b] Denotes ultimate strength.

**Table E–2   Values in SI Units**

| Material | Strength (kPa) (Yield values except where noted) | | | Modulus of Elasticity(E) (MPa) | Coefficient of Thermal Expansion ($°C^{-1}$)($10^{-6}$) |
| --- | --- | --- | --- | --- | --- |
| | Tension | Compression | Shear | | |
| Wood (dry)[a] | | | | | |
| Douglas fir | 40 000 | 25 000 | 3 500 | 12 000 | 4 |
| Redwood | 45 000 | 30 000 | 3 000 | 9 000 | 4 |
| Southern pine | 60 000 | 35 000 | 4 000 | 12 000 | 4 |
| Steel | | | | | |
| Mild, low-carbon | 345 000 | 345 000 | 200 000 | 200 000 | 11.7 |
| Cable, high-strength | 1 800 000[b] | — | — | 165 000 | 11.7 |
| Concrete | | | | | |
| Stone | 1 500[b] | 25 000[b] | 1 400[b] | 24 000 | 10 |
| Structural, lightweight | 1 100[b] | 25 000[b] | 900[b] | 15 000 | 10 |
| Brick masonry | 2 000[b] | 30 000[b] | 2 000[b] | 30 000 | 6 |
| Aluminum, structural | 200 000 | 200 000 | 140 000 | 70 000 | 22 |
| Iron, cast | 140 000[b] | 600 000[b] | 180 000[b] | 170 000 | 11 |
| Glass, plate | 70 000[b] | 250 000[b] | — | 70 000 | 8 |
| Polyester, glass-reinforced | 70 000[b] | 175 000[b] | 70 000[b] | 7 000 | 60 |

[a] Values given are for the parallel-to-grain direction.
[b] Denotes ultimate strength.

# Appendix F
# Properties of Areas

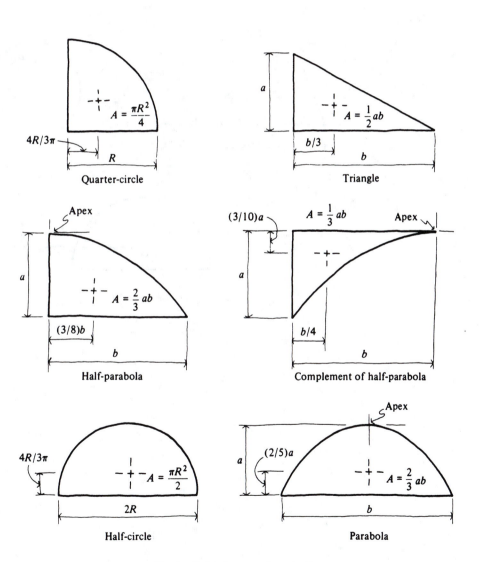

**Figure F–1** Areas and centroids.

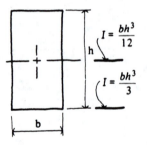

$$I = \frac{bh^3}{12}$$

$$I = \frac{bh^3}{3}$$

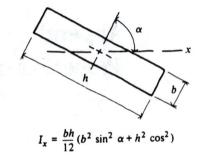

$$I_x = \frac{bh}{12}(b^2 \sin^2 \alpha + h^2 \cos^2 \alpha)$$

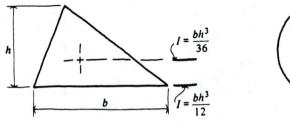

$$I = \frac{bh^3}{36}$$

$$I = \frac{bh^3}{12}$$

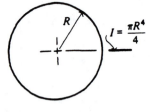

$$I = \frac{\pi R^4}{4}$$

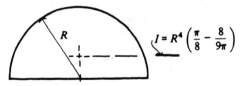

$$I = R^4 \left( \frac{\pi}{8} - \frac{8}{9\pi} \right)$$

**Figure F–2**   Moments of inertia.

# Appendix G
## Proof of Moment-Area Theorems

The radius of curvature at any point along a curve whose equation is $y = f(x)$ (Figure G–1) can be expressed as

$$R = \frac{[1 + (dy/dx)^2]^{3/2}}{d^2y/dx^2}$$

In beams, the slopes of elastic curves are always very small, and for this reason the term $(dy/dx)^2$ is insignificant and may be taken as zero. Therefore,

$$R = \frac{1}{d^2y/dx^2}$$

where $d^2y/dx^2$ is merely the rate of change of the slope. It is a measure of the change in slope between two points on the elastic curve as we shall use it. If the points $a$ and $b$ in Figure G–2 are allowed to approach each other, then $\theta_{ab}$ can be represented by $d^2y/dx^2$.

If we equate the two values obtained for $R$,

$$R = \frac{EI}{M} = \frac{1}{d^2y/dx^2}$$

we get

$$EI\frac{d^2y}{dx^2} = M \qquad \text{(G–1)}$$

Credit for the development of this equation is given to Leonhard Euler (1707–1783), a Swiss mathematician.

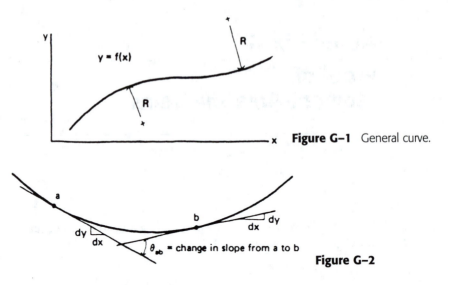

**Figure G–1** General curve.

**Figure G–2**

## FIRST MOMENT-AREA THEOREM

The basic equation developed above can be written as

$$\frac{d(dy/dx)}{dx} = \frac{M}{EI}$$

For the small slopes of elastic curves assumed previously, $dy/dx = \tan \theta = \theta$. Then

$$\frac{d^2y}{dx^2} = \frac{d\theta}{dx}$$

Also,

$$\frac{d\theta}{dx} = \frac{M}{EI}$$

which can be written as

$$d\theta = \frac{M}{EI}\,dx$$

which is valid for a small length of beam $dx$. It states that a small change in angle is equal to a small area. Integrating both sides of the equation with finite limits determined by the portion of the beam shown in Figure G–3, we get that the change in angle $\theta$ from point $A$ to point $B$ is equal to the shaded area under the $M/EI$ curve.

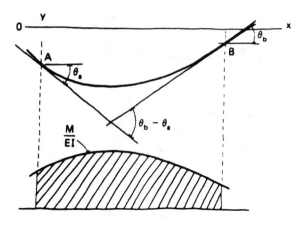

**Figure G–3** Portion of elastic curve and *M/EI* diagram.

# SECOND MOMENT-AREA THEOREM

Two tangent lines to the elastic curve at infinitesimally close points *c* and *d* will subtend a distance *dt* on the vertical line through *B*, as shown in Figure G–4.

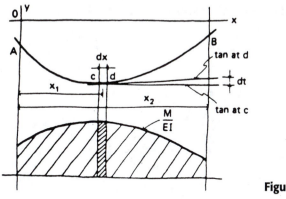

**Figure G–4**

$$\tan d\theta = \frac{dt}{x_2 - x_1}$$

Since for small values of $\theta$ the tangent function equals the angle itself, we can say that

$$d\theta = \frac{dt}{x_2 - x_1}$$

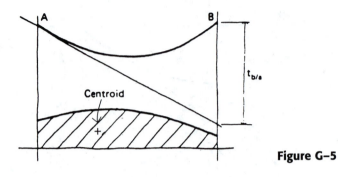

**Figure G–5**

Previously, $d\theta$ was declared equal to $(M/EI)dx$; therefore,

$$dt = (x_2 - x_1)\,\frac{M}{EI}\,dx$$

This states that a small vertical distance on a line through $B$ is equal to a small area, $(M/EI)dx$, times the distance from the line to the centroid of the area, $x_2 - x_1$. Integrating both sides of the equation with finite limits determined by the portion of the beam shown in Figure G–4, we get that the distance $t_{b/a}$ along a vertical line through $B$ is equal to the shaded area times the distance from the vertical line to the centroid of the shaded area (see Figure G–5).

# Appendix H

# Allowable Stresses* and Modulus of Elasticity Values for Selected Structural Sawn Lumber

* The allowable stress values in these tables may be modified when the design is governed by a temporary load condition below. The modulus of elasticity is *not* modified for a temporary load condition.

| | |
|---|---|
| Two months (snow) | 1.15 |
| Seven days (construction) | 1.25 |

| | |
|---|---|
| Wind or earthquake | 1.60 |
| Impact | 2.00 |

If the *full* load is to be applied, continuously or cumulatively, for more than 10 years, the $F_b$, $F_v$, and $F_c$ values shall be multiplied by 0.90.

**Table H–1   Values in Customary Units (psi)[a]**

| Nominal Size | Douglas fir-larch No. 2 | | | | Hem-fir No. 2 | | | | Southern pine No. 2 | | | |
|---|---|---|---|---|---|---|---|---|---|---|---|---|
| | $F_b{}^b$ | $F_v$ | $F_c$ | $E(\times 10^6)$ | $F_b{}^b$ | $F_v$ | $F_c$ | $E(\times 10^6)$ | $F_b{}^b$ | $F_v$ | $F_c$ | $E(\times 10^6)$ |
| 2 × 4 | 1313 | 95 | 1495 | 1.6 | 1275 | 75 | 1438 | 1.3 | 1500 | 90 | 1650 | 1.6 |
| 2 × 6 | 1138 | 95 | 1430 | 1.6 | 1105 | 75 | 1375 | 1.3 | 1250 | 90 | 1600 | 1.6 |
| 2 × 8 | 1050 | 95 | 1365 | 1.6 | 1020 | 75 | 1313 | 1.3 | 1200 | 90 | 1550 | 1.6 |
| 2 × 10 | 963 | 95 | 1300 | 1.6 | 935 | 75 | 1250 | 1.3 | 1050 | 90 | 1500 | 1.6 |
| 2 × 12 | 875 | 95 | 1300 | 1.6 | 850 | 75 | 1250 | 1.3 | 975 | 90 | 1450 | 1.6 |
| 4 × 4 | 1313 | 95 | 1495 | 1.6 | 1275 | 75 | 1438 | 1.3 | 1500 | 90 | 1650 | 1.6 |
| 4 × 6 | 1138 | 95 | 1430 | 1.6 | 1105 | 75 | 1375 | 1.3 | 1250 | 90 | 1600 | 1.6 |
| 4 × 8 | 1138 | 95 | 1365 | 1.6 | 1105 | 75 | 1313 | 1.3 | 1200 | 90 | 1550 | 1.6 |
| 4 × 10 | 1050 | 95 | 1300 | 1.6 | 1020 | 75 | 1250 | 1.3 | 1050 | 90 | 1500 | 1.6 |
| 4 × 12 | 963 | 95 | 1300 | 1.6 | 935 | 75 | 1250 | 1.3 | 975 | 90 | 1450 | 1.6 |
| 4 × 14 | 875 | 95 | 1170 | 1.6 | 850 | 75 | 1125 | 1.3 | 878 | 90 | 1305 | 1.6 |
| 4 × 16 | 875 | 95 | 1170 | 1.6 | 850 | 75 | 1125 | 1.3 | 878 | 90 | 1305 | 1.6 |
| 6 × 6 | 700 | 85 | 475 | 1.3 | 525 | 70 | 375 | 1.1 | 850 | 100 | 525 | 1.2 |
| 6 × 8 | 700 | 85 | 475 | 1.3 | 525 | 70 | 375 | 1.1 | 850 | 100 | 525 | 1.2 |
| 6 × 10 | 875 | 85 | 600 | 1.3 | 675 | 70 | 475 | 1.1 | 850 | 100 | 525 | 1.2 |
| 6 × 12 | 875 | 85 | 600 | 1.3 | 675 | 70 | 475 | 1.1 | 850 | 100 | 525 | 1.2 |
| 6 × 14 | 864 | 85 | 600 | 1.3 | 666 | 70 | 475 | 1.1 | 839 | 100 | 525 | 1.2 |
| 6 × 16 | 850 | 85 | 600 | 1.3 | 656 | 70 | 475 | 1.1 | 826 | 100 | 525 | 1.2 |
| 6 × 18 | 839 | 85 | 600 | 1.3 | 647 | 70 | 475 | 1.1 | 815 | 100 | 525 | 1.2 |
| 6 × 20 | 829 | 85 | 600 | 1.3 | 640 | 70 | 475 | 1.1 | 805 | 100 | 525 | 1.2 |
| 6 × 22 | 820 | 85 | 600 | 1.3 | 633 | 70 | 475 | 1.1 | 797 | 100 | 525 | 1.2 |
| 6 × 24 | 812 | 85 | 600 | 1.3 | 626 | 70 | 475 | 1.1 | 789 | 100 | 525 | 1.2 |

The values for nominal 8 × sizes are the same as those for their counterparts in the 6 × sizes.

[a] Dry-use conditions assumed. Size factor ($C_f$) values have been included.
[b] Bending design values ($F_b$) for dimension lumber (2 to 4 in. nominal thickness) may be increased by 15% when such members are *repetitive*. Repetitive members are joists, truss chords, rafters, or similar members which are spaced not more than 24 in. on center, are not less than three in number, and are joined by floor, roof, or other load-distributing elements adequate to support the design load.

Allowable stresses for 6 × 10 and larger Douglas fir-larch and hem-fir are based on Beams and Stringers size classification.

**Table H–2   Values in SI Units (kPa for stress, MPa for $E$)[a]**

| Standard Dressed Size | Douglas fir-larch No. 2 | | | | Hem-fir No. 2 | | | | Southern pine No. 2 | | | |
|---|---|---|---|---|---|---|---|---|---|---|---|---|
| | $F_b$[b] | $F_v$ | $F_c$ | $E$ | $F_b$[b] | $F_v$ | $F_c$ | $E$ | $F_b$[b] | $F_v$ | $F_c$ | $E$ |
| 38 × 89 | 9053 | 655 | 10 310 | 11 000 | 8791 | 517 | 9915 | 8960 | 10 340 | 620 | 11 380 | 11 030 |
| 38 × 139 | 7846 | 655 | 9859 | 11 000 | 7619 | 517 | 9480 | 8960 | 8618 | 620 | 11 030 | 11 030 |
| 38 × 185 | 7239 | 655 | 9411 | 11 000 | 7033 | 517 | 9052 | 8960 | 8274 | 620 | 10 690 | 11 030 |
| 38 × 235 | 6640 | 655 | 8963 | 11 000 | 6447 | 517 | 8618 | 8960 | 7239 | 620 | 10 340 | 11 030 |
| 38 × 285 | 6033 | 655 | 8963 | 11 000 | 5860 | 517 | 8618 | 8960 | 6722 | 620 | 9997 | 11 030 |
| 89 × 89 | 9053 | 655 | 10 310 | 11 000 | 8791 | 517 | 9915 | 8960 | 10 342 | 620 | 11 380 | 11 030 |
| 89 × 139 | 7846 | 655 | 9859 | 11 000 | 7619 | 517 | 9480 | 8960 | 8618 | 620 | 11 030 | 11 030 |
| 89 × 185 | 7846 | 655 | 9411 | 11 000 | 7619 | 517 | 9052 | 8960 | 8274 | 620 | 10 690 | 11 030 |
| 89 × 235 | 7239 | 655 | 8963 | 11 000 | 7033 | 517 | 8618 | 8 960 | 7239 | 620 | 10 340 | 11 030 |
| 89 × 285 | 6640 | 655 | 8963 | 11 000 | 6447 | 517 | 8618 | 8960 | 6722 | 620 | 9997 | 11 030 |
| 89 × 335 | 6033 | 655 | 8067 | 11 000 | 5860 | 517 | 7756 | 8960 | 6054 | 620 | 8998 | 11 030 |
| 89 × 385 | 6033 | 655 | 8067 | 11 000 | 5860 | 517 | 7756 | 8960 | 6054 | 620 | 8998 | 11 030 |
| 139 × 139 | 4826 | 586 | 3275 | 8960 | 3620 | 483 | 2586 | 7580 | 5860 | 689 | 3620 | 8270 |
| 139 × 190 | 4826 | 586 | 3275 | 8960 | 3620 | 483 | 2586 | 7580 | 5860 | 689 | 3620 | 8270 |
| 139 × 240 | 6033 | 586 | 4138 | 8960 | 4654 | 483 | 3275 | 7580 | 5860 | 689 | 3620 | 8270 |
| 139 × 290 | 6033 | 586 | 4138 | 8960 | 4654 | 483 | 3275 | 7580 | 5860 | 689 | 3620 | 8270 |
| 139 × 340 | 5957 | 586 | 4138 | 8960 | 4592 | 483 | 3275 | 7580 | 5785 | 689 | 3620 | 8270 |
| 139 × 390 | 5860 | 586 | 4138 | 8960 | 4523 | 483 | 3275 | 7580 | 5700 | 689 | 3620 | 8270 |
| 139 × 445 | 5785 | 586 | 4138 | 8960 | 4461 | 483 | 3275 | 7580 | 5619 | 689 | 3620 | 8270 |
| 139 × 495 | 5716 | 586 | 4138 | 8960 | 4413 | 483 | 3275 | 7580 | 5550 | 689 | 3620 | 8270 |
| 139 × 545 | 5654 | 586 | 4138 | 8960 | 4364 | 483 | 3275 | 7580 | 5495 | 689 | 3620 | 8270 |
| 139 × 595 | 5598 | 586 | 4138 | 8960 | 4317 | 483 | 3275 | 7580 | 5440 | 689 | 3620 | 8270 |

The values for 190 × sizes are the same as those for their counterparts in the 139 × sizes.

[a]  Dry-use conditions assumed. Size factor ($C_f$) values have been included.

[b]  Bending design values ($F_b$) for dimension lumber (38 to 89 mm in thickness) may be increased by 15% when such members are *repetitive*. Repetitive members are joists, truss chords, rafters, or similar members which are spaced not more than 600 mm on center, are not less than three in number, and are joined by floor, roof, or other load-distributing elements adequate to support the design load.

Allowable stresses for 139 × 240 and larger Douglas fir-larch and hem-fir are based on Beams and Stringers size classification.

# Appendix I
# Wood Section Properties

**Table I–1   Values in Customary Units**

| Nominal Size | Actual Size (in.) | Area (in.$^2$) | $S_x$ (in.$^3$) | $I_x$ (in.$^4$) | $I_y$ (in.$^4$) |
|---|---|---|---|---|---|
| 2 × 4 | 1½ × 3½ | 5.25 | 3.06 | 5.36 | 0.984 |
| 2 × 6 | 1½ × 5½ | 8.25 | 7.56 | 20.8 | 1.55 |
| 2 × 8 | 1½ × 7¼ | 10.9 | 13.1 | 47.6 | 2.04 |
| 2 × 10 | 1½ × 9¼ | 13.9 | 21.4 | 98.9 | 2.60 |
| 2 × 12 | 1½ × 11¼ | 16.9 | 31.6 | 178 | 3.16 |
| 4 × 4 | 3½ × 3½ | 12.3 | 7.15 | 12.5 | 12.5 |
| 4 × 6 | 3½ × 5½ | 19.3 | 17.6 | 48.5 | 19.7 |
| 4 × 8 | 3½ × 7¼ | 25.4 | 30.7 | 111 | 25.9 |
| 4 × 10 | 3½ × 9¼ | 32.4 | 49.9 | 231 | 33.0 |
| 4 × 12 | 3½ × 11¼ | 39.4 | 73.8 | 415 | 40.2 |
| 4 × 14 | 3½ × 13¼ | 46.4 | 102 | 678 | 47.3 |
| 4 × 16 | 3½ × 15¼ | 53.4 | 136 | 1034 | 54.5 |
| 6 × 6 | 5½ × 5½ | 30.3 | 27.7 | 76.3 | 76.3 |
| 6 × 8 | 5½ × 7½ | 41.3 | 51.6 | 193 | 104 |
| 6 × 10 | 5½ × 9½ | 52.3 | 82.7 | 393 | 132 |
| 6 × 12 | 5½ × 11½ | 63.3 | 121 | 697 | 159 |
| 6 × 14 | 5½ × 13½ | 74.3 | 167 | 1128 | 187 |
| 6 × 16 | 5½ × 15½ | 85.3 | 220 | 1707 | 215 |
| 6 × 18 | 5½ × 17½ | 96.3 | 281 | 2456 | 243 |
| 6 × 20 | 5½ × 19½ | 107 | 349 | 3398 | 270 |
| 6 × 22 | 5½ × 21½ | 118 | 424 | 4555 | 298 |
| 6 × 24 | 5½ × 23½ | 129 | 506 | 5948 | 326 |
| 8 × 8 | 7½ × 7½ | 56.3 | 70.3 | 264 | 264 |
| 8 × 10 | 7½ × 9½ | 71.3 | 113 | 536 | 334 |
| 8 × 12 | 7½ × 11½ | 86.3 | 165 | 951 | 404 |
| 8 × 14 | 7½ × 13½ | 101 | 228 | 1538 | 475 |
| 8 × 16 | 7½ × 15½ | 116 | 300 | 2327 | 545 |
| 8 × 18 | 7½ × 17½ | 131 | 383 | 3350 | 615 |
| 8 × 20 | 7½ × 19½ | 146 | 475 | 4634 | 685 |
| 8 × 22 | 7½ × 21½ | 161 | 578 | 6211 | 756 |
| 8 × 24 | 7½ × 23½ | 176 | 690 | 8111 | 826 |

**Table I–2   Values in SI Units**

| Size (mm) | Area $(mm^2)$ | $S_x$ $(10^3\ mm^3)$ | $I_x$ $(10^6\ mm^4)$ | $I_y$ $(10^6\ mm^4)$ |
|---|---|---|---|---|
| 38 × 89 | 3 380 | 50.1 | 2.23 | 0.407 |
| 38 × 139 | 5 280 | 122 | 8.51 | 0.636 |
| 38 × 185 | 7 030 | 217 | 20.1 | 0.846 |
| 38 × 235 | 8 930 | 350 | 41.1 | 1.07 |
| 38 × 285 | 10 800 | 514 | 73.3 | 1.30 |
| 89 × 89 | 7 920 | 118 | 5.23 | 5.23 |
| 89 × 139 | 12 400 | 287 | 19.9 | 8.17 |
| 89 × 185 | 16 500 | 508 | 47.0 | 10.9 |
| 89 × 235 | 20 900 | 819 | 96.3 | 13.8 |
| 89 × 285 | 25 400 | 1 210 | 172 | 16.7 |
| 89 × 335 | 29 800 | 1 660 | 279 | 19.7 |
| 89 × 385 | 34 300 | 2 200 | 423 | 22.6 |
| 139 × 139 | 19 300 | 448 | 31.1 | 31.1 |
| 139 × 190 | 26 400 | 836 | 79.4 | 42.5 |
| 139 × 240 | 33 400 | 1 330 | 160 | 53.7 |
| 139 × 290 | 40 300 | 1 950 | 283 | 64.9 |
| 139 × 340 | 47 300 | 2 680 | 455 | 76.1 |
| 139 × 390 | 54 200 | 3 520 | 687 | 87.3 |
| 139 × 445 | 61 900 | 4 590 | 1 020 | 99.6 |
| 139 × 495 | 68 800 | 5 680 | 1 400 | 111 |
| 139 × 545 | 75 800 | 6 880 | 1 880 | 122 |
| 139 × 595 | 82 700 | 8 200 | 2 440 | 133 |
| 190 × 190 | 36 100 | 1 140 | 109 | 109 |
| 190 × 240 | 45 600 | 1 820 | 219 | 137 |
| 190 × 290 | 55 100 | 2 660 | 386 | 166 |
| 190 × 340 | 64 600 | 3 660 | 622 | 194 |
| 190 × 390 | 74 100 | 4 820 | 939 | 223 |
| 190 × 445 | 84 600 | 6 270 | 1 400 | 254 |
| 190 × 495 | 94 100 | 7 760 | 1 920 | 283 |
| 190 × 545 | 104 000 | 9 400 | 2 560 | 312 |
| 190 × 595 | 113 000 | 11 200 | 3 340 | 340 |

# Appendix J

## Properties of Selected Steel Sections

**Table J–1   Values in Customary Units**

| | Designation | $A$ (in.$^2$) | $d$ (in.) | $b_f$ (in.) | $S_x$ (in.$^3$) | $I_x$ (in.$^4$) | $I_y$ (in.$^4$) |
|---|---|---|---|---|---|---|---|
| **Channels** | C6 × 10.5 | 3.09 | 6.00 | 2.034 | 5.06 | 15.2 | 0.866 |
| | C8 × 13.75 | 4.04 | 8.00 | 2.343 | 9.03 | 36.1 | 1.53 |
| | C10 × 25 | 7.35 | 10.00 | 2.886 | 18.2 | 91.2 | 3.36 |
| | C10 × 30 | 8.82 | 10.00 | 3.033 | 20.7 | 103 | 3.94 |
| **Columns** | W6 × 25 | 7.34 | 6.38 | 6.080 | 16.7 | 53.4 | 17.1 |
| | W8 × 35 | 10.3 | 8.12 | 8.020 | 31.2 | 127 | 42.6 |
| | W8 × 48 | 14.1 | 8.50 | 8.110 | 43.3 | 184 | 60.9 |
| | W8 × 67 | 19.7 | 9.00 | 8.280 | 60.4 | 272 | 88.6 |
| | W10 × 49 | 14.4 | 9.98 | 10.000 | 54.6 | 272 | 93.4 |
| | W10 × 88 | 25.9 | 10.84 | 10.265 | 98.5 | 534 | 179 |
| | W12 × 120 | 35.3 | 13.12 | 12.320 | 163 | 1 070 | 345 |
| | W12 × 190 | 55.8 | 14.38 | 12.670 | 263 | 1 890 | 589 |
| | W14 × 211 | 62.0 | 15.72 | 15.800 | 338 | 2 660 | 1 030 |
| | W14 × 370 | 109 | 17.92 | 16.475 | 607 | 5 440 | 1 990 |
| | W14 × 730 | 215 | 22.42 | 17.890 | 1 280 | 14 300 | 4 720 |
| **Beams** | **W36 × 300** | 88.3 | 36.74 | 16.655 | 1 110 | 20 300 | 1 300 |
| | W33 × 318 | 93.5 | 35.16 | 15.985 | 1 110 | 19 500 | 1 290 |
| | **W36 × 260** | 76.5 | 36.26 | 16.550 | 953 | 17 300 | 1 090 |
| | **W36 × 245** | 72.1 | 36.08 | 16.510 | 895 | 16 100 | 1 010 |
| | **W33 × 241** | 70.9 | 34.18 | 15.860 | 829 | 14 200 | 932 |
| | **W30 × 173** | 50.8 | 30.44 | 14.985 | 539 | 8 200 | 598 |
| | **W33 × 130** | 38.3 | 33.09 | 11.510 | 406 | 6 710 | 218 |
| | **W33 × 118** | 34.7 | 32.89 | 11.480 | 359 | 5 900 | 187 |
| | **W30 × 108** | 31.7 | 29.83 | 10.475 | 299 | 4 470 | 146 |
| | W21 × 132 | 38.8 | 21.83 | 12.440 | 295 | 3 220 | 333 |
| | W24 × 117 | 34.4 | 24.26 | 12.800 | 291 | 3 540 | 297 |
| | **W30 × 99** | 29.1 | 29.65 | 10.450 | 269 | 3 990 | 128 |
| | W27 × 102 | 30.0 | 27.09 | 10.015 | 267 | 3 620 | 139 |
| | W18 × 130 | 38.2 | 19.25 | 11.160 | 256 | 2 460 | 278 |

**Table J–1** *(Continued)*

| Designation | $A$ (in.$^2$) | $d$ (in.) | $b_f$ (in.) | $S_x$ (in.$^3$) | $I_x$ (in.$^4$) | $I_y$ (in.$^4$) |
|---|---|---|---|---|---|---|
| **W30 × 90** | 26.4 | 29.53 | 10.400 | 245 | 3 620 | 115 |
| W21 × 101 | 29.8 | 21.36 | 12.290 | 227 | 2 420 | 248 |
| **W27 × 84** | 24.8 | 26.71 | 9.960 | 213 | 2 850 | 106 |
| **W24 × 76** | 22.4 | 23.92 | 8.990 | 176 | 2 100 | 82.5 |
| W21 × 83 | 24.3 | 21.43 | 8.355 | 171 | 1 830 | 81.4 |
| **W24 × 68** | 20.1 | 23.73 | 8.965 | 154 | 1 830 | 70.4 |
| W18 × 76 | 22.3 | 18.21 | 11.035 | 146 | 1 330 | 152 |
| W14 × 90 | 26.5 | 14.02 | 14.520 | 143 | 999 | 362 |
| **W21 × 62** | 18.3 | 20.99 | 8.240 | 127 | 1 330 | 57.5 |
| W18 × 65 | 19.1 | 18.35 | 7.590 | 117 | 1 070 | 54.8 |
| W16 × 67 | 19.7 | 16.33 | 10.235 | 117 | 954 | 119 |
| **W21 × 50** | 14.7 | 20.83 | 6.530 | 94.5 | 984 | 24.9 |
| W16 × 57 | 16.8 | 16.43 | 7.120 | 92.2 | 758 | 43.1 |
| **W21 × 44** | 13.0 | 20.66 | 6.500 | 81.6 | 843 | 20.7 |
| W18 × 46 | 13.5 | 18.06 | 6.060 | 78.8 | 712 | 22.5 |
| W14 × 48 | 14.1 | 13.79 | 8.030 | 70.3 | 485 | 51.4 |
| **W18 × 40** | 11.8 | 17.90 | 6.015 | 68.4 | 612 | 19.1 |
| W12 × 50 | 14.7 | 12.19 | 8.080 | 64.7 | 394 | 56.3 |
| **W18 × 35** | 10.3 | 17.70 | 6.000 | 57.6 | 510 | 15.3 |
| **W16 × 31** | 9.12 | 15.88 | 5.525 | 47.2 | 375 | 12.4 |
| W12 × 35 | 10.3 | 12.50 | 6.560 | 45.6 | 285 | 24.5 |
| **W12 × 26** | 7.65 | 12.22 | 6.490 | 33.4 | 204 | 17.3 |
| W10 × 30 | 8.84 | 10.47 | 5.810 | 32.4 | 170 | 16.7 |
| **W14 × 22** | 6.49 | 13.74 | 5.000 | 29.0 | 199 | 7.00 |
| W12 × 16 | 4.71 | 11.99 | 3.990 | 17.1 | 103 | 2.82 |
| **W12 × 14** | 4.16 | 11.91 | 3.970 | 14.9 | 88.6 | 2.36 |
| **W10 × 12** | 3.54 | 9.87 | 3.960 | 10.9 | 53.8 | 2.18 |
| **W8 × 10** | 2.96 | 7.89 | 3.940 | 7.81 | 30.8 | 2.09 |
| W6 × 12 | 3.55 | 6.03 | 4.000 | 7.31 | 22.1 | 2.99 |

**Table J–2   Values in SI Units**

| | Designation | $A$ (mm$^2$) | $d$ (mm) | $b_f$ (mm) | $S_x$ ($10^3$ mm$^3$) | $I_x$ ($10^6$ mm$^4$) | $I_y$ ($10^6$ mm$^4$) |
|---|---|---|---|---|---|---|---|
| **Channels** | C150 × 16 | 1 990 | 152.0 | 51.7 | 83.3 | 6.33 | 0.360 |
| | C200 × 20 | 2 610 | 203.0 | 59.5 | 148 | 15.0 | 0.637 |
| | C250 × 37 | 4 740 | 254.0 | 73.3 | 299 | 38.0 | 1.40 |
| | C250 × 45 | 5 690 | 254.0 | 77.0 | 338 | 42.9 | 1.64 |
| **Columns** | W150 × 37 | 4 730 | 162 | 154.0 | 274 | 22.2 | 7.07 |
| | W200 × 52 | 6 640 | 206 | 204.0 | 512 | 52.7 | 17.8 |
| | W200 × 71 | 9 100 | 216 | 206.0 | 709 | 76.6 | 25.4 |
| | W200 × 100 | 12 700 | 229 | 210.0 | 987 | 113 | 36.6 |

**Table J–2 (*Continued*)**

| | Designation | *A* ($mm^2$) | *d* (mm) | $b_f$ (mm) | $S_x$ ($10^3$ $mm^3$) | $I_x$ ($10^6$ $mm^4$) | $I_y$ ($10^6$ $mm^4$) |
|---|---|---|---|---|---|---|---|
| | W250 × 73 | 9 310 | 253 | 254.0 | 893 | 113 | 38.8 |
| | W250 × 131 | 16 700 | 275 | 261.0 | 1 610 | 221 | 74.5 |
| | W310 × 179 | 22 800 | 333 | 313.0 | 2 670 | 445 | 144 |
| | W310 × 283 | 36 000 | 365 | 322.0 | 4 310 | 787 | 246 |
| | W360 × 314 | 40 000 | 399 | 401.0 | 5 510 | 1 100 | 426 |
| | W360 × 551 | 70 200 | 455 | 418.0 | 9 930 | 2 260 | 825 |
| | W360 × 1086 | 138 000 | 569 | 454.0 | 20 900 | 5 960 | 1 960 |
| **Beams** | **W920 × 446** | 57 000 | 933 | 423.0 | 18 200 | 8 470 | 540 |
| | W840 × 473 | 60 300 | 893 | 406.0 | 18 200 | 8 120 | 537 |
| | **W920 × 387** | 49 300 | 921 | 420.0 | 15 600 | 7 180 | 453 |
| | **W920 × 365** | 46 500 | 916 | 419.0 | 14 700 | 6 170 | 421 |
| | **W840 × 359** | 45 700 | 868 | 403.0 | 13 600 | 5 910 | 389 |
| | **W760 × 257** | 32 800 | 773 | 381.0 | 8 850 | 3 420 | 250 |
| | **W840 × 193** | 24 700 | 840 | 292.0 | 6 620 | 2 780 | 90.3 |
| | **W840 × 176** | 22 400 | 835 | 292.0 | 5 890 | 2 460 | 78.2 |
| | **W760 × 161** | 20 500 | 758 | 266.0 | 4 910 | 1 860 | 60.7 |
| | W530 × 196 | 25 000 | 558 | 316.0 | 4 840 | 1 340 | 139 |
| | W610 × 174 | 22 200 | 616 | 325.0 | 4 770 | 1 470 | 124 |
| | **W760 × 147** | 18 800 | 753 | 265.0 | 4 410 | 1 660 | 52.9 |
| | W690 × 152 | 19 400 | 688 | 254.0 | 4 390 | 1 510 | 57.8 |
| | W460 × 193 | 24 700 | 489 | 283.0 | 4 170 | 1 020 | 115 |
| | **W760 × 134** | 17 000 | 750 | 264.0 | 4 000 | 1 500 | 47.7 |
| | W530 × 150 | 19 200 | 543 | 312.0 | 3 720 | 1 010 | 103 |
| | **W690 × 125** | 16 000 | 678 | 253.0 | 3 510 | 1 190 | 44.1 |
| | **W610 × 113** | 14 400 | 608 | 228.0 | 2 880 | 875 | 34.3 |
| | W530 × 123 | 15 700 | 544 | 212.0 | 2 800 | 761 | 33.8 |
| | **W610 × 101** | 12 900 | 603 | 228.0 | 2 530 | 764 | 29.5 |
| | W460 × 113 | 14 400 | 463 | 280.0 | 2 400 | 556 | 63.3 |
| | W360 × 134 | 17 100 | 356 | 369.0 | 2 330 | 415 | 151 |
| | **W530 × 92** | 11 800 | 533 | 209.0 | 2 080 | 554 | 23.9 |
| | W410 × 100 | 12 700 | 415 | 260.0 | 1 920 | 398 | 49.5 |
| | W460 × 97 | 12 300 | 466 | 193.0 | 1 910 | 445 | 22.8 |
| | **W530 × 74** | 9 490 | 529 | 166.0 | 1 550 | 410 | 10.4 |
| | W410 × 85 | 10 800 | 417 | 181.0 | 1 510 | 315 | 18.0 |
| | **W530 × 66** | 8 370 | 525 | 165.0 | 1 330 | 350 | 8.57 |
| | W460 × 68 | 8 730 | 459 | 154.0 | 1 290 | 297 | 9.41 |
| | W360 × 72 | 9 130 | 350 | 204.0 | 1 150 | 201 | 21.4 |
| | **W460 × 60** | 7 590 | 455 | 153.0 | 1 120 | 255 | 7.96 |
| | W310 × 74 | 9 480 | 310 | 205.0 | 1 060 | 165 | 23.4 |
| | **W460 × 52** | 6 640 | 450 | 152.0 | 942 | 212 | 6.34 |

**Table J–2** *(Continued)*

| Designation | $A$ $(mm^2)$ | $d$ $(mm)$ | $b_f$ $(mm)$ | $S_x$ $(10^3\ mm^3)$ | $I_x$ $(10^6\ mm^4)$ | $I_y$ $(10^6\ mm^4)$ |
|---|---|---|---|---|---|---|
| **W410 × 46** | 5 890 | 403 | 140.0 | 774 | 156 | 5.14 |
| W310 × 52 | 6 670 | 318 | 167.0 | 748 | 119 | 10.3 |
| **W310 × 39** | 4 930 | 310 | 165.0 | 547 | 84.8 | 7.23 |
| W250 × 45 | 5 700 | 266 | 148.0 | 535 | 71.1 | 7.03 |
| **W360 × 33** | 4 190 | 349 | 127.0 | 475 | 82.9 | 2.91 |
| W310 × 24 | 3 040 | 305 | 101.0 | 281 | 42.8 | 1.16 |
| **W310 × 21** | 2 680 | 303 | 101.0 | 244 | 37.0 | 0.986 |
| **W250 × 18** | 2 280 | 251 | 101.0 | 179 | 22.5 | 0.919 |
| **W200 × 15** | 1 910 | 200 | 100.0 | 128 | 12.8 | 0.870 |
| W150 × 18 | 2 290 | 153 | 102.0 | 120 | 9.19 | 1.26 |

# Appendix K

## Shear, Moment, and Deflection Equations

①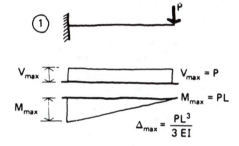

$V_{max} = P$

$M_{max} = PL$

$\Delta_{max} = \dfrac{PL^3}{3\,EI}$

②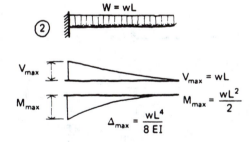

$V_{max} = wL$

$M_{max} = \dfrac{wL^2}{2}$

$\Delta_{max} = \dfrac{wL^4}{8\,EI}$

③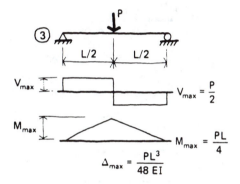

$V_{max} = \dfrac{P}{2}$

$M_{max} = \dfrac{PL}{4}$

$\Delta_{max} = \dfrac{PL^3}{48\,EI}$

④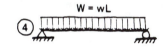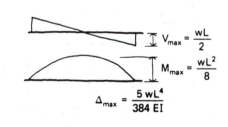

$V_{max} = \dfrac{wL}{2}$

$M_{max} = \dfrac{wL^2}{8}$

$\Delta_{max} = \dfrac{5\,wL^4}{384\,EI}$

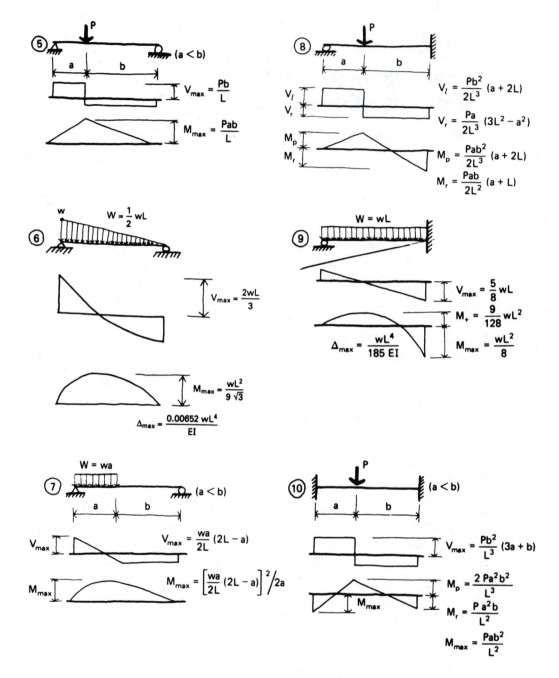

**⑤** $(a < b)$

$$V_{max} = \frac{Pb}{L}$$

$$M_{max} = \frac{Pab}{L}$$

**⑧**

$$V_l = \frac{Pb^2}{2L^3}(a + 2L)$$

$$V_r = \frac{Pa}{2L^3}(3L^2 - a^2)$$

$$M_p = \frac{Pab^2}{2L^3}(a + 2L)$$

$$M_r = \frac{Pab}{2L^2}(a + L)$$

**⑥** $W = \frac{1}{2}wL$

$$V_{max} = \frac{2wL}{3}$$

$$M_{max} = \frac{wL^2}{9\sqrt{3}}$$

$$\Delta_{max} = \frac{0.00652\, wL^4}{EI}$$

**⑨** $W = wL$

$$V_{max} = \frac{5}{8}wL$$

$$M_+ = \frac{9}{128}wL^2$$

$$\Delta_{max} = \frac{wL^4}{185\, EI}$$

$$M_{max} = \frac{wL^2}{8}$$

**⑦** $W = wa$ $(a < b)$

$$V_{max} = \frac{wa}{2L}(2L - a)$$

$$M_{max} = \left[\frac{wa}{2L}(2L - a)\right]^2 \Big/ 2a$$

**⑩** $(a < b)$

$$V_{max} = \frac{Pb^2}{L^3}(3a + b)$$

$$M_p = \frac{2\,Pa^2b^2}{L^3}$$

$$M_r = \frac{P\,a^2b}{L^2}$$

$$M_{max} = \frac{Pab^2}{L^2}$$

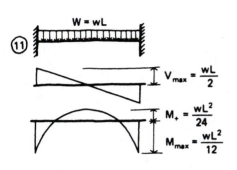

(11)

$$V_{max} = \frac{wL}{2}$$

$$M_+ = \frac{wL^2}{24}$$

$$M_{max} = \frac{wL^2}{12}$$

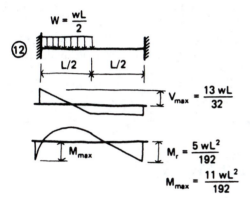

(12)

$$V_{max} = \frac{13\,wL}{32}$$

$$M_r = \frac{5\,wL^2}{192}$$

$$M_{max} = \frac{11\,wL^2}{192}$$

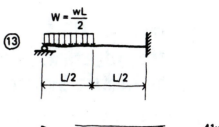

(13)

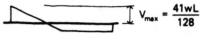

$$V_{max} = \frac{41wL}{128}$$

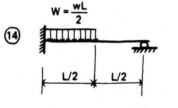

$$M_{max} = \frac{7wL^2}{128}$$

(14)

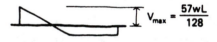

$$V_{max} = \frac{57wL}{128}$$

$$M_{max} = \frac{9wL^2}{128}$$

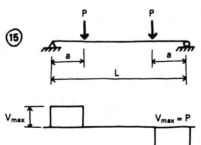

(15)

$$V_{max}$$

$$V_{max} = P$$

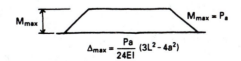

$$M_{max}$$

$$M_{max} = P_a$$

$$\Delta_{max} = \frac{Pa}{24EI}\,(3L^2 - 4a^2)$$

# Appendix L
## Introduction to the SI Metric System

## WHAT IS SI METRIC?

SI metric is the name given to the new measurement system which is being adopted on a worldwide basis. It differs somewhat from the long-standing European metric system. SI stands for Le Système International d'Unités, a name generated by the 36 nations meeting at the 11th General Conference on Weights and Measures (CGPM) in 1960.

SI is a coherent means of measurement based on the meter–kilogram–second–ampere system of fundamental units. Conversions within the system (e.g., ounces to pounds and inches to feet, as we do in the customary system are never necessary.

## HOW NEW IS THE METRIC SYSTEM TO THE UNITED STATES?

In 1866, Congress made the metric system a legal system of units for U.S. use. In 1875, the United States and 16 other nations formed the General Conference on Weights and Measures (CGPM). The United States has been active in the periodic meetings of this group.

In 1893, an Executive Order made the meter and the kilogram fundamental standards from which the pound and the yard would henceforth be derived.

In 1960, the CGPM established the SI system and has subsequently modified it in several meetings.

## WHO IS COORDINATING THE CONVERSION TO SI METRIC IN THE UNITED STATES?

The Omnibus Trade and Competitiveness Act of 1988 and its amendments declared the metric system as the preferred system of measurement in the United States and required its use in all federal activities to the extent feasible. Federal agencies formed the Construction Metrication Council within the National Institute of Building Sciences (NIBS) in Washington, D.C. The council is responsible for coordinating activities and distributing information and metric resources.

## WHAT ARE THE PRINCIPAL UNITS USED IN STRUCTURES THAT WILL BE OF CONCERN TO THE ARCHITECT?

| *Name* | *Symbol* | *Use* |
|---|---|---|
| meter | m | site plan dimensions, building plans |
| millimeter | mm | building plans and details |
| square millimeter | $mm^2$ | small areas |
| square meter | $m^2$ | large areas |
| hectare | ha | very large areas (1 hectare equals $10^4$ $m^2$) |
| cubic millimeter | $mm^3$ | small volumes |
| cubic meter | $m^3$ | large volumes |
| section modulus | $mm^3$ | property of cross section |
| moment of inertia | $mm^4$ | property of cross section |
| kilogram | kg | mass of all building materials |
| Newton | N | force (all structural computations) |
| Pascal | Pa | stress or pressure (all structural computations; one pascal equals one newton per square meter) |
| mass density | $kg/m^3$ | density of materials |
| degree Celsius | °C | temperature measurement |

## WHAT ARE THE COMMON PREFIXES?

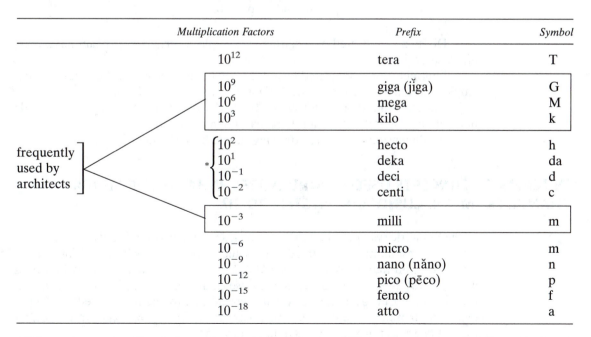

| | *Multiplication Factors* | *Prefix* | *Symbol* |
|---|---|---|---|
| | $10^{12}$ | tera | T |
| | $10^9$ | giga (jĭga) | G |
| | $10^6$ | mega | M |
| | $10^3$ | kilo | k |
| | $10^2$ | hecto | h |
| | * $10^1$ | deka | da |
| frequently used by architects | $10^{-1}$ | deci | d |
| | $10^{-2}$ | centi | c |
| | $10^{-3}$ | milli | m |
| | $10^{-6}$ | micro | m |
| | $10^{-9}$ | nano (năno) | n |
| | $10^{-12}$ | pico (pēco) | p |
| | $10^{-15}$ | femto | f |
| | $10^{-18}$ | atto | a |

* To be consistent and avoid confusion, prefixes should change in steps of $10^3$; therefore, these four should be avoided if at all possible.

## WHAT ARE SOME RULES OF "GRAMMAR"?

Double prefixes should never be used; for instance, use Gm (gigameter), not Mkm (megakilometer).

Base units are not capitalized unless you are writing a *symbol* that is derived from a proper name; for example, 12 meters or 12 m, 60 newtons or 60 N.

Plurals are accomplished normally, except for quantities less than 1. In such cases the "s" is deleted; for instance, 2.6 meters, 0.6 meter.

Prefix symbols are not capitalized except for M (mega), G (giga), and T (tera). This avoids confusion with m (meter), g (gram), and t (metric ton). One metric ton (t) is equal to one megagram (Mg).

Periods are not used after symbols except at the end of a sentence. Commas should not be used to clarify groups of digits; instead, use spaced groups of three on each side of the decimal point.

$$832 \quad 604.789 \quad 06 \quad \text{not} \quad 832{,}604.78906$$

$$20\ 800 \quad \text{not} \quad 20{,}800$$

Exception: The space is optional in groups of four digits.

$$1486 \quad \text{or} \quad 1\ 486$$

$$0.3248 \quad \text{or} \quad 0.324\ 8$$

Division is indicated by a slash; for example, a certain steel beam has a mass of 100 kg/m.

Multiplication is indicated by a dot placed at mid-height of the letters; for example, a certain moment or torque might be given as 100 kN · m.

Decimals (not dual units) should be used; for instance, the length of one side of a building lot, state 118.6 m rather than 118 m, 600 mm.

All dimensions on a given drawing should have the same units.

## IN TERMS OF CONCEPTS USED IN STRUCTURES, WHAT IS THE MAJOR CHANGE FROM THE CUSTOMARY SYSTEM TO SI?

In the SI metric system of units, there is a clear distinction made between mass and force. The customary system treated mass and force as if they both had force units; that is, we said that a beam weighed 100 *lb* per foot and that the force in a truss member was 1800 *lb*. It was correct to use pounds for force but not for weight (mass). In the European metric system, we said that the beam weighed 149 *kg* per meter and that the force in a truss member was 818 *kg of force* (kgf). It was correct to use kilograms for weight (mass) but not for force.

$$F \text{ is equal to } MA \qquad F \text{ is not equal to } M$$

SI units do not confuse the two terms. We can "weigh" items such as cubic meters of concrete by establishing their mass in kilograms, but force is expressed in newtons and pressure (stress) is in newtons per square meter (pascals).

A commonly used illustration to explain the difference between force and mass is to look at what happens in different fields of gravity. Assume that you are holding a 1-kg mass in the palm of your hand. On earth, the kg would exert a force of 9.8 N downward on your hand. (This would vary slightly, depending upon whether you were located at sea level or on top of Mt. Everest.) On our moon it would push with a force of 1.6 N, and on the surface of Jupiter you would feel a force of 24.1 N!

For structural engineering purposes on earth, we can multiply our loads (if given in kg) by 9.8 to get the number of newtons of force we must design for.

## HOW WILL THE NEW UNITS MODIFY DESIGN DRAWINGS?

Conceptually, the square meter is the new unit of plan area replacing the square foot. Length measurements may be in meters or millimeters, except that it is desirable to express all the measurements on a single drawing in the same units. (Among other advantages, this obviates the need for placing m or mm as a suffix to each dimension.) The millimeter is preferred for all detail, section, and plan drawings up through the scale of 1 : 200. On plans, this results in small numbers for wall thicknesses and large numbers for room dimensions, but eliminates the need for fractions. Even on details, the millimeter is small enough that, with few exceptions, fractions can be avoided.

The basic building modules recommended are 100 mm and 300 mm. The 300-mm dimension is very close to 12 inches and is an easy concept to adopt. At the same time, it is much more flexible than the foot, in that it is *evenly* divisible by 2, 3, 4, 5, 6, 10, 15, 20, 25, 30, 50, 60, 100, and 150.

## WHAT ARE COMMONLY USED SCALES?

| Customary | Nearest Convenient Ratio | Metric Equivalent |
|---|---|---|
| $\frac{1}{16}$" = 1'–0" | 1 : 200 | 5 mm = 1 m |
| $\frac{1}{8}$" = 1'–0" | 1 : 100 | 10 mm = 1 m |
| $\frac{1}{4}$" = 1'–0" | 1 : 50 | 20 mm = 1 m |
| $\frac{1}{2}$" = 1'–0" | 1 : 20 | 50 mm = 1 m |
| $\frac{3}{4}$" = 1'–0" | 1 : 10 | 100 mm = 1 m |
| $1\frac{1}{2}$" = 1'–0" | 1 : 10 | 100 mm = 1 m |
| 3" = 1'–0" | 1 : 5 | 200 mm = 1 m |
| 1" = 20' | 1 : 200 | 5 mm = 1 m |
| 1" = 50' | 1 : 500 | 2 mm = 1 m |

## ARE THERE SOME CONVERSION FACTORS THAT WE CAN USE WHILE BECOMING FAMILIAR WITH THE NEW UNITS?

There are endless tables of conversion factors available. However, the more we use conversion factors, the longer it will be before we are able to "think" in metric. In any event, one must keep the desired level of accuracy in mind when making conversions. For example, suppose that you are working with a reinforced concrete beam $12 \times 20$ inches in cross section. You wish to convert its area in square inches to millimeters squared. (The dimensions imply an accuracy of plus or minus 1 square inch or about 0.4%.) Following the table below, you could convert to metric by multiplying 240 by 6.451 600 E+02 to get an area of 154 838 mm$^2$. To use this quantity would be deceiving in terms of accuracy because it is subject to the same $\pm 0.4\%$ tolerance level, or in this case about 600 mm$^2$. In other words, the area could range from approximately 154 200 to about 155 400. Expressing the converted area as simply 155 000 mm$^2$ would be much more consistent.

A few conversion factors that may prove useful in structural analysis are presented below.

| *To Convert From* | *To* | *Multiply By* |
|---|---|---|
| inches | mm | 2.540 000 E+01 |
| feet | m | 3.048 000 E−01 |
| in$^2$ | mm$^2$ | 6.451 600 E+02 |
| ft$^2$ | m$^2$ | 9.290 304 E−02 |
| in$^3$ | mm$^3$ | 1.638 706 E+04 |
| ft$^3$ | m$^3$ | 2.831 685 E−02 |
| in$^4$ | mm$^4$ | 4.162 314 E+05 |
| °F | °C | $t°C = (t°F + 32)/1.8$ |
| lb (mass) per foot | kg/m | 1.488 163 E+00 |
| lb (force) per foot | N/m | 1.459 390 E+01 |
| strain/°F (thermal expansion) | strain/°C | 1.800 000 E+00 |
| lb (force) | N | 4.448 222 E+00 |
| kip (force) | kN | 4.448 222 E+00 |
| lb-ft (moment) | N · m | 1.355 818 E+00 |
| kip-ft (moment) | kN · m | 1.355 818 E+00 |
| psi (stress) | kPa | 6.894 757 E+00 |
| ksi (stress) | MPa | 6.894 757 E+00 |
| psf (uniform load) | kN/m$^2$ | 4.788 026 E−02 |

## REFERENCES

ASTM E380, Standard Practice for Use of the International System of Units (SI).

ASTM E621, Standard Practice for the Use of Metric (SI) Units in Building Design and Construction.

Metric Guide for Federal Construction. NIBS.

# Answers to Problems

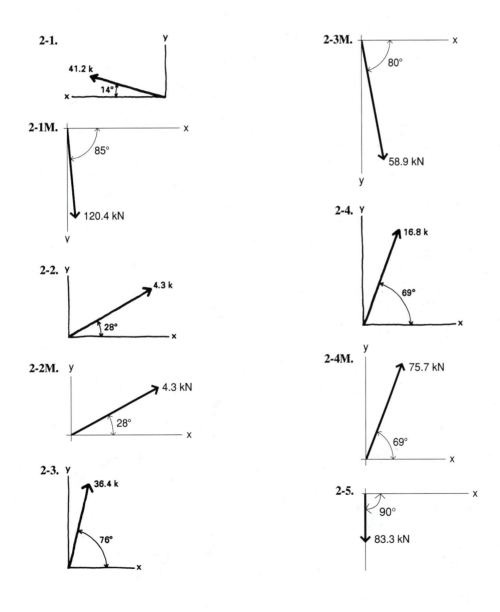

**2-1.** 41.2 k, 14°

**2-1M.** 85°, 120.4 kN

**2-2.** 4.3 k, 28°

**2-2M.** 4.3 kN, 28°

**2-3.** 36.4 k, 76°

**2-3M.** 80°, 58.9 kN

**2-4.** 16.8 k, 69°

**2-4M.** 75.7 kN, 69°

**2-5.** 90°, 83.3 kN

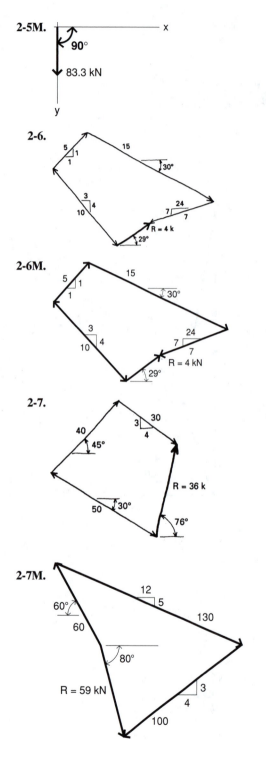

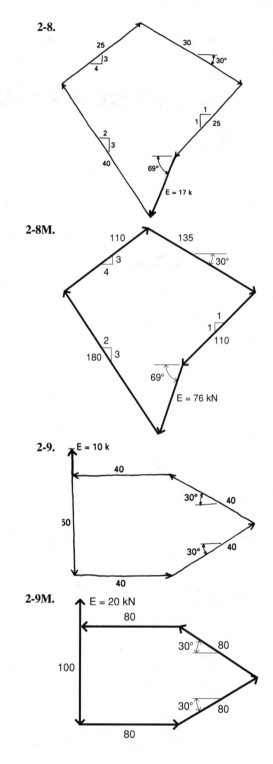

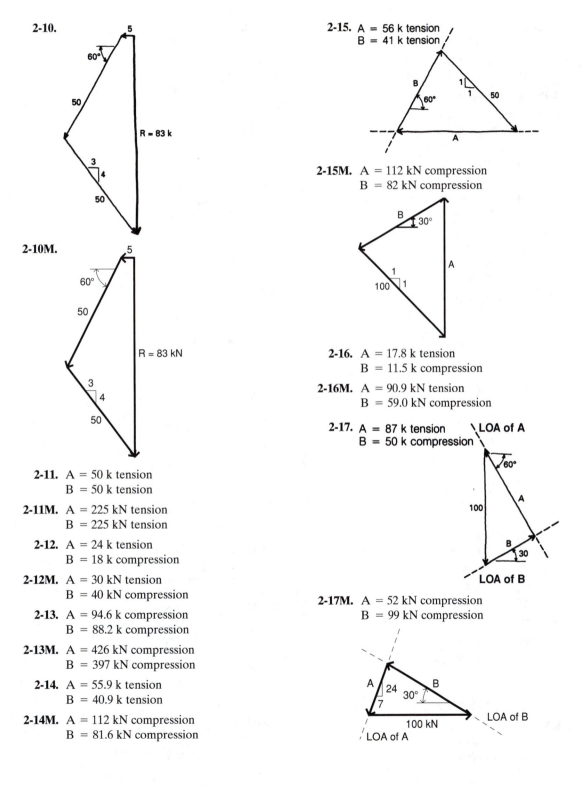

**2-10.**

**2-10M.**

**2-11.** A = 50 k tension
B = 50 k tension

**2-11M.** A = 225 kN tension
B = 225 kN tension

**2-12.** A = 24 k tension
B = 18 k compression

**2-12M.** A = 30 kN tension
B = 40 kN compression

**2-13.** A = 94.6 k compression
B = 88.2 k compression

**2-13M.** A = 426 kN compression
B = 397 kN compression

**2-14.** A = 55.9 k tension
B = 40.9 k tension

**2-14M.** A = 112 kN compression
B = 81.6 kN compression

**2-15.** A = 56 k tension
B = 41 k tension

**2-15M.** A = 112 kN compression
B = 82 kN compression

**2-16.** A = 17.8 k tension
B = 11.5 k compression

**2-16M.** A = 90.9 kN tension
B = 59.0 kN compression

**2-17.** A = 87 k tension
B = 50 k compression

**2-17M.** A = 52 kN compression
B = 99 kN compression

**2-18.** A = 198 k tension
B = 100 k tension

**2-18M.** A = 198 kN compression
B = 100 kN compression

**2-19.** A and B approach infinity

**2-19M.** A and B approach infinity

**2-20.** $M_O$ = 460 k-ft $\circlearrowright$
$M_A$ = 255 k-ft $\circlearrowright$
$M_B$ = 260 k-ft $\circlearrowleft$

**2-20M.** $M_O$ = 110 kN•m $\circlearrowright$
$M_A$ = 187 kN•m $\circlearrowleft$
$M_B$ = 250 kN•m $\circlearrowright$

**2-21.** $M_O$ = 704 k-ft $\circlearrowleft$
$M_A$ = 0
$M_B$ = 1248 k-ft $\circlearrowright$
$M_C$ = 480 k-ft $\circlearrowright$

**2-21M.** $M_O$ = 660 kN•m $\circlearrowleft$
$M_A$ = 0
$M_B$ = 1170 kN•m $\circlearrowright$
$M_C$ = 450 kN•m $\circlearrowright$

**2-24.** $A_x$ = 0
$A_y$ = 44 k ↑
$M_A$ = 408 k-ft $\circlearrowleft$

**2-24M.** $B_x$ = 0
$B_y$ = 36 kN ↑
$M_B$ = 162 kN•m $\circlearrowright$

**2-25.** $A_x$ = 10 k →
$A_y$ = 20 k ↑
$M_A$ = 60 k-ft $\circlearrowright$

**2-25M.** $A_x$ = 16 kN ←
$A_y$ = 20 kN ↑
$M_A$ = 32 kN•m $\circlearrowright$

**2-26.** $A_x$ = 1 k →
$A_y$ = 10 k ↑
$M_A$ = 19 k-ft $\circlearrowleft$

**2-26M.** $B_x$ = 20 k ←
$B_y$ = 10 kN ↑
$M_B$ = 30 kN•m $\circlearrowleft$

**2-27.** $A_x$ = 0
$A_y$ = 135 k ↑
$M_A$ = 810 k-ft $\circlearrowleft$

**2-27M.** $A_x$ = 0
$A_y$ = 36 kN ↑
$M_A$ = 60 kN•m $\circlearrowleft$

**2-28.** $A_x$ = 0
$A_y$ = 16.4 k ↑
$B_y$ = 17.6 k ↑

**2-28M.** $A_x$ = 0
$A_y$ = 62 kN ↑
$B_y$ = 68 kN ↑

**2-29.** $A_y$ = 7.1 k ↑
$B_x$ = 32 k ←
$B_y$ = 7.1 k ↓

**2-29M.** $A_x$ = 50 kN ←
$B_y$ = 0
$B_x$ = 50 kN ←

**2-30.** $A_y$ = 66 k ↑
$B_x$ = 0
$B_y$ = 50 k ↑

**2-30M.** $A_x$ = 0
$A_y$ = 63 kN ↑
$B_y$ = 45 kN ↑

**2-31.** $A_x$ = 6 k ←
$A_y$ = 0
$B_y$ = 8 k ↑

**2-31M.** $A_x$ = 60 kN ←
$A_y$ = 0
$B_y$ = 80 kN ↑

**2-32.** $A_y$ = 0
$B_x$ = 0
$B_y$ = 30 k ↑

**2-32M.** $A_x$ = 0
$A_y$ = 100 kN ↑
$B_y$ = 0

**2-33.** $A_x$ = 0
$A_y$ = 24 k ↑
$B_x$ = 32 k →

**2-33M.** $A_x$ = 2350 kN ←
$A_y$ = 1200 kN ↑
$B_x$ = 2000 kN →

**2-34.** $A_x$ = 0
$A_y$ = 28.3 k ↑
$B_x$ = 28.3 k →

**2-34M.** $A_y$ = 136 kN ↑
$B_x$ = 60 kN ←
$B_y$ = 36 kN ↓

**2-35.** (a) $A_y = 75$ k ↑
    $B_x = 0$
    $B_y = 25$ k ↓
(b) $A_x = 150$ k →
    $A_y = 75$ k ↑
    $B_x = 150$ k ←
    $B_y = 25$ k ↓

**2-35M.** (a) $A_x = 0$
    $A_y = 25$ kN ↓
    $B_x = 0$
    $B_y = 75$ kN ↑
(b) $A_x = 150$ kN ←
    $A_y = 25$ kN ↓
    $B_x = 150$ kN →
    $B_y = 75$ kN ↑

**2-36.** $A_x = 60$ k →
    $A_y = 30$ k ↑
    $B_x = 60$ k ←
    $B_y = 10$ k ↑

**2-36M.** $A_x = 50$ kN →
    $A_y = 50$ kN ↑
    $B_x = 50$ kN ←
    $B_y = 50$ kN ↑

**2-37.** $A_x = 45$ k →
    $A_y = 67.5$ k ↑
    $B_x = 45$ k ←
    $B_y = 22.5$ k ↑

**2-37M.** $A_x = 27$ kN →
    $A_y = 15$ kN ↑
    $B_x = 27$ kN ←
    $B_y = 45$ kN ↑

**2-38.** $A_x = 48$ k ←
    $A_y = 108$ k ↑
    $B_x = 48$ k →
    $B_y = 48$ k ↓

**2-38M.** $A_x = 50$ kN ←
    $A_y = 25$ kN ↓
    $B_x = 50$ kN →
    $B_y = 175$ kN ↑

**2-39.** $A_x = 36$ k →
    $A_y = 9$ k ↑
    $B_x = 36$ k ←
    $B_y = 24$ k ↑

**2-39M.** $A_x = 66.7$ kN →
    $A_y = 50$ kN ↑
    $B_x = 66.7$ kN ←
    $B_y = 50$ kN ↑

**2-40.** $A_x = 30$ k →
    $A_y = 30$ k ↑
    $B_x = 30$ k ←

**2-40M.** $A_x = 60$ kN →
    $A_y = 60$ kN ↑
    $B_x = 60$ kN ←

**2-41.** $A_x = 30$ k ←
    $A_y = 15$ k ↑
    $B_x = 30$ k ←
    $B_y = 15$ k ↑

**2-41M.** $A_x = 60$ kN →
    $A_y = 30$ kN ↑
    $B_x = 60$ kN ←
    $B_y = 30$ kN ↑

**2-42.** $A_x = 9$ k →
    $A_y = 6.75$ k ↓
    $B_y = 6.75$ k ↑

**2-42M.** $A_x = 10$ kN ←
    $A_y = 10$ kN ↓
    $B_x = 10$ kN ←
    $B_y = 10$ kN ↑

**2-43.** $A_x = 24$ k →
    $A_y = 6$ k ↓
    $B_x = 24$ k ←
    $B_y = 12$ k ↑

**2-43M.** $A_x = 77$ kN ←
    $A_y = 33$ kN ↑
    $B_x = 77$ kN →
    $B_y = 9$ kN ↓

**2-44.** $A_x = 20$ k ←
    $A_y = 24$ k ↓
    $B_x = 20$ k →
    $B_y = 40$ k ↑

**2-44M.** $A_x = 28.2$ kN →
    $A_y = 84.7$ kN ↓
    $B_x = 88.2$ kN ←
    $B_y = 84.7$ kN ↑

**2-45.** $A_x = 4$ k $\rightarrow$
$A_y = 4$ k $\downarrow$
$B_x = 4$ k $\leftarrow$
$B_y = 4$ k $\uparrow$

**2-45M.** $A_x = 11$ kN $\rightarrow$
$A_y = 11$ kN $\downarrow$
$B_x = 11$ kN $\leftarrow$
$B_y = 11$ kN $\uparrow$

**2-46.** 1.15 k/ft

**2-46M.** 6.5 kN/m

**2-47.** 1070 lb

**2-47M.** 35.2 kN

**2-48.** 21 600 lb

**2-48M.** (a) 8 kN
(b) 2.67 kN

**2-49.** (a) stable, determinate
(b) stable, indeterminate, first degree
(c) stable, determinate
(d) stable, determinate
(e) unstable
(f) stable, indeterminate, second degree

**2-51.**

**2-51M.**

**2-52.**

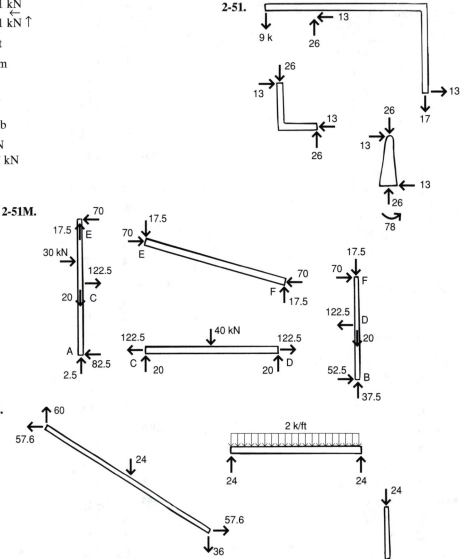

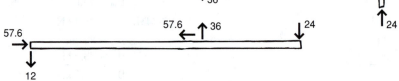

**2-52M.**

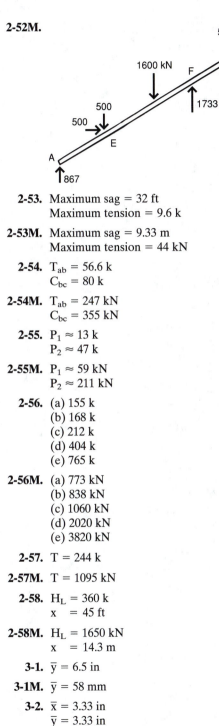

**2-53.** Maximum sag = 32 ft
Maximum tension = 9.6 k

**2-53M.** Maximum sag = 9.33 m
Maximum tension = 44 kN

**2-54.** $T_{ab}$ = 56.6 k
$C_{bc}$ = 80 k

**2-54M.** $T_{ab}$ = 247 kN
$C_{bc}$ = 355 kN

**2-55.** $P_1 \approx 13$ k
$P_2 \approx 47$ k

**2-55M.** $P_1 \approx 59$ kN
$P_2 \approx 211$ kN

**2-56.** (a) 155 k
(b) 168 k
(c) 212 k
(d) 404 k
(e) 765 k

**2-56M.** (a) 773 kN
(b) 838 kN
(c) 1060 kN
(d) 2020 kN
(e) 3820 kN

**2-57.** T = 244 k

**2-57M.** T = 1095 kN

**2-58.** $H_L$ = 360 k
x   = 45 ft

**2-58M.** $H_L$ = 1650 kN
x   = 14.3 m

**3-1.** $\bar{y}$ = 6.5 in

**3-1M.** $\bar{y}$ = 58 mm

**3-2.** $\bar{x}$ = 3.33 in
$\bar{y}$ = 3.33 in

**3-2M.** $\bar{x}$ = 36 mm
$\bar{y}$ = 36 mm

**3-3.** $\bar{y}$ = 6.5 in

**3-3M.** $\bar{y}$ = 85 mm

**3-4.** $\bar{y}$ = 4.82 in

**3-4M.** $\bar{y}$ = 86 mm

**3-5.** $\bar{y}$ = 5.36 in

**3-5M.** $\bar{y}$ = 65 mm

**3-6.** $\bar{y}$ = 7.46 in

**3-6M.** $\bar{y}$ = 190 mm

**3-7.** $\bar{x}$ = 28.6 ft
$\bar{y}$ = 18.3 ft

**3-7M.** $\bar{x}$ = 9.5 m
$\bar{y}$ = 6.1 m

**3-8.** $\bar{y}$ = 22.5 m

**3-8M.** $\bar{y}$ = 560 mm

**3-9.** $\bar{y}$ = 0.7 in

**3-9M.** $\bar{y}$ = 19 mm

**3-10.** $I_{xx}$ = 981 in$^4$
$I_{yy}$ = 469 in$^4$

**3-10M.** $I_{xx}$ = 1.22 (10)$^6$ mm$^4$
$I_{yy}$ = 0.62 (10)$^6$ mm$^4$

**3-11.** $I_{xx}$ = 122 in$^4$
$I_{yy}$ = 72 in$^4$

**3-11M.** $I_{yy}$ = 11.4 (10)$^6$ mm$^4$

**3-12.** $I_{xx}$ = 1460 in$^4$

**3-12M.** $I_{xx}$ = 2090 (10)$^6$ mm$^4$

**3-13.** $I_{xx}$ = 1980 in$^4$

**3-13M.** $I_{xx}$ = 772 (10)$^6$ mm$^4$

**3-14.** $I_{xx}$ = 5510 in$^4$

**3-14M.** $I_{xx}$ = 2170 (10)$^6$ mm$^4$

**3-15.** (a) $I_{yy}$ = 22 500 in$^4$
(b) $I_{xx}$ = 5000 in$^4$

**3-15M.** (a) $I_{yy}$ = 9150 (10)$^6$ mm$^4$
(b) $I_{xx}$ = 1980 (10)$^6$ mm$^4$

**3-16.** $\bar{y}$ = 6.34 in
$I_{xx}$ = 139.5 in

**3-16M.** $\bar{y}$   = 160 mm
$I_{xx}$ = 69.9 (10)$^6$ mm$^4$

**3-17M.** $I_{xx}$ = 134 m$^4$

**3-19.** $I_{xx}$ = 2910 in$^4$

**3-19M.** $I_{xx} = 14.9 \ (10)^6 \ mm^4$

**3-20.** $\bar{y} = 5.1$ in
$I_{xx} = 240 \ in^4$

**3-20M.** $\bar{y} = 129$ mm
$I_{xx} = 99.6 \ (10)^6 \ mm^4$

**3-22.** $I_{xx} = 139.5 \ in^4$

**3-22M.** $I_{xx} = 69.9 \ (10)^6 \ mm^4$

**3-23.** $I_{xx} = 529.6 \ in^4$

**3-23M.** $I_{xx} = 209 \ (10)^6 \ mm^4$

**3-24.** (a) 0.6%
(b) 35.4%

**3-25M.** 6%

**3-26.** $\bar{y} = 13$ in.
$I_{xx} = 25 \ 500 \ in^4$

**3-27.** $r_x = 4.19$ in
$r_y = 0.96$ in

**3-28.** $r_x = 3.92$ in
$r_y = 2.71$ in

**3-28M.** $r_x = 22$ mm
$r_y = 16$ mm

**3-29M.** $r_x = 136$ mm

**3-30.** $r_x = 1.16$ in

**3-31.** $r_x = 4.08$ in
$r_y = 8.66$ in

**3-31M.** $r_x = 102$ mm
$r_y = 219$ mm

**3-32.** $s = 3.06$ in

**3-32M.** $s = 77$ mm

**4-1.** $f_a = 33.9$ ksi

**4-1M.** $f_a = 90$ MPa

**4-2.** $f_a = 330$ psi

**4-2M.** $f_a = 2330$ kPa

**4-3.** $D = 3$ in

**4-3M.** $D = 78$ mm

**4-4.** $f_s = 9.55$ ksi

**4-4M.** $f_s = 76$ MPa

**4-5.** $f_s = 12.7$ ksi

**4-5M.** $f_s = 91.6$ MPa

**4-6.** $f = 18.3$ ksi

**4-6M.** $f = 136$ MPa

**4-7.** $\epsilon = 0.000 \ 69$

**4-7M.** $\epsilon = 0.0006$

**4-8.** $\epsilon = 0.001 \ 04$

**4-8M.** $\epsilon = 0.001$

**4-9.** $\delta = 1.2$ ft

**4-9M.** $\delta = 0.4$ m

**4-10.** 11.964 in

**4-10M.** 299.1 mm

**4-11.** $E = 3.77 \ (10)^6$ psi

**4-11M.** $E = 20$ GPa

**4-12.** $f = 100$ ksi

**4-12M.** $f = 640$ MPa

**4-13.** $E = 29 \ 000$ ksi

**4-13M.** $E = 200$ GPa

**4-14.** $f = 1350$ psi

**4-14M.** $F = 9000$ kPa

**4-15.** $E = 3 \ (10)^6$ psi

**4-15M.** $E = 28.9$ GPa

**4-17.** $\delta = 12.2$ in

**4-17M.** $\delta = 0.359$ m

**4-18.** $\delta = 0.15$ in

**4-18M.** $\delta = 4$ mm

**4-19.** $D = 2$ in

**4-19M.** $D = 55$ mm

**4-20.** $D = 4.1$ in

**4-20M.** $D = 111$ mm

**4-21.** $L = 31.1$ ft

**4-21M.** $L = 8.8$ m

**4-22.** $L_1 = 399.84$ ft
$L_2 = 400.16$ ft

**4-22M.** $L_1 = 124.95$ m
$L_2 = 125.06$ m

**4-23.** $\delta = 0.276$ in

**4-23M.** $\delta = 8$ mm

**4-24.** width = 0.4 in

**4-24M.** width = 10 mm

**4-25.** $\delta = 0.122$ in

**4-25M.** $\delta = 3.6$ mm

**4-26.** f = 33.9 ksi

**4-26M.** f = 234 MPa

**4-27.** L = 5.65 ft

**4-27M.** L = 1.9 m

**4-28M.** f = 46.5 MPa

**6-1.**

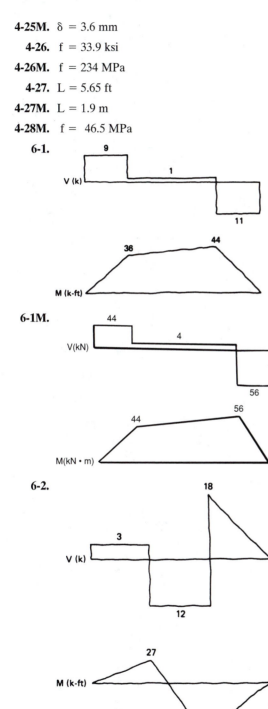

**6-1M.**

**6-2.**

**6-2M.**

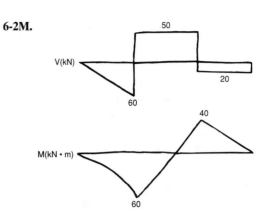

**6-3.**

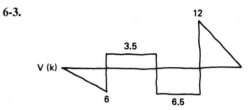

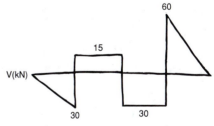

**6-3M.**

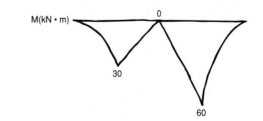

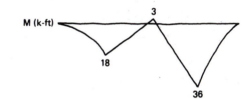

**432**

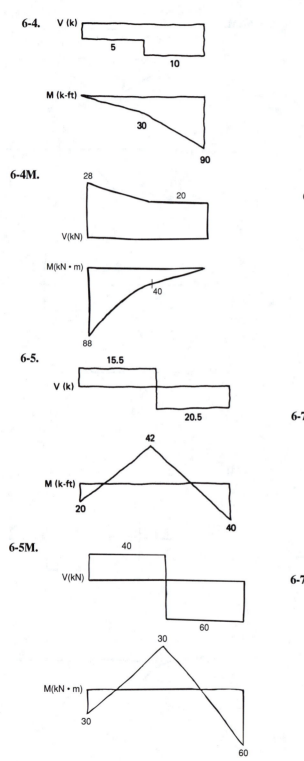

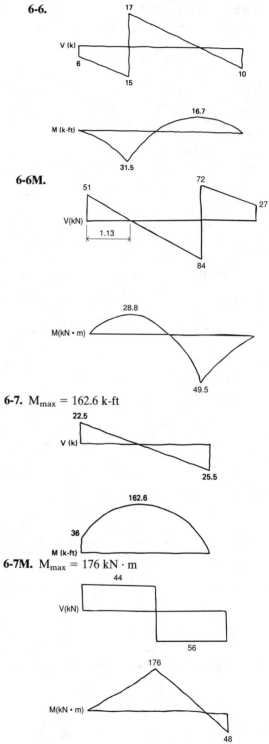

**6-7.** $M_{max} = 162.6$ k-ft

**6-7M.** $M_{max} = 176$ kN · m

**6-8.**

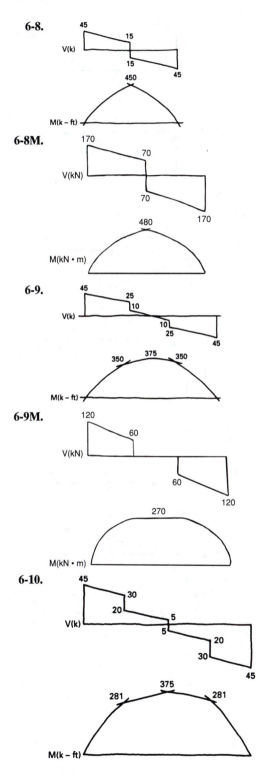

**6-8M.**

**6-9.**

**6-9M.**

**6-10.**

**6-10M.**

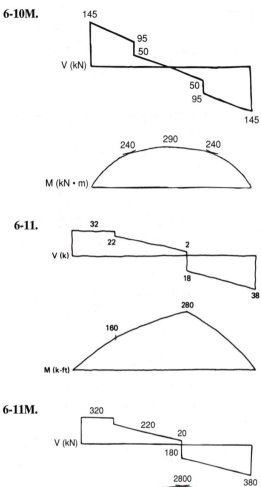

**6-11.**

**6-11M.**

**6-12.**

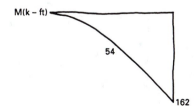

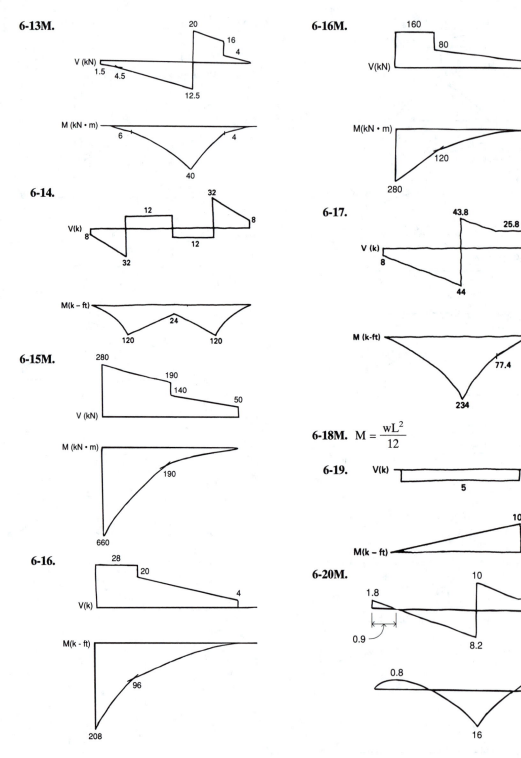

**6-13M.**

**6-14.**

**6-15M.**

**6-16.**

**6-16M.**

**6-17.**

**6-18M.** $M = \dfrac{wL^2}{12}$

**6-19.**

**6-20M.**

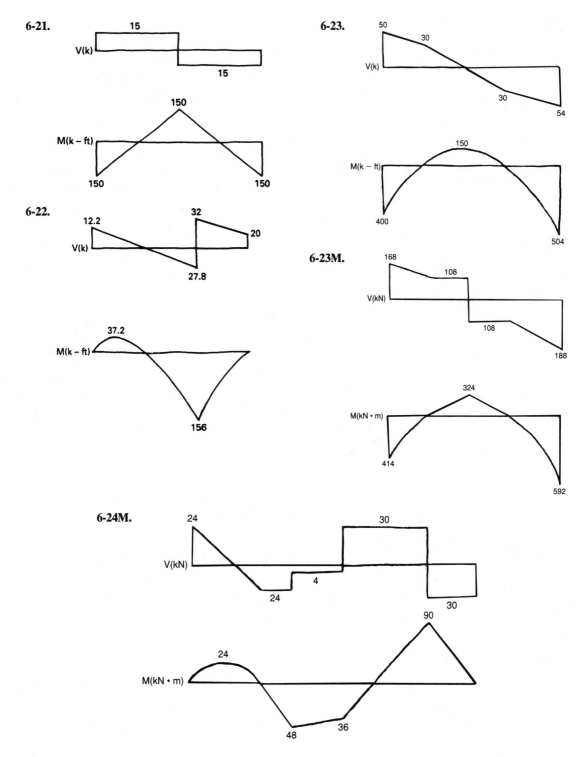

**6-21.**

V(k)   15 / 15

M(k – ft)   150 ... 150 / 150

**6-22.**

V(k)   12.2 / 32 / 20 / 27.8

M(k – ft)   37.2 / 156

**6-23.**

V(k)   50 / 30 / 30 / 54

M(k – ft)   150 / 400 / 504

**6-23M.**

V(kN)   168 / 108 / 108 / 188

M(kN • m)   324 / 414 / 592

**6-24M.**

V(kN)   24 / 30 / 24 / 4 / 30

M(kN • m)   90 / 24 / 48 / 36

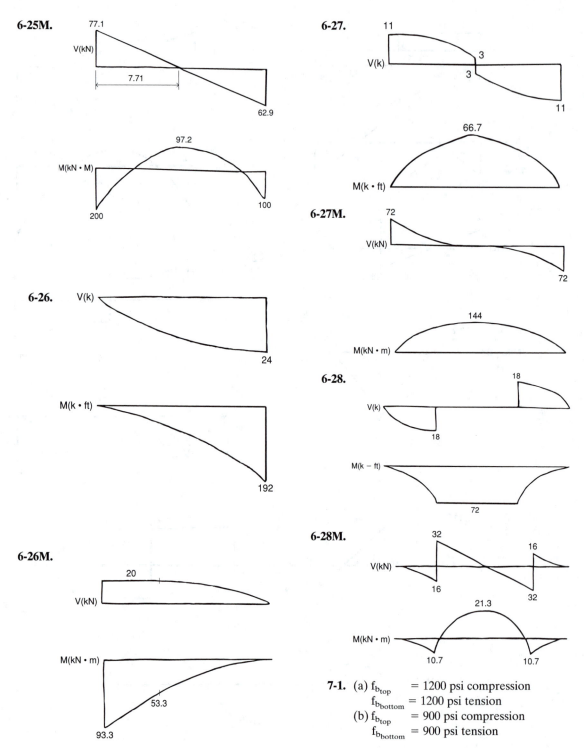

**6-25M.**

V(kN)

77.1

7.71

62.9

M(kN · M)

97.2

200

100

**6-26.**

V(k)

24

M(k · ft)

192

**6-26M.**

V(kN)

20

M(kN · m)

53.3

93.3

**6-27.**

V(k)

11

3

3

11

M(k · ft)

66.7

**6-27M.**

V(kN)

72

72

M(kN · m)

144

**6-28.**

V(k)

18

18

M(k – ft)

72

**6-28M.**

V(kN)

32

16

16

32

M(kN · m)

21.3

10.7

10.7

**7-1.** (a) $f_{b_{top}}$   = 1200 psi compression
$f_{b_{bottom}}$ = 1200 psi tension
(b) $f_{b_{top}}$   = 900 psi compression
$f_{b_{bottom}}$ = 900 psi tension

**7-1M.** (a) $f_{b_{top}}$ = 8930 kPa compression
$f_{b_{bottom}}$ = 8930 kPa tension
(b) $f_{b_{top}}$ = 5720 kPa compression
$f_{b_{bottom}}$ = 5720 kPa tension

**7-2.** $f_{b_{top}}$ = 1100 psi tension
$f_{b_{bottom}}$ = 1100 psi compression

**7-2M.** $f_{b_{top}}$ = 8280 kPa tension
$f_{b_{bottom}}$ = 8280 kPa compression

**7-3.** $f_b$ = 16.5 ksi

**7-3M.** $f_b$ = 134 MPa

**7-4.** (a) $f_b$ = 14.3 ksi
(b) $f_b$ = 0.838 ksi

**7-4M.** (a) $f_b$ = 108 MPa
(b) $f_b$ = 7.74 MPa

**7-5.** w = 94.9 lb/ft

**7-5M.** w = 1.48 kN/m

**7-6.** L = 17.9 ft

**7-6M.** L = 5.42 m

**7-7.** No, 37.4 ksi > 33 ksi

**7-7M.** Yes, $f_b$ = 220 MPa < 230 MPa

**7-8.** $f_b$ = 1520 psi

**7-8M.** $f_b$ = 13 000 kPa

**7-9.** $f_b$ = 21.6 ksi

**7-9M.** $f_b$ = 172 MPa

**7-10.** $f_{b_{top}}$ = 1610 psi compression
$f_{b_{bottom}}$ = 1930 psi tension

**7-10M.** $f_{b_{top}}$ = 13 800 kPa compression
$f_{b_{bottom}}$ = 16 500 kPa tension

**7-11.** $f_b$ = 1125 psi

**7-11M.** $f_b$ = 8190 kPa

**7-12.** $f_{b_{compression}}$ = 9.24 ksi
$f_{b_{tension}}$ = 22.4 ksi

**7-12M.** $f_{b_{compression}}$ = 70 MPa
$f_{b_{tension}}$ = 143 MPa

**7-13.** $f_{b_{compression}}$ = 441 psi
$f_{b_{tension}}$ = 525 psi

**7-13M.** $f_{b_{compression}}$ = 3360 kPa
$f_{b_{tension}}$ = 3990 kPa

**7-14M.** $f_{b_{top}}$ = 16 000 kPa compression
$f_{b_{bottom}}$ = 13 100 kPa tension

**7-15.** $f_b$ = 31.6 ksi

**7-15M.** $f_b$ = 223 MPa

**7-16.** W 36 × 160

**7-16M.** W 920 × 238

**7-17.** (a) 2 × 12
(b) 2 × 10
(c) 2 × 10
(d) 2 × 8

**7-17M.** (a) 38 × 285
(b) 38 × 235
(c) 38 × 235
(d) 38 × 185

**7-18.** 285 ft

**7-18M.** 87.5 m

**7-19.** (a) $f_b$ = 122 psi
(b) $f_b$ = 840 psi

**7-19M.** (a) $f_b$ = 1440 kPa
(b) $f_b$ = 7560 kPa

**7-20.** (a) 1500 psi
(b) 200 psi

**7-20M.** (a) 12 100 kPa
(b) 1610 kPa

**7-21.** 14.1 ksi

**7-21M.** 102 MPa

**8-1.**

**8-2M.**

**8-4.** b = 2.47 in.

**8-4M.** b = 69 mm

**8-5.** Shearing stress maximizes just above the notch

**8-6M.**

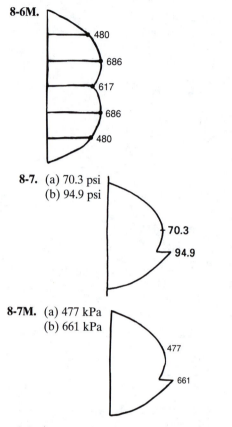

480
686
617
686
480

**8-7.** (a) 70.3 psi
(b) 94.9 psi

70.3
94.9

**8-7M.** (a) 477 kPa
(b) 661 kPa

477
661

**8-8.** $f_{v_{MAX}} = 110$ psi, larger than the allowable values.

**8-8M.** $f_{v_{MAX}} = 766$ kPa, larger than the allowable values.

**8-9M.** (a) 120 mm
(b) 52 mm

**8-10.** $f_{v_{MAX}} = 57.8$ psi

**8-10M.** $f_{v_{MAX}} = 420$ kPa

**8-11.** $f_{v_{MAX}} = 27.5$ psi

**8-11M.** $f_{v_{MAX}} = 192$ kPa

**8-12.** 57.9 psi < 90 psi, OK in shear.

**8-12M.** 336 kPa < 620 kPa, OK in shear.

**8-13.** Yes, 93 psi < 95 psi

**8-13M.** No, 725 kPa > 655 kPa

**8-14.** $f_{v_{MAX}} = 86$ psi

**8-14M.** No, 800 kPa > 655 kPa

**8-15.** $f_{v_{MAX}} = 45.3$ psi

**8-15M.** $f_{v_{MAX}} = 378$ kPa

**8-16.** $f_{v_{MAX}} = 55$ psi

**8-16M.** $f_{v_{MAX}} = 431$ kPa

**8-17.** $f_{v_{MAX}} = 126$ psi, larger than the allowable values.

**8-17M.** $f_{v_{MAX}} = 840$ kPa, larger than the allowable values.

**8-18.** $f_{v_{MAX}} \approx 2.5$ ksi

**8-18M.** $f_{v_{MAX}} \approx 19.9$ MPa

**8-19.** $f_{v_{MAX}} \approx 2.9$ ksi

**8-19M.** $f_{v_{MAX}} \approx 22.3$ MPa

**8-20.** 12.5%

**8-20M.** 11%

**8-21.** Yes, 18.8 ksi < 20.0 ksi

**8-22M.** Yes, 126 MPa < 140 MPa

**9-1.** $\Theta = \dfrac{wL^3}{6EI}, \Delta = \dfrac{wL^4}{8EI}$

**9-2.** $\Delta = \dfrac{PL^3}{48EI}$

**9-3.** 0.5 in. < 0.8 in., deflection is OK

**9-3M.** 8 mm < 17 mm, deflection is OK

**9-4.** $\Delta_{MAX} = \dfrac{6156 \text{ k-ft}^3}{EI}$

**9-4M.** $\Delta_{MAX} = \dfrac{460 \text{ kN} \cdot \text{m}^3}{EI}$

**9-5.** $\Delta_{FE} = 0.83$ in.

**9-5M.** $\Delta_{FE} = 20$ mm

**9-6.** $\Delta_{MAX} = 2.31$ in.

**9-6M.** $\Delta_{MAX} = 57$ mm

**9-7.** $\Delta_M = 0.75$ in.

**9-7M.** $\Delta_M = 40$ mm

**9-8.** $\Delta = 0.18$ in

**9-8M.** $\Delta = 4$ mm

**9-9.** $\Delta_{FE} = 0.75$ in.

**9-9M.** $\Delta_{FE} = 65$ mm

**9-10.** $\Delta_M = 1.26$ in.

**9-10M.** $\Delta_M = 27$ mm

**9-11M.** $\Delta_M = 47$ mm

**9-12.** $\Delta_M = 2.15$ in.

**9-12M.** $\Delta_M = 52$ mm

**9-13.** 85%

**9-13M.** 85%

**9-14.**

4.5

V (k)

7.5

10.1

M (k-ft)

18

**9-14M.**

7.5

V(kN)

12.5

5.625

M(kN · m)

10

**9-15.** $R_A = R_C = 0.3125\ P$, $R_B = 1.375\ P$

**9-16.** $R_B = 18$ kips $\downarrow$

**9-16M.** $R_B = 100$ kN $\downarrow$

**9-17.** $P_A = 16$ kips, $P_B = 27$ kips

**9-17M.** $P_A = 32$ kN, $P_B = 54$ kN

**9-18.** $A_y = 3$ kips

4.5

M(k – ft)

36

18

M(k – ft)

**9-18M.** $A_y = 5$ kN

2.5

M(kN · m)

20

10

M(kN · m)

**9-19.** $A_y = 1.5$ kips

1.125

M(k – ft)

54

9

M(k – ft)

**9-19M.** $A_y = 2.5$ kN

0.625

M(kN · m)

30

5

M(kN · m)

**9-20.** (a) The moment will gradually increase.
(b) $\Delta_A = 1/3$ in.

**9-20M.** (a) The moment will gradually increase.
(b) $\Delta_A = 7$ mm

**9-21.** $\Delta_A = \Delta_C = 0.69$ in.

**9-22M.** hinge shear force at B = 11.25 kN

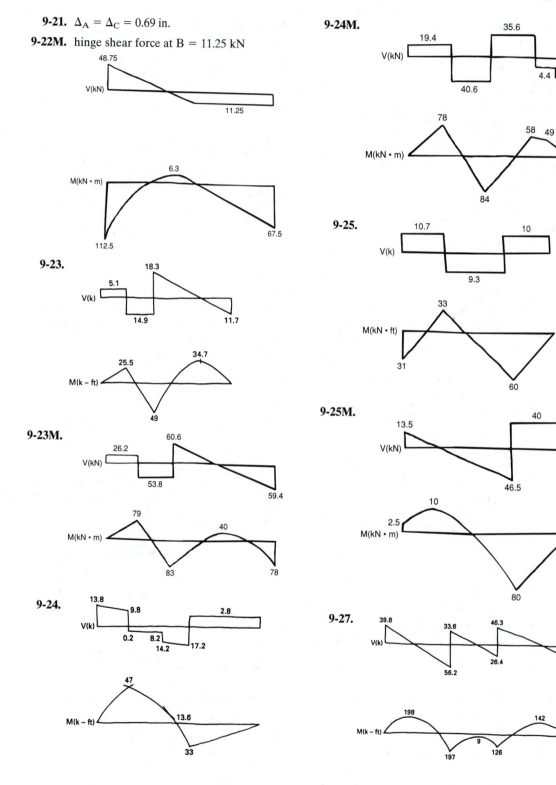

**9-23.**

**9-23M.**

**9-24.**

**9-24M.**

**9-25.**

**9-25M.**

**9-27.**

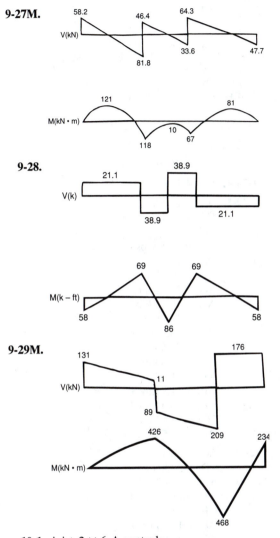

**9-27M.**

V(kN)

58.2    46.4    64.3

33.6    47.7

81.8

M(kN · m)

121    81

118    10    67

**9-28.**

V(k)

21.1    38.9

38.9    21.1

M(k – ft)

69    69

58    58

86

**9-29M.**

V(kN)

131    176

11

89

M(kN · m)

426    209    234

468

**10-1.** joist: $2 \times 6$, $\Delta$ controls
header: $2 - 2 \times 10$'s, M controls
lintel: $2 - 2 \times 8$'s, M controls
(Note: For construction reasons, $2 \times 10$'s
may be desirable for the lintel.)

**10-1M.** joist: $38 \times 139$, M controls
header: $3 - 38 \times 235$'s, M controls
lintel: $3 - 38 \times 235$'s, M controls

**10-2.** beam: $4 \times 6$, M controls
girder: $6 \times 10$, M controls

**10-2M.** beam: $89 \times 385$, M controls
girder: $139 \times 240$, M controls

**10-3.** joist: $2 \times 10$, M controls
beam: $2 - 2 \times 10$'s, M controls

**10-3M.** joist: $38 \times 285$, $\Delta$ controls
beam: $2 - 38 \times 285$'s, $\Delta$ controls

**10-5.** beam AB, $W24 \times 68$, M controls
beam AC, $W18 \times 35$, $\Delta$ controls

**10-5M.** beam AB, $W610 \times 113$, M controls
beam AC, $W460 \times 60$, $\Delta$ controls

**10-6.** beams: $W27 \times 84$, $\Delta$ controls
girder: $W36 \times 260$, $\Delta$ controls

**10-6M.** beams: $W610 \times 101$, $\Delta$ controls
girder: $W840 \times 359$, $\Delta$ controls

**10-7.** $W21 \times 132$, $\Delta$ controls

**10-7M.** $W530 \times 196$, M controls

**11-1.** $\left(\dfrac{P}{A}\right)_{cr} = 783$ psi, $P_{cr} = 23\ 700$ lb

**11-1M.** $P_{cr} = 80.6$ kN

**11-2.** $\left(\dfrac{P}{A}\right)_{cr} = 573$ psi, $P_{cr} = 7050$ lb

**11-2M.** $\left(\dfrac{P}{A}\right)_{cr} = 3900$ kPa, $P_{cr} = 30.9$ kN

**11-3.** $\left(\dfrac{P}{A}\right)_{cr} = 573$ psi, $P_{cr} = 11\ 060$ lb

**11-3M.** $\left(\dfrac{P}{A}\right)_{cr} = 3900$ kPa, $P_{cr} = 48.4$ kN

**11-4.** $\left(\dfrac{P}{A}\right)_{cr} = 1060$ psi, $P_{cr} = 32\ 120$ lb

**11-4M.** $\left(\dfrac{P}{A}\right)_{cr} = 7130$ kPa, $P_{cr} = 138$ kN

**11-6.** $\left(\dfrac{P}{A}\right)_{cr} = 11.6$ ksi, $P_{cr} = 85.1$ kips

**11-6M.** $\left(\dfrac{P}{A}\right)_{cr} = 82.2$ MPa, $P_{cr} = 389$ kN

**11-7.** $\left(\dfrac{P}{A}\right)_{cr} = 17.5$ ksi

**11-7M.** $\left(\dfrac{P}{A}\right)_{cr} = 116$ MPa

**11-8.** $P_{cr} = 1370$ lb

**11-8M.** $P_{cr} = 6.6$ kN

**11-9.** $\left(\dfrac{P}{A}\right)_{cr} = 12.2$ ksi, $P_{cr} = 179$ kips

**11-9M.** $\left(\dfrac{P}{A}\right)_{cr} = 67.5$ MPa, $P_{cr} = 640$ kN

**11-10.** $\left(\dfrac{P}{A}\right)_{cr} = 23.7$ ksi, $P_{cr} = 174$ kips

**11-10M.** $\left(\dfrac{P}{A}\right)_{cr} = 168$ MPa, $P_{cr} = 793$ kN

**11-11.** $\left(\dfrac{P}{A}\right)_{cr} = 2.9$ ksi, $P_{cr} = 21.3$ kips

**11-11M.** $\left(\dfrac{P}{A}\right)_{cr} = 20.4$ MPa, $P_{cr} = 96.5$ kN

**11-12.** $\left(\dfrac{P}{A}\right)_{cr} = 24.1$ ksi

**11-12M.** $P_{cr} = 160$ MPa

**11-13.** $L = 26.6$ ft

**11-13M.** $L = 8.2$ m

**11-14.** (a) $\left(\dfrac{P}{A}\right)_{cr} = 5.6$ ksi, $P_{cr} = 110$ kips

**11-14M.** (a) $\left(\dfrac{P}{A}\right)_{cr} = 39.7$ MPa, $P_{cr} = 504$ kN

**11-15.** $P_{cr} = 40.3$ kips

**11-15M.** $\left(\dfrac{P}{A}\right)_{cr} = 7340$ kPa, $P_{cr} = 91$ kN

**11-16M.** $\left(\dfrac{P}{A}\right)_{cr} = 128$ MPa

**11-17.** $\left(\dfrac{P}{A}\right)_{cr} = 25.5$ ksi, $P_{cr} = 263$ kips

**11-17M.** $\left(\dfrac{P}{A}\right)_{cr} = 129$ MPa, $P_{cr} = 826$ kN

**11-18.** (a) $P_{cr} = 7470$ lb
(b) $w_{cr} = 5600$ lb/ft

**11-18M.** (a) $P_{cr} = 36.1$ kN
(b) $w_{cr} = 90.3$ kN/m

**11-19.** $\left(\dfrac{P}{A}\right)_{cr} = 17.3$ ksi, $P_{cr} = 254$ kips

**11-19M.** $\left(\dfrac{P}{A}\right)_{cr} = 99.3$ MPa, $P_{cr} = 941$ kN

**11-20.** $\left(\dfrac{P}{A}\right)_{cr} = 16.7$ ksi

**11-20M.** $\left(\dfrac{P}{A}\right)_{cr} = 36$ MPa

**11-21.** $\left(\dfrac{P}{A}\right)_{cr} = 519$ psi, $P_{cr} = 4280$ lb

**11-21M.** $\left(\dfrac{P}{A}\right)_{cr} = 4170$ kPa, $P_{cr} = 22$ kN

**11-22.** $x = 0.20$

**11-22M.** $x = 0.20$

**11-23.** Yes, 2530 plf > 2000 plf

**11-24M.** $L = 4.8$ m

**12-1.** AC = 30 kips C
CD = 30 kips C
DE = 112.5 kips C
EF = 97.5 kips C
FG = 0
EI = 0
BG = 30 kips C
AH = 112.5 kips T
HI = 135 kips T
IJ = 135 kips T
JB = 97.5 kips T
FJ = 37.5 kips T
AD = 116.7 kips C
HE = 31.8 kips C
EJ = 53.0 kips C
FB = 137.9 kips C
DH = 22.5 kips T

**12-1M.** AC = 30 kN C
CD = 30 kN C
AD = 116.7 kN C
AH = 112.5 kN T
DH = 22.5 kN T
DE = 112.5 kN C
EH = 31.8 kN C
HI = 135 kN T
EI = 0
EF = 97.5 kN C
EJ = 53.0 kN C
IJ = 135 kN T
FJ = 37.5 kN T
JB = 97.5 kN T
FB = 138 kN C
FB = 0
GB = 30 kN C

**12-2.** AC = BE = 40 kips C
AF = BF = 0

CF = EF = 33.6 kips T
CD = ED = 30.4 kips C
DF = 30 kips C

**12-2M.** AC = BE = 40 kN C
AF = BF = 0
CF = EF = 33.6 kN T
CE = DE = 30.4 kN C
DF = 30 kN C

**12-3.** AC = 9.4 kips C
AD = 7.2 kips T
DF = 4.8 kips T
DE = 3.2 kips T
CD = 3.2 kips C
CE = 7.8 kips C

**12-3M.** AC = 20 kN C
CD = 0
AD = 48.1 kN C
AE = 26.7 kN T
DF = 33.3 kN C
DE = 20.0 kN T
EG = 53.3 kN T
FG = 0
EF = 33.3 kN C
FB = 66.7 kN C
GB = 53.3 kN T

**12-4.** FB = 10 kips T
GB = 8 kips C
EG = 8 kips C
FG = 0
EF = 10 kips C
DF = 20 kips T
DE = 6 kips T
AE = 16 kips C
CD = 30 kips T
AC = 12 kips T
AD = 14.4 kips C

**12-4M.** CG = BH = 20 kN C
CD = BF = 16 kN T
DG = FH = 24 kN C
GE = HE = 20 kN T
GA = HA = 40 kN C
AE = 48 kN C

**12-5.** AB = 30 kips T
AC = BE = DE = 0
CD = 60 kips C
AD = 42.4 kips T
BD = 42.4 kips C

**12-5M.** AB = 150 kN T
AC = DE = EB = 0
CD = 300 kN C
AD = 212 kN T
BD = 212 kN C

**12-6.** AC = 12 kips C
AE = DE = 0
CD = BE = 16 kips T
CD = BD = 20 kips C

**12-6M.** AC = 60 kN C
AE = DE = 0
CE = BE = 80 kN T
CD = BD = 100 kN C

**12-7.** DB = 22.5 kips C
EB = 18 kips T
AE = 5 kips T
AC = 13.5 kips C
CD = 22.5 kips C
CE = 14 kips T
DE = 3 kips T

**12-7M.** AC = 67.5 kN C
AE = 25 kN T
CE = 70 kN T
EB = 90 kN T
CD = BD = 112.5 kN C
DE = 15 kN T

**12-8.** AC = BG = 40 kips C
AH = BJ = HI = JI = 120 kips T
CD = GF = DH = FJ = 0
AD = BF = 170 kips C
DI = FI = 56.6 kips T

**12-8M.** AC = BG = 400 kN C
CD = FG = DH = FJ = 0
EI = 800 kN C
AD = BF = 1700 kN C
AH = HI = JI = BJ = 1200 kN T
DI = FI = 566 kN T
DE = FE = 1600 kN C

**12-9.** DB = 42.4 kips C

**12-9M.** DB = 212 kN C

**12-10.** DE = 0

**12-10M.** DE = 0

**12-11.** DF = 20 kips T
EG = 8 kips C
EF = 10 kips C

**12-11M.** DF = 33.3 kN C
EG = 53.3 kN T
EF = 33.3 kN C

**12-12.** AC = 285 kips T
BC = 212 kips C
BD = 115 kips C
CD = 100 kips T
CE = 135 kips T
DE = 141 kips C
DF = 15 kips C
EG = 49.5 kips T
FG = 21.2 kips C
EF = 15 kips T

**12-12M.** (See answers for 12-4M.)

**12-13.** AB = 15 kips T
AC = 22.5 kips T
BE = 22.5 kips C
AD = 27 kips T
BD = 27 kips C
CD = 20 kips C
ED = 10 kips T
CF = 7.5 kips T
EH = 7.5 kips C
CG = 18 kips T
EG = 18 kips C
FG = 15 kips C
HG = 5 kips T

FI = HK = JK = 0
FJ = 9 kips T
HJ = 9 kips C
IJ = 10 kips C

**12-13M.** IJ = 30 kN C
IF = KJ = KH = 0
JH = 27 kN C
JF = 27 kN T
FG = 45 kN C
FC = 22.5 kN T
GH = 15 kN T
HE = 22.5 kN C
GE = 54 kN C
GC = 54 kN T
CD = 60 kN C
CA = 67.5 kN T
DE = 30 kN T
EB = 67.5 kN C
DB = 81 kN C
DA = 81 kN T
AB = 45 kN T

**12-14.** AB = 26.7 kips

**12-14M.** HI = 48 kN T

**12-15.** (a) 1125 lb tension
(b) 562.5 lb tension

**12-15M.** EF = 6.7 kN C

# Index

## A

Aalto, Alvar, 341
Accuracy of Computations, 16
Allowable Stress. *See* Stress
Allowable stress design (ASD), 159
American Forest and Paper Association, 331
American Institute of Steel Construction
    (AISC), 331
Answer free-body diagram (AFBD), 54
Answers to Problems, 423–444
Area
    first moment of, 108
    properties of (*table*), 401–402
    second moment of, 118
    statical moment of, 108
    tributary, 212, 287–289
Average unit strain. *See* Strain
Axes, principal, 118
Axial stress. *See* Stress

## B

Bartning, Otto, 162
Bay, structural, 333
Beams
    collar, 80, 305
    continuous, 271
    cross-sectional shape, 117, 284
    deflection, 249–259
    formulas for (*table*), 415–417
    "ideal," 284
    lateral bracing, 226

    lateral buckling, 226–228
    repetitive, 293, 407, 408
    selection of steel, 312
    selection of wood, 292
    types of, 8
Beam-columns, 363
Bending, inefficiency of, 5, 199
Bending stress, 284. *See also* Stress
Bracing
    of beams, 226
    of columns, 353
Brittania Bridge, 131
Buckling
    of beams, 226–228
    of columns, 333–358
    diagonal compression, 229
    effect of compression, 5
    inelastic, 363
Building material weights (*table*), 398

## C

Cables, 91–102, 149
Center of gravity, 107
Centroid, 107–112
Centroid location (*table*), 401
Codes, 11
Collar beam, 80, 305
Cologne Cathedral, 144
Columns
    bracing of, 353–358
    buckling of, 333–358
    end conditions, 343

Columns *(Contd.)*
   failure modes, 334
   inelastic buckling, 363
   slenderness ratio, 336
Components, 20
Compression
   diagonal, 229
   stress, 5
Concrete, nature of, 165, 286
Connections, symbols, 48–51
Continuous beams, 271
   moment coefficients for, 280
Contributary area. *See* Tributary area
Counters, 368
Couple, 43, 47
Creep, 168
Crown Hall, 228
Cuvier, Georges, 9

**D**

Dead loads, 4
Deflection
   formulas, use of, 262
   formulas *(table)*, 415–417
   limitations *(table)*, 250
   theory, 249–259
Determinacy, 83–85
   of trusses, 237
Diagonal compression, 229
Dimension lumber, 286
Duration of load, 297, 408

**E**

Eames, Charles, 367
Eccentric loading, 336, 363
Economy, 2
Eiffel, Gustave, 17
Efficiency, 5, 159
Elastic buckling, columns, 333–358
Elastic curve, 250
Equilibrant, 28
Equilibrium, 4
   cantilevers, 52–56
   concurrent forces, 30–37

equations of, 31, 52
   moment, 52
   simple beams and frames, 58–63
   simple cables, 91–102
   truss joints, 368–375
   two-force members, 66–73
Euler equation, 336
   derivation, 396
Euler, Leonhard, 335

**F**

Factor of safety, 159
Ferris, George W., 17
First moment. *See* Area
Flexural stress. *See* Stress
Flexure formula, 202, 284
   derivation, 389–392
Force
   components, 20
   definition, 19
   direction of, 19
   equilibrant, 28
   moment of, 40
   polygon, 28
   redundant, 84
   resultant, 20
   sense of, 19
Framing direction, 289
Free-body diagram (FBD), 51
Fujisawa Municipal Gymnasium, 99
Furness, Frank, 144

**G**

Gable members, 77
Goldsmith, Myron, 178
Greeley Laboratory, 178

**H**

Hooke, Robert, 147
Horizontal projection, 77, 300
Horizontal shear. *See* Stress

## I

Imaginary span, 274
Indeterminacy, 84
Inertia, moment of, 116–124
Inflection point, 170
Ingalls Hockey Rink, 99
Isozaki, Arata, 131

## J

Joint equilibrium (trusses), 368–375

## K

Kahn, Louis I., 131
Kansas City Regency Hyatt Hotel, 143
Kitakyushu City Museum, 131
Krier, Leon, 178

## L

Ladovsky, Nikolai, 178
Lateral bracing
  beams, 226
    columns, 353–358
Lateral buckling, beams, 226–228
Laugier, Abbe, 6
LeCorbusier, 6, 291
Limit states design, 159
Live loads, 4
Load and resistance factor design (LRFD),
      159
Loads
  duration of, 297, 408
  eccentric, 336, 363
  types, 4
Loading patterns, 280
Lumber, dimension, 286

## M

Maisons Dom-ino, 6
Majowiecki, Massimo, 131

Maki, Fumikiko, 99
*Manual of Steel Construction,* 331
*Manual of Timber Construction,* 331
Masonry, 168
Materials
  characteristics of, 161–168
  properties of, 286
  properties (*table*), 399, 400
  weights (*table*), 398
Maybeck, Bernard, 367
Mayer Tower, 178
Mechanical system, 290
Mechanics, 19
Mechanics of materials, 19
Method of joints, 368–375
Method of sections, 379–383
Mies van der Rohe, 228
Modulus of elasticity, 149
Moment, definition of, 40
Moment-area theorems, 251
  proof of, 403–406
Moment arm, 40
Moment coefficients, 280
Moment diagrams, 169–196
  sign convention, 169
  *table,* 415–417
Moment of inertia, 107, 116–124, 199
  formulas (*table*), 402
Munday, Richard, 162

## N

National Design Specification, 295, 331
Navier, Claude L. M. H., 149, 201
Neutral axis, 200
Notre-Dame du Raincy Church, 162

## O

Openings in floor decks, 288
Otto, Frei, 341

## P

Panel points, 366
Parallel-axis theorem, 127–132

Parent, Antoine, 201
Perret, Auguste, 162
Point of inflection, 170
Ponding, 249
Principal axes, 118
Problem answers, 423–444
Projection, horizontal, 77, 301
Properties of areas (*table*), 401
Properties of materials, 286
Properties of materials (*table*), 399, 400
Properties of sections
    steel (*table*), 411–414
    wood (*table*), 409–410

# R

Radius of gyration, 134–136, 336
Reinforced concrete, nature of, 165
Repetitive members, 293, 407, 408
Resultant, 20
Rigid body, 51
Rome Olympic Stadium, 131
Rudolph, Paul, 178

# S

Saarinen, Eero, 99
Ste. Chapelle Church, 162
Saint-Venant, Barré de, 250
Salk Institute Research Building, 131
Second moment. *See* Area
Section modulus, 218, 285, 292, 312
    self-weight, 202, 220, 222
Shear diagrams, 170–189
    sign convention, 169
    table, 415–417
Shear formula, 232, 284
    shed members, 77
SI Metric System, 418
Simbirchev, 178
Slenderness ratio, 336
    sloped members, 77
Span/depth ratio (*table*), 12–15
Span, imaginary, 274
Stability, 2, 83
Statical moment, 108, 232
Statics, 19, 52
    of cables, 91–102
    procedures, 105

Steel
    design aids, 331
    nature of, 166, 286
    section properties (*table*), 411–414
Steel Joist Institute, 331
Stephenson, Robert, 131
Stiffness, 2, 148
Stirling, James, 367
Stirrups, 230
Strain
    average unit, 142
    flexural, 200, 389
    shearing, 141
    tangential, 141
    thermal, 153–157
Strength of structure, 2
Stress
    allowable in steel, 220, 222, 314
    allowable in wood (*table*), 407–408
    axial, 139
    bending, 5, 200, 284
    compressive, 5, 335
    in connections, 141
    critical buckling, 336
    definition, 139
    flexural, 199–214, 218–223
        derivation of formula, 389–392
    horizontal shearing, 229–238
        derivation of formula, 393–395
    normal, 139, 141
    shearing, 5
    tangential, 139, 141
    tensile, 5
    thermal, 153–157
    yield, 147
Structural bay, 288, 333
Structural design, 3
Structural planning, 3
Structural systems, 12–16
Structure
    in building, 8–16
    definition, 1
    in nature, 5–8
Superposition
    indeterminate structures, 265–268
    principle of, 260
Support conditions
    fixed, 51

Support conditions (*Contd.*)
  hanger, 48
  pin, 49
  roller, 50
Support moment, 271

# T

Tangential deviation, 251
Temperature, effects of, 153–157
Tension, diagonal, 229. *See also* Stress
Theorem of three moments, 271–277
Timber. *See* Wood
Tributary area, 212, 287–289
Trinity Church, 162
Truss
  analysis, 368–383
    method of joints, 368–375
    method of sections, 379–383
  definition, 365
  open-web bar joist, 387
  types, 366, 386
  Vierendeel, 387
Two-force member, 66–73, 365

# U

United Airlines Hangar, 178

# V

Varignon, Pierre, 41
Varignon's theorem, 41

# W

Weights, building materials (*table*), 398
Wood
  allowable stresses (*table*), 407, 408
  beams, selection of, 292
  design aids, 331
  nature of, 161, 286
  section properties (*table*), 409, 410
Wright, Frank Lloyd, 367

Vierendeel, Arthur, 387
Vierendeel "truss," 387
Villa Savoye, 291

# Y

Yield point, 147
Young, Thomas, 149
Young's modulus, 149

# Z

Zero members, in trusses, 368, 370, 373
Zero shear, in beams, 182

mm      millimeter (a unit of length, $10^{-3}$ meter)

M      bending moment

MN      meganewton (a unit of force, $10^6$ newtons)

MPa      megapascal (a unit of stress, $10^6$ pascals)

n.a.      neutral axis (really a plane)

N      newton (a unit of force)

psi      pound per square inch (a unit of stress)

P      force, concentrated (point) load (lb, kip, kN, MN)

Pa      pascal (a unit of stress)

$P_{cr}$      critical (failure) load in column calculations (lb, kip, kN, MN)

$\left(\dfrac{P}{A}\right)_{cr}$      critical (failure) stress in column calculations (psi, ksi, kPa, MPa)

Q      statical moment of that area of cross section between the horizontal plane under investigation and the near edge of the beam, taken with respect to the neutral axis, in shearing stress calculations ($in^3$, $mm^3$, $m^3$)

r      radius of gyration (in, mm, m)

$r_x$      radius of gyration with respect to an x-axis

$r_y$      radius of gyration with respect to a y-axis

R      force, reaction or resultant (lb, kip, kN)

s.w.      self-weight

S      section modulus, I/c ($in^3$, $mm^3$, $m^3$)

$S_r$      section modulus required ($in^3$, $mm^3$, $m^3$)

$S_x$      section modulus with respect to the x-centroidal axis

$S_y$      section modulus with respect to the y-centroidal axis

t      thickness, often of the web of a steel beam cross section (in, mm, m)